METHODS IN
ENZYMOLOGY

Cryo-EM, Part C

Analyses, Interpretation, and
Case studies

METHODS IN ENZYMOLOGY

Editors-in-Chief

JOHN N. ABELSON AND MELVIN I. SIMON

Division of Biology
California Institute of Technology
Pasadena, California

Founding Editors

SIDNEY P. COLOWICK AND NATHAN O. KAPLAN

VOLUME FOUR HUNDRED AND EIGHTY-THREE

METHODS IN
ENZYMOLOGY

Cryo-EM, Part C

Analyses, Interpretation, and
Case studies

EDITED BY

GRANT J. JENSEN
Division of Biology and Howard Hughes Medical Institute
California Institute of Technology
Pasadena, California, USA

AMSTERDAM • BOSTON • HEIDELBERG • LONDON
NEW YORK • OXFORD • PARIS • SAN DIEGO
SAN FRANCISCO • SINGAPORE • SYDNEY • TOKYO
ELSEVIER   Academic Press is an imprint of Elsevier

Academic Press is an imprint of Elsevier
525 B Street, Suite 1900, San Diego, CA 92101-4495, USA
30 Corporate Drive, Suite 400, Burlington, MA 01803, USA
32 Jamestown Road, London NW1 7BY, UK

First edition 2010

For information on all Academic Press publications
visit our website at elsevierdirect.com

ISBN: 978-0-12-384993-9
ISSN: 0076-6879

Printed and bound in United States of America
10   11   12   10 9 8 7 6 5 4 3 2 1

Contents

CONTRIBUTORS

Francisco Asturias
Department of Cell Biology, The Scripps Research Institute, La Jolla, California, USA

Mariah R. Baker
National Center for Macromolecular Imaging, Verna and Marrs McLean Department of Biochemistry and Molecular Biology, Baylor College of Medicine, Houston, Texas, USA

Matthew L. Baker
National Center for Macromolecular Imaging, Verna and Marrs McLean Department of Biochemistry and Molecular Biology, Baylor College of Medicine, Houston, Texas, USA

Martin Beck
European Molecular Biology Laboratory, Heidelberg, Germany

Edward J. Brignole
Department of Cell Biology, The Scripps Research Institute, La Jolla, California, USA

Bridget Carragher
National Resource for Automated Molecular Microscopy, Department of Cell Biology, The Scripps Research Institute, La Jolla, California, USA

Anchi Cheng
National Resource for Automated Molecular Microscopy, Department of Cell Biology, The Scripps Research Institute, La Jolla, California, USA

Frank DiMaio
Department of Biochemistry, University of Washington, Seattle, Washington, USA

Kenneth H. Downing
Life Sciences Division, Donner Laboratory, Lawrence Berkeley National Laboratory, Berkeley, California, USA

Nadav Elad
Department of Life Sciences, and The National Institute for Biotechnology in the Negev, Ben Gurion University of the Negev, Beer-Sheva, Israel

Andreas Engel
Department of Pharmacology, Case Western Reserve University, Cleveland, Ohio, USA

Friedrich Förster
Max-Planck Institute of Biochemistry, Department of Structural Biology, Martinsried, Germany

Lauren Fisher
The Scripps Research Institute, La Jolla, California, USA

Yoshinori Fujiyoshi
Department of Biophysics, Kyoto University, Oiwake, Kitashirakawa, Sakyo-ku, Kyoto, Japan

Jan Giesebrecht
Institut für medizinische Physik und Biophysik, Charité, Universitätsmedizin Berlin, Berlin, Germany

John Paul Glaves
Department of Biochemistry, School of Molecular and Systems Medicine, and National Institute for Nanotechnology, University of Alberta, Edmonton, Alberta, Canada

Bong-Gyoon Han
Life Sciences Division, Lawrence Berkeley National Laboratory, University of California, Berkeley, California, USA

Dorit Hanein
Sanford-Burnham Medical Research Institute, La Jolla, California, USA

Amber Herold
National Resource for Automated Molecular Microscopy, Department of Cell Biology, The Scripps Research Institute, La Jolla, California, USA

Richard K. Hite
Department of Cell Biology, Harvard Medical School, Boston, Massachusetts, USA

Eric Hou
National Resource for Automated Molecular Microscopy, Department of Cell Biology, The Scripps Research Institute, La Jolla, California, USA

Corey F. Hryc
National Center for Macromolecular Imaging, Verna and Marrs McLean Department of Biochemistry and Molecular Biology, Baylor College of Medicine, Houston, Texas, USA

Christopher Irving

National Resource for Automated Molecular Microscopy, Department of Cell Biology, The Scripps Research Institute, La Jolla, California, USA

Erica L. Jacovetty

National Resource for Automated Molecular Microscopy, Department of Cell Biology, The Scripps Research Institute, La Jolla, California, USA

Pick-Wei Lau

National Resource for Automated Molecular Microscopy, Department of Cell Biology, The Scripps Research Institute, La Jolla, California, USA

Catherine L. Lawson

Department of Chemistry and Chemical Biology and Research Collaboratory for Structural Bioinformatics, Rutgers, The State University of New Jersey, USA

Jun Liu

Department of Pathology and Laboratory Medicine, University of Texas Medical School at Houston, Houston, Texas, USA

Justus Loerke

Institut für medizinische Physik und Biophysik, Charité, Universitätsmedizin Berlin, Berlin, Germany

Dmitry Lyumkis

National Resource for Automated Molecular Microscopy, Department of Cell Biology, The Scripps Research Institute, La Jolla, California, USA

Asaf Mader

Department of Life Sciences, and The National Institute for Biotechnology in the Negev, Ben Gurion University of the Negev, Beer-Sheva, Israel

Ohad Medalia

Department of Life Sciences, and The National Institute for Biotechnology in the Negev, Ben Gurion University of the Negev, Beer-Sheva, Israel

Arne Moeller

National Resource for Automated Molecular Microscopy, Department of Cell Biology, The Scripps Research Institute, La Jolla, California, USA

Anke M. Mulder

National Resource for Automated Molecular Microscopy, Department of Cell Biology, The Scripps Research Institute, La Jolla, California, USA

Eva Nogales

Life Sciences Division, Donner Laboratory, and Life Sciences Division, Lawrence Berkeley National Laboratory, and Department of Molecular and Cell Biology, Howard Hughes Medical Institute, UC Berkeley, Berkeley, California, USA

Clinton S. Potter
National Resource for Automated Molecular Microscopy, Department of Cell Biology, The Scripps Research Institute, La Jolla, California, USA

James Pulokas
National Resource for Automated Molecular Microscopy, Department of Cell Biology, The Scripps Research Institute, La Jolla, California, USA

Joel D. Quispe
National Resource for Automated Molecular Microscopy, Department of Cell Biology, The Scripps Research Institute, La Jolla, California, USA

Andreas D. Schenk
Department of Cell Biology, Harvard Medical School, Boston, Massachusetts, USA

Christian M. T. Spahn
Institut für medizinische Physik und Biophysik, Charité, Universitätsmedizin Berlin, Berlin, Germany

Elizabeth Villa
Max-Planck Institute of Biochemistry, Department of Structural Biology, Martinsried, Germany

Niels Volkmann
Sanford-Burnham Medical Research Institute, La Jolla, California, USA

Neil R. Voss
National Resource for Automated Molecular Microscopy, Department of Cell Biology, The Scripps Research Institute, La Jolla, California, USA

Thomas Walz
Department of Cell Biology, and Howard Hughes Medical Institute, Harvard Medical School, Boston, Massachusetts, USA

Andrew Ward
The Scripps Research Institute, La Jolla, California, USA

Hanspeter Winkler
Institute of Molecular Biophysics, Florida State University, Tallahassee, Florida, USA

Elizabeth R. Wright
Department of Pediatrics, Emory University School of Medicine, Atlanta, Georgia, USA

Howard S. Young
Department of Biochemistry, School of Molecular and Systems Medicine, and National Institute for Nanotechnology, University of Alberta, Edmonton, Alberta, Canada

PREFACE

In this, the fifty-fourth year of *Methods in Enzymology*, we celebrate the discovery and initial characterization of thousands of individual enzymes, the sequencing of hundreds of whole genomes, and the structure determination of tens of thousands of proteins. In this context, the architectures of multienyzme/multiprotein complexes and their arrangement within cells have now come to the fore. A uniquely powerful method in this field is electron cryomicroscopy (cryo-EM), which in its broadest sense, is all those techniques that image cold samples in the electron microscope. Cryo-EM allows individual enzymes and proteins, macromolecular complexes, assemblies, cells, and even tissues to be observed in a "frozen-hydrated," near-native state free from the artifacts of fixation, dehydration, plastic-embedding, or staining typically used in traditional forms of EM (Chapter 3, Vol. 481). This series of volumes is therefore dedicated to a description of the instruments, samples, protocols, and analyses that belong to the growing field of cryo-EM.

The material could have been organized well by two schemes. The first is by the symmetry of the sample. Because the fundamental limitation in cryo-EM is radiation damage (Chapter 15, Vol. 481), a defining characteristic of each method is whether and how low-dose images of identical copies of the specimen can be averaged. In the most favorable case, large numbers of identical copies of the specimen of interest, like a single protein, can be purified and crystallized within thin "two-dimensional" crystals (Chapter 1, Vol. 481). In this case, truly *atomic* resolution reconstructions have been obtained through averaging very low dose images of millions of copies of the specimen (Chapter 11, Vol. 481; Chapter 4, Vol. 482; Chapters 5 and 6, Vol. 483). The next most favorable case is helical crystals, which present a range of views of the specimen within a single image (Chapter 2, Vol. 481 and Chapter 7, Vol. 483) and can also deliver atomically interpretable reconstructions, although through quite different data collection protocols and reconstruction mathematics (Chapters 5 and 6, Vol. 482). At an intermediate level of (60-fold) symmetry, icosahedral viruses have their own set of optimal imaging and reconstruction protocols, and are just now also reaching atomic interpretability (Chapters 7 and 14, Vol. 482). Less symmetric particles, such as many multienyzme/multiprotein complexes, invite yet another set of challenges and methods (Chapters 3, 5, and 6, Vol. 481; Chapters 8–10, Vol. 482). Many are conformationally heterogeneous, requiring that images of different particles be first classified and then averaged (Chapters 10 and 12, Vol. 482; Chapters 8 and 9, Vol. 483). Heterogeneity and the precision to which these images can be aligned have limited most such reconstructions to "sub-nanometer" resolution, where the folds of proteins are clear but not much more (Chapter 1, Vol. 483). Finally, the most challenging samples are those which are simply unique (Chapter 8,Vol. 481), eliminating any chance of improving the clarity of reconstructions through averaging. For these, tomographic methods are required (Chapter 12, Vol. 481; Chapter 13, Vol. 482), and only nanometer resolutions can be obtained (Chapters 10–13, Vol. 483).

But instead of organizing topics according to symmetry, following a wonderful historical perspective by David DeRosier (Historical Perspective,Vol. 481), I chose to order the topics in experimental sequence: Sample preparation and data collection/microscopy (Vol. 481); 3-D reconstruction (Vol. 482); and analyses and interpretation, including case studies (Vol. 483). This organization emphasizes how the relatedness of the mathematics (Chapter 1, Vol. 482), instrumentation (Chapters 10 and 14, Vol. 482), and methods (Chapter 15, Vol. 482; Chapter 9, Vol. 481) underlying all cryo-EM approaches allow practictioners to easily move between them. It further highlights how in a growing number of recent cases, the methods are being mixed (Chapter 13, Vol. 481), for instance, through the application of "single particle-like" approaches to "unbend" and average 2-D and helical crystals (Chapter 6, Vol. 482), but also average subvolumes within tomograms. Moreover, different samples are always more-or-less well-behaved, so the actual resolution achieved may be less than theoretically possible for a particular symmetry, or to the opposite effect; extensively known constraints may allow a more specific interpretation than usual for a given resolution (Chapters 2-4 and 6, Vol. 483). Nevertheless, within each section, the articles are ordered as much as possible according to the symmetry of the sample as described above (i.e. methods for preparing samples proceed from 2-D and helical crystals to sectioning of high-pressure-frozen tissues; Chapter 8, Vol. 481). The cryo-EM beginner with a new sample must then first recognize its symmetry and then identify the relevant chapters within each volume.

As a final note, our field has not yet reached a consensus on the placement of the prefix "cryo" and other details of the names of cryo-EM techniques. Thus, "cryo-electron microscopy" (CEM), "electron cryo-microscopy" (ECM), and "cryo-EM" should all be considered synonyms here. Likewise, "single particle reconstruction" (SPR) and "single particle analysis" (SPA) refer to a single technique, as do "cryo-electron tomography" (CET), "electron cryo-tomography" (ECT), and cryo-electron microscope tomography (cEMT).

GRANT J. JENSEN

METHODS IN ENZYMOLOGY

VOLUME XVI. Fast Reactions
Edited by KENNETH KUSTIN

VOLUME XVII. Metabolism of Amino Acids and Amines (Parts A and B)
Edited by HERBERT TABOR AND CELIA WHITE TABOR

VOLUME XVIII. Vitamins and Coenzymes (Parts A, B, and C)
Edited by DONALD B. MCCORMICK AND LEMUEL D. WRIGHT

VOLUME XIX. Proteolytic Enzymes
Edited by GERTRUDE E. PERLMANN AND LASZLO LORAND

VOLUME XX. Nucleic Acids and Protein Synthesis (Part C)
Edited by KIVIE MOLDAVE AND LAWRENCE GROSSMAN

VOLUME XXI. Nucleic Acids (Part D)
Edited by LAWRENCE GROSSMAN AND KIVIE MOLDAVE

VOLUME XXII. Enzyme Purification and Related Techniques
Edited by WILLIAM B. JAKOBY

VOLUME XXIII. Photosynthesis (Part A)
Edited by ANTHONY SAN PIETRO

VOLUME XXIV. Photosynthesis and Nitrogen Fixation (Part B)
Edited by ANTHONY SAN PIETRO

VOLUME XXV. Enzyme Structure (Part B)
Edited by C. H. W. HIRS AND SERGE N. TIMASHEFF

VOLUME XXVI. Enzyme Structure (Part C)
Edited by C. H. W. HIRS AND SERGE N. TIMASHEFF

VOLUME XXVII. Enzyme Structure (Part D)
Edited by C. H. W. HIRS AND SERGE N. TIMASHEFF

VOLUME XXVIII. Complex Carbohydrates (Part B)
Edited by VICTOR GINSBURG

VOLUME XXIX. Nucleic Acids and Protein Synthesis (Part E)
Edited by LAWRENCE GROSSMAN AND KIVIE MOLDAVE

VOLUME XXX. Nucleic Acids and Protein Synthesis (Part F)
Edited by KIVIE MOLDAVE AND LAWRENCE GROSSMAN

VOLUME XXXI. Biomembranes (Part A)
Edited by SIDNEY FLEISCHER AND LESTER PACKER

VOLUME XXXII. Biomembranes (Part B)
Edited by SIDNEY FLEISCHER AND LESTER PACKER

VOLUME XXXIII. Cumulative Subject Index Volumes I–XXX
Edited by MARTHA G. DENNIS AND EDWARD A. DENNIS

VOLUME XXXIV. Affinity Techniques (Enzyme Purification: Part B)
Edited by WILLIAM B. JAKOBY AND MEIR WILCHEK

VOLUME XXXV. Lipids (Part B)
Edited by JOHN M. LOWENSTEIN

VOLUME XXXVI. Hormone Action (Part A: Steroid Hormones)
Edited by BERT W. O'MALLEY AND JOEL G. HARDMAN

VOLUME XXXVII. Hormone Action (Part B: Peptide Hormones)
Edited by BERT W. O'MALLEY AND JOEL G. HARDMAN

VOLUME XXXVIII. Hormone Action (Part C: Cyclic Nucleotides)
Edited by JOEL G. HARDMAN AND BERT W. O'MALLEY

VOLUME XXXIX. Hormone Action (Part D: Isolated Cells, Tissues,
and Organ Systems)
Edited by JOEL G. HARDMAN AND BERT W. O'MALLEY

VOLUME XL. Hormone Action (Part E: Nuclear Structure and Function)
Edited by BERT W. O'MALLEY AND JOEL G. HARDMAN

VOLUME XLI. Carbohydrate Metabolism (Part B)
Edited by W. A. WOOD

VOLUME XLII. Carbohydrate Metabolism (Part C)
Edited by W. A. WOOD

VOLUME XLIII. Antibiotics
Edited by JOHN H. HASH

VOLUME XLIV. Immobilized Enzymes
Edited by KLAUS MOSBACH

VOLUME XLV. Proteolytic Enzymes (Part B)
Edited by LASZLO LORAND

VOLUME XLVI. Affinity Labeling
Edited by WILLIAM B. JAKOBY AND MEIR WILCHEK

VOLUME XLVII. Enzyme Structure (Part E)
Edited by C. H. W. HIRS AND SERGE N. TIMASHEFF

VOLUME XLVIII. Enzyme Structure (Part F)
Edited by C. H. W. HIRS AND SERGE N. TIMASHEFF

VOLUME XLIX. Enzyme Structure (Part G)
Edited by C. H. W. HIRS AND SERGE N. TIMASHEFF

VOLUME L. Complex Carbohydrates (Part C)
Edited by VICTOR GINSBURG

VOLUME LI. Purine and Pyrimidine Nucleotide Metabolism
Edited by PATRICIA A. HOFFEE AND MARY ELLEN JONES

VOLUME LII. Biomembranes (Part C: Biological Oxidations)
Edited by SIDNEY FLEISCHER AND LESTER PACKER

VOLUME 71. Lipids (Part C)
Edited by JOHN M. LOWENSTEIN

VOLUME 72. Lipids (Part D)
Edited by JOHN M. LOWENSTEIN

VOLUME 73. Immunochemical Techniques (Part B)
Edited by JOHN J. LANGONE AND HELEN VAN VUNAKIS

VOLUME 74. Immunochemical Techniques (Part C)
Edited by JOHN J. LANGONE AND HELEN VAN VUNAKIS

VOLUME 75. Cumulative Subject Index Volumes XXXI, XXXII, XXXIV–LX
Edited by EDWARD A. DENNIS AND MARTHA G. DENNIS

VOLUME 76. Hemoglobins
Edited by ERALDO ANTONINI, LUIGI ROSSI-BERNARDI, AND EMILIA CHIANCONE

VOLUME 77. Detoxication and Drug Metabolism
Edited by WILLIAM B. JAKOBY

VOLUME 78. Interferons (Part A)
Edited by SIDNEY PESTKA

VOLUME 79. Interferons (Part B)
Edited by SIDNEY PESTKA

VOLUME 80. Proteolytic Enzymes (Part C)
Edited by LASZLO LORAND

VOLUME 81. Biomembranes (Part H: Visual Pigments and Purple Membranes, I)
Edited by LESTER PACKER

VOLUME 82. Structural and Contractile Proteins (Part A: Extracellular Matrix)
Edited by LEON W. CUNNINGHAM AND DIXIE W. FREDERIKSEN

VOLUME 83. Complex Carbohydrates (Part D)
Edited by VICTOR GINSBURG

VOLUME 84. Immunochemical Techniques (Part D: Selected Immunoassays)
Edited by JOHN J. LANGONE AND HELEN VAN VUNAKIS

VOLUME 85. Structural and Contractile Proteins (Part B: The Contractile Apparatus and the Cytoskeleton)
Edited by DIXIE W. FREDERIKSEN AND LEON W. CUNNINGHAM

VOLUME 86. Prostaglandins and Arachidonate Metabolites
Edited by WILLIAM E. M. LANDS AND WILLIAM L. SMITH

VOLUME 87. Enzyme Kinetics and Mechanism (Part C: Intermediates, Stereo-chemistry, and Rate Studies)
Edited by DANIEL L. PURICH

VOLUME 88. Biomembranes (Part I: Visual Pigments and Purple Membranes, II)
Edited by LESTER PACKER

VOLUME 157. Biomembranes (Part Q: ATP-Driven Pumps and Related Transport: Calcium, Proton, and Potassium Pumps)
Edited by SIDNEY FLEISCHER AND BECCA FLEISCHER

VOLUME 158. Metalloproteins (Part A)
Edited by JAMES F. RIORDAN AND BERT L. VALLEE

VOLUME 159. Initiation and Termination of Cyclic Nucleotide Action
Edited by JACKIE D. CORBIN AND ROGER A. JOHNSON

VOLUME 160. Biomass (Part A: Cellulose and Hemicellulose)
Edited by WILLIS A. WOOD AND SCOTT T. KELLOGG

VOLUME 161. Biomass (Part B: Lignin, Pectin, and Chitin)
Edited by WILLIS A. WOOD AND SCOTT T. KELLOGG

VOLUME 162. Immunochemical Techniques (Part L: Chemotaxis and Inflammation)
Edited by GIOVANNI DI SABATO

VOLUME 163. Immunochemical Techniques (Part M: Chemotaxis and Inflammation)
Edited by GIOVANNI DI SABATO

VOLUME 164. Ribosomes
Edited by HARRY F. NOLLER, JR., AND KIVIE MOLDAVE

VOLUME 165. Microbial Toxins: Tools for Enzymology
Edited by SIDNEY HARSHMAN

VOLUME 166. Branched-Chain Amino Acids
Edited by ROBERT HARRIS AND JOHN R. SOKATCH

VOLUME 167. Cyanobacteria
Edited by LESTER PACKER AND ALEXANDER N. GLAZER

VOLUME 168. Hormone Action (Part K: Neuroendocrine Peptides)
Edited by P. MICHAEL CONN

VOLUME 169. Platelets: Receptors, Adhesion, Secretion (Part A)
Edited by JACEK HAWIGER

VOLUME 170. Nucleosomes
Edited by PAUL M. WASSARMAN AND ROGER D. KORNBERG

VOLUME 171. Biomembranes (Part R: Transport Theory: Cells and Model Membranes)
Edited by SIDNEY FLEISCHER AND BECCA FLEISCHER

VOLUME 172. Biomembranes (Part S: Transport: Membrane Isolation and Characterization)
Edited by SIDNEY FLEISCHER AND BECCA FLEISCHER

VOLUME 173. Biomembranes [Part T: Cellular and Subcellular Transport: Eukaryotic (Nonepithelial) Cells]
Edited by SIDNEY FLEISCHER AND BECCA FLEISCHER

VOLUME 174. Biomembranes [Part U: Cellular and Subcellular Transport: Eukaryotic (Nonepithelial) Cells]
Edited by SIDNEY FLEISCHER AND BECCA FLEISCHER

VOLUME 175. Cumulative Subject Index Volumes 135–139, 141–167

VOLUME 176. Nuclear Magnetic Resonance (Part A: Spectral Techniques and Dynamics)
Edited by NORMAN J. OPPENHEIMER AND THOMAS L. JAMES

VOLUME 177. Nuclear Magnetic Resonance (Part B: Structure and Mechanism)
Edited by NORMAN J. OPPENHEIMER AND THOMAS L. JAMES

VOLUME 178. Antibodies, Antigens, and Molecular Mimicry
Edited by JOHN J. LANGONE

VOLUME 179. Complex Carbohydrates (Part F)
Edited by VICTOR GINSBURG

VOLUME 180. RNA Processing (Part A: General Methods)
Edited by JAMES E. DAHLBERG AND JOHN N. ABELSON

VOLUME 181. RNA Processing (Part B: Specific Methods)
Edited by JAMES E. DAHLBERG AND JOHN N. ABELSON

VOLUME 182. Guide to Protein Purification
Edited by MURRAY P. DEUTSCHER

VOLUME 183. Molecular Evolution: Computer Analysis of Protein and Nucleic Acid Sequences
Edited by RUSSELL F. DOOLITTLE

VOLUME 184. Avidin-Biotin Technology
Edited by MEIR WILCHEK AND EDWARD A. BAYER

VOLUME 185. Gene Expression Technology
Edited by DAVID V. GOEDDEL

VOLUME 186. Oxygen Radicals in Biological Systems (Part B: Oxygen Radicals and Antioxidants)
Edited by LESTER PACKER AND ALEXANDER N. GLAZER

VOLUME 187. Arachidonate Related Lipid Mediators
Edited by ROBERT C. MURPHY AND FRANK A. FITZPATRICK

VOLUME 188. Hydrocarbons and Methylotrophy
Edited by MARY E. LIDSTROM

VOLUME 189. Retinoids (Part A: Molecular and Metabolic Aspects)
Edited by LESTER PACKER

VOLUME 399. Ubiquitin and Protein Degradation (Part B)
Edited by RAYMOND J. DESHAIES

VOLUME 400. Phase II Conjugation Enzymes and Transport Systems
Edited by HELMUT SIES AND LESTER PACKER

VOLUME 401. Glutathione Transferases and Gamma Glutamyl Transpeptidases
Edited by HELMUT SIES AND LESTER PACKER

VOLUME 402. Biological Mass Spectrometry
Edited by A. L. BURLINGAME

VOLUME 403. GTPases Regulating Membrane Targeting and Fusion
Edited by WILLIAM E. BALCH, CHANNING J. DER, AND ALAN HALL

VOLUME 404. GTPases Regulating Membrane Dynamics
Edited by WILLIAM E. BALCH, CHANNING J. DER, AND ALAN HALL

VOLUME 405. Mass Spectrometry: Modified Proteins and Glycoconjugates
Edited by A. L. BURLINGAME

VOLUME 406. Regulators and Effectors of Small GTPases: Rho Family
Edited by WILLIAM E. BALCH, CHANNING J. DER, AND ALAN HALL

VOLUME 407. Regulators and Effectors of Small GTPases: Ras Family
Edited by WILLIAM E. BALCH, CHANNING J. DER, AND ALAN HALL

VOLUME 408. DNA Repair (Part A)
Edited by JUDITH L. CAMPBELL AND PAUL MODRICH

VOLUME 409. DNA Repair (Part B)
Edited by JUDITH L. CAMPBELL AND PAUL MODRICH

VOLUME 410. DNA Microarrays (Part A: Array Platforms and Web-Bench Protocols)
Edited by ALAN KIMMEL AND BRIAN OLIVER

VOLUME 411. DNA Microarrays (Part B: Databases and Statistics)
Edited by ALAN KIMMEL AND BRIAN OLIVER

VOLUME 412. Amyloid, Prions, and Other Protein Aggregates (Part B)
Edited by INDU KHETERPAL AND RONALD WETZEL

VOLUME 413. Amyloid, Prions, and Other Protein Aggregates (Part C)
Edited by INDU KHETERPAL AND RONALD WETZEL

VOLUME 414. Measuring Biological Responses with Automated Microscopy
Edited by JAMES INGLESE

VOLUME 415. Glycobiology
Edited by MINORU FUKUDA

VOLUME 416. Glycomics
Edited by MINORU FUKUDA

VOLUME 417. Functional Glycomics
Edited by MINORU FUKUDA

VOLUME 418. Embryonic Stem Cells
Edited by IRINA KLIMANSKAYA AND ROBERT LANZA

VOLUME 419. Adult Stem Cells
Edited by IRINA KLIMANSKAYA AND ROBERT LANZA

VOLUME 420. Stem Cell Tools and Other Experimental Protocols
Edited by IRINA KLIMANSKAYA AND ROBERT LANZA

VOLUME 421. Advanced Bacterial Genetics: Use of Transposons and Phage for Genomic Engineering
Edited by KELLY T. HUGHES

VOLUME 422. Two-Component Signaling Systems, Part A
Edited by MELVIN I. SIMON, BRIAN R. CRANE, AND ALEXANDRINE CRANE

VOLUME 423. Two-Component Signaling Systems, Part B
Edited by MELVIN I. SIMON, BRIAN R. CRANE, AND ALEXANDRINE CRANE

VOLUME 424. RNA Editing
Edited by JONATHA M. GOTT

VOLUME 425. RNA Modification
Edited by JONATHA M. GOTT

VOLUME 426. Integrins
Edited by DAVID CHERESH

VOLUME 427. MicroRNA Methods
Edited by JOHN J. ROSSI

VOLUME 428. Osmosensing and Osmosignaling
Edited by HELMUT SIES AND DIETER HAUSSINGER

VOLUME 429. Translation Initiation: Extract Systems and Molecular Genetics
Edited by JON LORSCH

VOLUME 430. Translation Initiation: Reconstituted Systems and Biophysical Methods
Edited by JON LORSCH

VOLUME 431. Translation Initiation: Cell Biology, High-Throughput and Chemical-Based Approaches
Edited by JON LORSCH

VOLUME 432. Lipidomics and Bioactive Lipids: Mass-Spectrometry–Based Lipid Analysis
Edited by H. ALEX BROWN

CHAPTER ONE

ANALYSES OF SUBNANOMETER RESOLUTION CRYO-EM DENSITY MAPS

Matthew L. Baker,* Mariah R. Baker,* Corey F. Hryc,* and Frank DiMaio†

Contents

* National Center for Macromolecular Imaging, Verna and Marrs McLean Department of Biochemistry and Molecular Biology, Baylor College of Medicine, Houston, Texas, USA
† Department of Biochemistry, University of Washington, Seattle, Washington, USA

Methods in Enzymology, Volume 483
ISSN 0076-6879, DOI: 10.1016/S0076-6879(10)83001-0

1

Abstract

Today, electron cryomicroscopy (cryo-EM) can routinely achieve subnanometer resolutions of complex macromolecular assemblies. From a density map, one can extract key structural and functional information using a variety of computational analysis tools. At subnanometer resolution, these tools make it possible to isolate individual subunits, identify secondary structures, and accurately fit atomic models. With several cryo-EM studies achieving resolutions beyond 5 Å, computational modeling and feature recognition tools have been employed to construct backbone and atomic models of the protein components directly from a density map. In this chapter, we describe several common classes of computational tools that can be used to analyze and model subnanometer resolution reconstructions from cryo-EM. A general protocol for analyzing subnanometer resolution density maps is presented along with a full description of steps used in analyzing the 4.3 Å resolution structure of Mm-cpn.

1. Introduction

Electron microscopy has played an increasingly important role in understanding the structure and function of macromolecular assemblies that contribute to numerous biological processes. It offers an advantage over other structural techniques, like X-ray crystallography, by imaging macromolecular assemblies in near-native conditions (Baumeister and Steven, 2000; Frank, 2002). Even at non-atomic resolutions, three-dimensional (3D) reconstructions (volumetric density maps) from electron microscopy can describe the size, shape, and composition of a macromolecular assembly.

In 1975, the first subnanometer resolution electron microscopy data was derived from regularly arrayed 2D crystals of bacteriorhodopsin by Henderson and Unwin (1975). At 7 Å resolution, the seven transmembrane helices of bacteriorhodopsin were clearly visible. Fifteen years later, Henderson *et al.* (1990) reported the first atomic model constructed directly from an electron microscopy density map. Another milestone was achieved in 2005, as water molecules were clearly resolved in the 1.9 Å resolution structure of aquaporin-0 (Gonen *et al.*, 2005).

In single-particle electron cryomicroscopy (cryo–EM), a macromolecular assembly does not need to form a regular array. Rather, images of randomly oriented particles are processed to generate a 3D density map

(Cong and Ludtke, 2010). Early 3D reconstructions achieved relatively low resolutions due to several factors including small data sets, sample heterogeneity, and technical limitations of the microscopes. In 1997, a significant milestone was achieved in single-particle cryo-EM; two reconstructions of the Hepatitis B virus obtained subnanometer resolutions (9 and 7.4 Å resolution; Böttcher *et al.*, 1997; Conway *et al.*, 1997). From these density maps, it was possible for the first time to clearly identify the core capsid protein and visualize rod-like structures corresponding to α-helices. Using the connectivity of the helices, the overall fold of the 183 amino acid core capsid protein was proposed. Several years later, a reconstruction of the rice dwarf virus (6.8 Å resolution) was the first cryo-EM reconstruction to clearly resolve β-sheets as flat planes of density (Zhou *et al.*, 2001). Like the density maps from the Hepatitis B reconstructions, this resolution allowed for the description of the protein fold, though no models were constructed. Subsequent atomic models confirmed the proposed folds for both Hepatitis B (Wynne *et al.*, 1999) and rice dwarf virus (Nakagawa *et al.*, 2003). Today, technical advances in computing, specimen preparation and data acquisition have made it possible to routinely achieve subnanometer resolutions on a wide variety of specimens with single-particle cryo-EM (Chiu *et al.*, 2005).

A decade after the first subnanometer resolution cryo-EM structures, another milestone was reached; the first near-atomic resolution structures were produced by single-particle cryo-EM. In reconstructions of rotavirus (3.88 Å; Zhang *et al.*, 2008), GroEL (4.2 Å; Ludtke *et al.*, 2008), cytoplasmic polyhedrosis virus (4.0 Å; Yu *et al.*, 2008), and bacteriophage ε15 (4.5 Å; Jiang *et al.*, 2008), the pitch of α-helices and the separation of β-strands were visualized. Though these structures did not have the resolution to utilize standard X-ray crystallographic tools for model construction (typically starting at ~3.5 Å resolution), *de novo* Cα backbone models were built from the cryo-EM density maps of cytoplasmic polyhedrosis virus, GroEL, and bacteriophage ε15 using a combination of computational and geometric tools (Baker *et al.*, 2007, Ju *et al.*, 2007). The *de novo* models built directly from these density maps relied almost entirely on visual interpretation of the density and manual structure assignment. Several recent state-of-the-art reconstructions have now resolved sidechain densities and allowed for the construction of complete atomic models directly from the density map (Cong *et al.*, 2010; Yu *et al.*, 2008; Zhang *et al.*, 2010).

2. FEATURES IN A SUBNANOMETER RESOLUTION DENSITY MAP

Regardless of the resolution, computational tools are critical in analyzing, interpreting, and annotating structural information in cryo-EM density maps. As such, a number of specialized computational tools have been developed. However, before describing these tools and the general protocol for analyzing a cryo-EM density map, it is important to establish an understanding of the features visible in a density map as a function of resolution (Fig. 1.1). Apparent in many of the density maps deposited in the EMDB (http://emdatabank.org), subnanometer resolution structures have distinct boundaries that allow for subunits and individual domains to be identified and segmented from the entire map (Fig. 1.1A). These well-defined densities help to accurately fit known structural models to a density map (Fig. 1.1B). α-Helices appear as long rod-like densities, and β-sheets appear as thin, continuous planes at ~ 8 Å resolution (Fig. 1.1C). At slightly higher resolutions, connectivity between the secondary structure elements (SSEs) becomes evident (Fig. 1.1D). α-Helices begin to develop features and the pitch becomes visible at ~ 5 Å resolution. Beyond 4.5 Å resolution, the thin, flat planes of β-sheets become broken, as individual strands are resolved (Fig. 1.1E). By 4 Å resolution, sidechain densities become recognizable (Fig. 1.1F) and a relatively unambiguous trace of a protein backbone can be seen (Fig. 1.1G).

3. Tools: Analyzing a Subnanometer Resolution Density Map

While subnanometer resolutions span a wide range of detectable features, tools for analyzing their structure can be grouped into two classes: feature recognition and data integration. In principle, many of these tools are not restricted to subnanometer resolutions, although higher resolution features do provide significant advantages in validation. Feature recognition tools, however, are generally resolution-specific. The following briefly describes - some "standard" tools and techniques for analyzing subnanometer resolution cryo-EM density maps (Table 1.1).

3.1. Fitting atomic models

Perhaps the most common method for analyzing a subnanometer resolution density map is fitting a known atomic model into a density map. There are a number of different approaches to fit atomic models to the density map, though an exhaustive rotational and translational search of the model within the map is generally used (reviewed in Rossmann *et al.*, 2005). "Rigid-body" fitting attempts to identify the maximum overlap of the model with the density and minimize the amount of model unaccounted for within the density. Various fitting programs report scores differently, so it is important to consult the program documentation, as well as visually inspect the fit of a model to the density map. Aside from program-specific fitting scores, tools

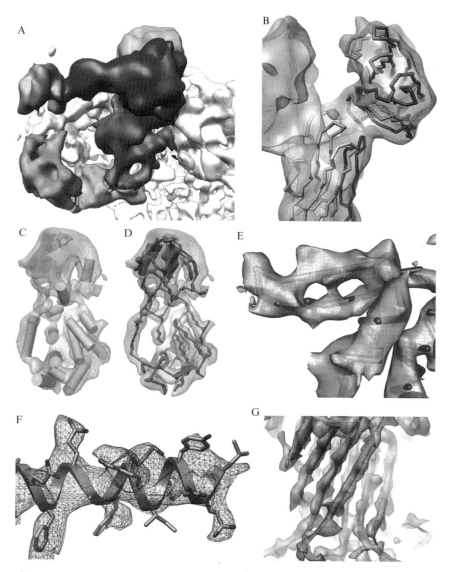

Figure 1.1 Features at subnanometer resolutions. A gallery of structural features from cryo-EM reconstructions is shown. (A) Domains in the clamp region of the 9.5 Å resolution reconstruction of RyR1 can be observed (Serysheva *et al.*, 2008). (B) The atomic models of VP5★ (lower left) and VP8★ (upper right) are fit to the density map corresponding to the VP4 spikes in the 9.5 Å resolution structure of rotavirus (Li *et al.*, 2009). (C) At slightly higher resolutions, secondary structures (α-helices are depicted as cylinders and β-sheets are depicted as planes) can be clearly seen in the capsid proteins of rice dwarf virus at 6.8 Å resolution (Zhou *et al.*, 2001). (D) Around this resolution, possible connections between secondary structure elements can be identified computationally using density skeletonization (red), again as seen in the 6.8 Å resolution rice dwarf virus capsid protein. (E) Increasing resolution reveals the separation of β-strands in GroEL at 4.2 Å resolution (Ludtke *et al.*, 2008). (F) Large, bulky sidechains begin to appear in TriC reconstruction at 4.0 Å resolution (Cong *et al.*, 2010). (G) An unambiguous backbone is apparent in VP6 of rotavirus at ∼3.8 Å resolution (Zhang *et al.*, 2008). (See Color Insert.)

Table 1.1   Programs for analyzing subnanometer resolution cryo-EM density maps

Segmentation	*Manual*
	Amira (Visage Imaging, Gmbh)
	Avizo (VSG, France)
	Chimera (Pettersen *et al.*, 2004)
	Automatic
	CoDiv (Volkmann, 2002)
	EMAN (Ludtke *et al.*, 1999; Tang *et al.*, 2007)
	Segger (Pintilie *et al.*, 2010)
	VolRover (Baker *et al.*, 2006b)
Fitting atomic models	*Rigid body*
	Chimera (Pettersen *et al.*, 2004)
	CoFi (Volkmann and Hanein, 1999)
	Coot (Emsley and Cowtan, 2004)
	DockEM (Roseman, 2000)
	EMFit (Rossmann, 2000)
	Foldhunter (Jiang *et al.*, 2001)
	Mod-EM (Topf *et al.*, 2005)
	O (Jones *et al.*, 1991)
	Situs (Wriggers *et al.*, 1999)
	UROX (Siebert and Navaza, 2009)
	Flexible
	DireX (Schröder *et al.*, 2007)
	Flex-EM (Topf *et al.*, 2008)
	MDFF (Trabuco *et al.*, 2009)
	NMFF (Tama *et al.*, 2004)
	NORMA (Suhre *et al.*, 2006)
	Yup.scx (Tan *et al.*, 2008)
	Situs (Rusu *et al.*, 2008)
	Validation
	FH-stat (Serysheva *et al.*, 2005)
Secondary structure	*Helixhunter* (Jiang *et al.*, 2001)
identification	Sheetminer/Sheetracer (Kong and Ma, 2003;
	Kong *et al.*, 2004)
	SSEHunter (Baker *et al.*, 2007)
Modeling	EM-IMO (Zhu *et al.*, 2010)
	Gorgon (http://gorgon.wustl.edu)
	Modeller (Topf *et al.*, 2005, 2006)
	Rosetta (Baker *et al.*, 2006a; DiMaio *et al.*, 2009)
Visualization	Amira (Visage Imaging, Gmbh)
	Avizo (VSG, France)
	Chimera (Pettersen *et al.*, 2004)
	PyMol (DeLano Scientific LLC, USA)
	VMD (Humphrey *et al.*, 1996)

Common computational tools used in the analysis of subnanometer resolution cryo-EM density maps are listed. Numerous other algorithms have also been published, though only currently downloadable tools are listed.

are available for independently reporting the quality of fit for a model within a density map (Serysheva *et al.*, 2005; Volkmann, 2009).

In addition to fitting an atomic model as a "rigid-body" within the density map, atomic models can be morphed, or flexibly fit, into the density map (Schröder *et al.*, 2007; Suhre *et al.*, 2006; Tama *et al.*, 2004; Tan *et al.*, 2008; Topf *et al.*, 2008; Trabuco *et al.*, 2009). In these types of programs an atomic model is allowed to relax or bend at certain points to better fit the density map. This is particularly useful when fitting structures of different conformations or homologous structures.

3.2. Constrained modeling with cryo-EM density

If a structure for one or more of the protein subunits in the assembly is not known, computational modeling approaches can be used to generate homology models or *ab initio* models for proteins. In these cases, the cryo-EM density map can be used directly to facilitate the construction and evaluation of a protein/domain structural model (Baker *et al.*, 2006a; DiMaio *et al.*, 2009; Topf *et al.*, 2005, 2006; Zhu *et al.*, 2010). In a constrained comparative modeling approach, an initial sequence–structure alignment is allowed to evolve, simultaneously improving the homology model and its fit to the density map (Topf *et al.*, 2005, 2006). This type of approach can also be used to improve local regions of a model within the density map. When a template structure is not known, constrained *ab initio* modeling may be used to build domains or small proteins (Baker *et al.*, 2006a). In this approach, a gallery of models are built computationally and then later evaluated by their fit to the density map. The resolution of the density map is key in determining the accuracy of the models (Topf *et al.*, 2005). While not restricted to subnanometer resolutions, models built within a subnanometer resolution density map will likely have the correct fold, though atom placement may only be approximate.

3.3. Extracting protein subunits from a density map

Like fitting atomic models to a density map, density segmentation, the process of identifying and isolating a single protein or domain from the cryo-EM density map, is not exclusive to subnanometer resolutions. However, at subnanometer resolutions, cryo-EM density maps move from being "blob-like" to having distinct features, including sharp drop-offs in the density that characterizes the edges and boundaries between subunits (Chiu *et al.*, 2005). The actual process of segmenting a subunit can be done using interactive tools commonly found in visualization programs such as Chimera (Pettersen *et al.*, 2004), Amira (Visage Imaging, GmbH), and Avizo (VSG, France) or computational techniques such as those based on the watershed transform (Ludtke *et al.*, 1999; Pintilie *et al.*, 2010; Volkmann, 2002).

3.4. Secondary structure identification

In cryo-EM, feature recognition tools have been typically used to identify SSEs within a subnanometer resolution density map. At subnanometer resolutions, helices are clearly resolved as long cylinders with relatively high density. Helixhunter (Jiang *et al.*, 2001) was the first tool designed to computationally detect α-helices in a density map using a simple cross-correlation search with a prototypical helix template over three translational and two rotational degrees of freedom. β-Sheets, which resemble thin planes, are more diverse in their structure and not amenable to correlation-based approaches. Rather, morphological analysis of the density was used to first localize β-sheets and strands in subnanometer resolution density maps (Kong and Ma, 2003; Kong *et al.*, 2004). Later developments lead to SSEHunter, a single tool to detect both α-helices and β-sheets (Baker *et al.*, 2007). With SSEHunter, the helix correlation routine is paired with a local geometry analysis and a density skeleton to detect SSEs. With any of these feature recognition tools, it is advisable to operate on segmented density maps rather than the entire map when detecting SSEs.

3.5. *De novo* modeling

In conjunction with development of SSEHunter, a new density skeletonization routine (Ju *et al.*, 2007) was developed that preserves both the features and the topology in a density map. This density skeleton provided clear connections between observable SSEs. Coupling density skeletonization with the aforementioned feature recognition tools has given rise to a new protocol for *de novo* structural modeling (Jiang *et al.*, 2008; Ludtke *et al.*, 2008). The *de novo* method attempts to construct a model directly from a density map without the aid of an existing structural template. Borrowing methods for constructing atomic models from X-ray crystallographic density maps, SSEs in the sequence and the density map are correlated, providing initiation points for model construction (Abeysinghe *et al.*, 2008). This sequence-to-structure correspondence assigns residue positions to SSEs in the map.

Once SSE anchor points are established, models can be constructed using various X-ray crystallographic model building toolkits, such as O (Jones *et al.*, 1991) and Coot (Emsley and Cowtan, 2004), both of which offer tools for placing atoms within a cryo-EM density map. Gorgon (http://gorgon.wustl.edu), a molecular modeling toolkit tailored to subnanometer resolution cryo-EM density maps, provides a comprehensive suite of utilities to analyze subnanometer resolution density maps, including the ability to generate secondary structure assignments, a sequence-to-structure correspondence routine and Cα model construction utilities.

4. Protocol: From Density Map to Atomic Model

With an extensive software library (Table 1.1), analyzing a subnanometer resolution density map can be an overwhelming task. Simply identifying the proper tools and understanding the expected results can be
difficult. Figure 1.2 provides a simplified flowchart describing the assorted

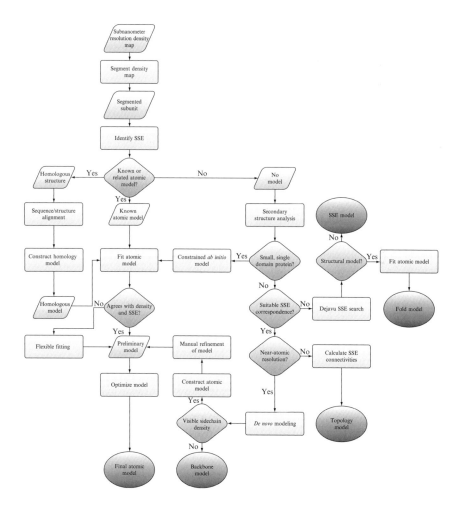

Figure 1.2 Analyzing a subnanometer resolution cryo-EM density map. A general
scheme for analyzing subnanometer resolution cryo-EM density maps is depicted.
Different projects may take advantage of additional information during the analysis
process and thus deviate from the overall scheme.

paths and decisions one may encounter in analyzing a subnanometer resolution density map.

In the following sections, we describe a general protocol (based on Fig. 1.2) for analyzing a subnanometer resolution density map. In a subsequent section, we also detail the analysis of the 4.3-Å resolution structure of *Methanococcus maripaludis* chaperonin (Mm-cpn), a group II chaperonin. It should be noted that this protocol is not restricted to single-particle cryo-EM and can be used to analyze any type of subnanometer resolution density map.

The steps described in this protocol illustrate a complete pathway for constructing an atomic model from a cryo-EM density map; however, most of the steps can be used independently depending on numerous constraints, such as resolution and available structural information. As such, consulting the various user-guides and documentation for the individual software packages will help the user to optimally select parameters for each particular project.

In this protocol, several image processing, visualization and modeling software packages are utilized, including Coot (Emsley and Cowtan, 2004), EMAN/EMAN2 (Ludtke *et al.*, 2005; Tang *et al.*, 2007), Gorgon (http://gorgon.wustl.edu), Rosetta (Bradley *et al.*, 2005), and UCSF Chimera (Pettersen *et al.*, 2004). Other programs may be substituted, though their exact usage, including inputs and outputs, will vary. Additionally, the amount of time and computational resources required by each of these tools will vary considerably based on the project and experience of the user.

4.1. Segmentation

To begin modeling an individual protein subunit, one subunit/domain must be extracted from the macromolecular assembly. Generally, individual protein densities within subnanometer resolution reconstructions have relatively sharp fall-offs at their boundaries. Identifying the boundaries between subunits can be accomplished using a variety of different approaches. Manual segmentation requires the user to visually identify and demarcate the subunits. By adjusting the isosurface display threshold in visualization programs, such as Chimera and Gorgon, these boundaries can usually be seen clearly. The user can then create a mask as either a set of 2D slices or a single 3D volume, or "erase" spurious density to produce an initial segmentation of a protein subunit. The initial segmentation is usually relatively crude to avoid removing too much density but still adequate for analysis. Once all subunits have been segmented, the user can then check to see that all density has been assigned a segment and no overlapping segments are present. This may require several iterations to improve the initial segmentation. Alternatively, automated approaches like EMAN's segment3d (Ludtke *et al.*, 1999) and Segger (Pintilie *et al.*, 2010) will cluster every voxel in a density map based on a set of user-provided parameters (e.g., number of subunits, symmetry, *etc.*). Visual inspection and manual adjustment of the subunits may be required.

4.2. Identifying secondary structure elements

Once the map has been segmented, SSEs are then identified using SSE-Hunter (Baker *et al.*, 2007). This step requires a cubic density map with an even number of voxels. Additionally, to maximize compatibility with other modeling programs, the origin of the segmented subunit should be reset to zero. To accomplish this, *proc3d* from EMAN or *e2proc3d.py* from EMAN2 can be used with the "clip" and "origin" options.

SSEHunter can be executed in three different ways: (1) directly from the command line (*ssehunter3.py* in EMAN and *e2ssehunter.py* in EMAN2), (2) as a plug-in to UCSF's Chimera, or (3) within Gorgon. The first two options require the installation of EMAN or EMAN2, while Gorgon's version of SSEHunter is built-in. In each of these cases, the user is required to provide the resolution in Å, the sampling of the map (Å/pixel) and a threshold corresponding to the highest isosurface value at which all density of the segmented subunit appears to be connected (obtained visually, ~ 2–4σ above the mean density).

Map sizes between 48^3 and 160^3 generally require less than 15 min to run on a modern desktop and return a set of pseudoatoms, typically corresponding to $\sim 50\%$ of the total number of amino acids represented in the segmented volume. Encoded in the B-factor column of the SSEHunter PDB file are the per-pseudoatom SSEHunter scores, which range between -3 and 3. These values represent the likelihood of a density region to be either α-helix (0 -3) or β-sheet (-3 -0). In Gorgon, the pseudoatoms are automatically colored from blue (-3, β-sheet) to white (0) to red, (3, α-helix) where the intensity of the color reflects the score and confidence of the prediction.

In addition to these pseudoatoms, a density skeleton is calculated with SSEHunter. As mentioned previously, this skeleton is a simplified geometrical representation of the density map that preserves both features and topology. Gorgon offers an improved threshold-free, grayscale density skeleton that can be substituted place of the SSEHunter skeleton in later steps for identifying the connectivity between SSEs.

4.3. Secondary structure annotation

Next, pseudoatoms of similar values are grouped into their respective SSEs using SSEBuilder (Baker *et al.*, 2007; Fig. 1.3). Like SSEHunter, SSE-Builder can be accessed from within Chimera or Gorgon. In SSEBuilder,

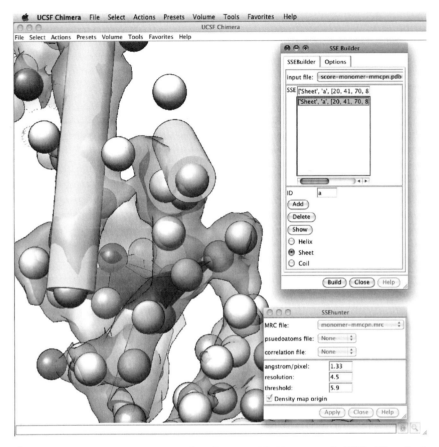

Figure 1.3 Secondary structure identification. SSEHunter and SSEBuilder, both EMAN programs, can be run as plug-ins to UCSF's Chimera. The results for the apical domain of Mm-cpn are shown: red spheres represent helix like regions and the blue spheres represent sheet like regions. Regions of similar scoring pseudoatoms from SSEHunter are grouped and built using SSEBuilder. Helices are depicted as cylinders and sheets are depicted as planes. (See Color Insert.)

pseudoatoms with similar scores are manually grouped into individual SSEs and then automatically constructed as VRML objects. The process continues until all visible SSEs are assigned. When selecting SSEs, generally, the minimum size is three pseudoatoms for an α-helix and five pseudoatoms for a β-sheet. False positives may occur; therefore, only groups of pseudoatoms that resemble SSEs in the density (α-helices appear as long cylinders, β-sheets resemble thin, curved surfaces) should be annotated.

4.4. Structural homologues from sequence

Until this point, analysis of the density map has focused on observing features within the density map itself. However, it is possible that an atomic model for one or more of the protein components in the density map is known. This may be a single protein, a domain from the macromolecular assembly or a related structure.

While it is relatively straightforward to search and retrieve a protein or domain with a known structure from the Protein Data Bank (http://pdb.org), identifying and building structural models from related structures is more complicated. Homologous structures can be identified using a number of sequence-based search methods including BLAST (Altschul *et al.*, 1990) and FASTA (Lipman and Pearson, 1985). These types of searches generally return a sequence alignment between the sequence of interest and a sequence with a known structure from which a homologous model can be constructed. A number of tools are available for this procedure, ranging from the fairly automated Swiss-Model web service (Arnold *et al.*, 2006) to more flexible modeling suites like Modeller (Sali *et al.*, 1995) and Rosetta (Bradley *et al.*, 2005).

In addition to these tools, there are several web-based tools that integrate the search and model building steps including Phyre (Kelley and Sternberg, 2009), Fugue (Shi *et al.*, 2001), and 3D Jigsaw (Bates *et al.*, 2001). Simplifying the process even further, meta-prediction servers, like BioInfoBank (Ginalski *et al.*, 2003), provide a convenient way to submit a sequence to multiple prediction servers and view the results. Regardless of the method or service used, fairly reliable atomic models can be produced when a suitable template structure is identified. As each method reports potential models differently, it is important to consult the documentation for each of the tools in determining the validity of the model.

4.5. Identifying structural homologues from SSEs

If structural homologues cannot be identified from the sequence, it is still possible to detect homologues based on the locations and orientations of the SSEs (Baker *et al.*, 2003, 2005). The identified SSEs can be used as inputs into two structural similarity search programs, DejaVu (Kleywegt and Jones, 1997) and COSEC (Mizuguchi and Go, 1995). Results obtained from SSEHunter and SSEBuilder are compatible with either program; an additional format conversion utility is provided with EMAN (*dejavu2sse.py*). While a sequence-based search will return a sequence alignment, a structure-based search does not. Rather, a structure-based search returns a list of possible folds that match a set of SSEs regardless of sequence.

4.6. Fitting atomic models in cryo-EM density maps

Once a suitable atomic model has been found or constructed, the model can be fit into the density map with any of the aforementioned tools. Fitting can be done either as a rigid body or flexibly fit, where the model is allowed to morph to better fit the density. To avoid mis- or overinterpretation, it is important to note the features in the density map. The shape and overall fold are likely to be recognizable in the segmented density map at subnanometer resolutions. More specifically, the atomic model should correspond well to the observed SSEs. Independent validation tools (*fh-stat.py* in EMAN; Serysheva *et al.*, 2005) and confidence intervals (Volkmann, 2009) may also be used to assess the statistical likelihood of the model positions in the map.

4.7. Predicting SSEs from sequence

When no known or homologous structure is available, it is still possible to construct structural models for individual proteins or subunits *de novo*. In this case, a combination of secondary structure prediction and feature recognition can assign sequence elements, such as helices and strands, to SSEs identified within the density map, a process termed SSE correspondence (Abeysinghe *et al.*, 2008; Ludtke *et al.*, 2008).

Having identified the positions of the SSEs within a segmented density map, the next step is to define the SSEs within the sequence of interest. A number of web-based programs can be used to predict secondary structure including SSPro (Pollastri *et al.*, 2002), JPred (Cole *et al.*, 2008), and PsiPred (McGuffin *et al.*, 2000). These programs generally provide a secondary structure assignment and confidence score to each amino acid. Due to errors in prediction, a consensus alignment built from multiple predictions may be better than the results from a single secondary structure prediction server. For convenience, Gorgon contains a tool that will remotely run the sequence prediction, retrieve the predictions, and format the results for the subsequent SSE correspondence routine.

4.8. SSE correspondence

Once SSEs have been identified in both the sequence and density map, a correspondence search can be preformed within Gorgon (Fig. 1.4A). This step provides the initial anchor points to place Cα atoms and construct a protein backbone. Determining the SSE correspondence requires four inputs: helix and sheet locations produced by SSEBuilder, the cryo-EM skeleton from SSEHunter or Gorgon, and the sequence/prediction from the previous step. Once entered, a sequence–structure correspondence is calculated and displayed graphically. In Gorgon, the SSE correspondence results are shown as a list ranked in order from best to worst correspondence.

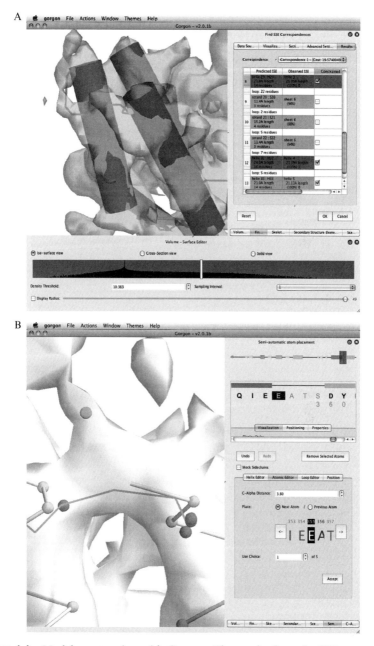

Figure 1.4 Model construction with Gorgon. The results from the SSE correspondence search on the apical domain of the 4.2-Å resolution structure of GroEL in Gorgon are shown in (A). Helices are shown as cylinders, while sheets are shown as planes. Potential connectivity is depicted as solid lines. A corresponding color scheme for these elements is shown in the SSE correspondence window on the right. (B) Gorgon contains several methods for assigning atoms to the density in the semi-automated atom placement tool. The atomic editor function illustrates the addition of Cα atoms along a density skeleton (not shown). The user can cycle through the possible locations, select the desired position and proceed to the next residue.

To assess individual correspondences, the lengths of the SSEHunter/SSE-Builder helices are compared to the sequence-predicted helices. Ideally, the lengths of correctly matched SSEs should not differ by more than three amino acids. In a correspondence, one or more SSEs may have been incorrectly assigned. If this occurs, the user may constrain the correct individual SSE correspondences by selecting only correctly paired SSEs and re-running the correspondence search. This process is relatively quick (<5 s for most cases) and can be done repeatedly until the user is satisfied with the correspondence.

4.9. Cα placement

From the previous step, an initial topology for the protein structure is established. Starting from this topology assignment, placement of the Cα backbone atoms begins with helices, followed by sheets and loops. The following steps are described using Gorgon.

4.10. Assigning Cα positions in helices

Based on the chosen SSE correspondence, Cα positions of the helices are registered with the corresponding sequence. This process and the following modeling steps can be done with the "Semi-Automatic Atom Placement" tool found in Gorgon. Possible errors in the SSE correspondence may require the user to adjust helix length and directionality when assigning the helix residues. Helices are initially represented as cylinders; manual adjustment of the helix position may be necessary to best fit the density. Bulky sidechains in the helices provide visual cues and help anchor the position and pitch of the helix within the density (Jiang *et al.*, 2008; Ludtke *et al.*, 2008; Yu *et al.*, 2008). Again, Gorgon contains a variety of mechanisms for optimizing helix position, including an option to show relative sidechain size at Cα atom positions. Once the assignment of helices is complete, strands and loops can be assigned.

4.11. Assigning Cα positions in sheets and loops

For the purposes of modeling a Cα backbone, β-strands are treated as loops. Extending from the assigned residues in the α-helices, the remaining residues can be assigned with two unique options found in Gorgon's "Semi-Automatic Atom Placement" tool. With the "Atomic Editor" the last assigned residue before an unassigned section of sequence is selected. The possible positions for the next unassigned amino acid are shown in the density map along the skeleton for which the Cα–Cα distance is satisfied (Fig. 1.4B). The user then interactively selects a position for the unassigned amino acid. The model is updated and possible positions for the next amino acid are shown. This process continues until a previously assigned Cα is joined. Alternatively, loops may be assigned using the "Loop Editor." With the endpoints of the loops selected, the "Loop Editor" allows the user to sketch out the approximate path through the density. Unlike the "Atomic Editor," no Cα–Cα distance constraints are enforced, though Gorgon provides a convenient way to visualize Cα–Cα distances. Red bond distances are too long, blue bonds are too short, and white bonds are approximately the correct length (3.8 ± 0.5 Å). To aid in the placement of Cα atoms, the density skeleton, calculated by SSEHunter or in Gorgon, can provide possible paths between the SSEs. Using these two techniques, Cα atoms are placed for any unassigned amino acids.

4.12. Fixing an atomic model

In models utilizing a known or homologous structure, adjustment of the Cα positions begins with the SSEs. Entire SSEs are first moved to register with the corresponding features in the density map using modeling programs such as Coot, Gorgon, or Chimera. Once the SSEs of the model have been fit to the density map, the remaining Cα atoms are moved individually to best fit the density. It is possible that some portions of the atomic model structure may be absent. If this is the case, the previous steps can be used to build any missing residues before proceeding.

4.13. Cα optimization

After all Cα atoms have been assigned, the next step is to adjust the atom positions to optimally fit the density while maintaining reasonable Cα–Cα bond distances (~3.8 Å) and angles (~60–120° between three consecutive Cαs), proper secondary structure features and no atom/bond clashes. Optimization of Cα positions begins with helices; the pitch of the model helices should register well with the pitch of the helix observed in the density. In the event that β-strands are resolved, the distance between neighboring strand Cα atoms should be between 4.5 and 5 Å. Once Cα positions in SSEs have been optimized Cα positions in the loops can be adjusted.

4.14. Building a macromolecular model

Following the initial construction of a Cα model, the models are then placed back into the context of the full cryo-EM density map with all other models. As previously mentioned, a number of fitting routines may be used for this step.

At this point, the Cα models are assessed by visually examining how well they fit into the density map. A "good model" will occupy the entire subunit density map and have no clashes with neighboring subunits. Residues in neighboring subunit models should also be readjusted so that the minimum distance between any Cα atoms is ∼4.5 Å. The last two steps will likely be iterated multiple times to improve the model until no inter- and intrasubunit clashes are evident and the models account for all the density in the reconstruction. The final refined backbone model is then saved as a PDB file.

4.15. Map rescaling

To enhance the high-resolution information in the density maps, such as sidechains, for subsequent map building, the density map can be rescaled in Fourier space using the Cα backbone model (Fernández *et al.*, 2008; Zhang *et al.*, 2010). The model is first blurred to approximately the same resolution as the original density map. Structure factors can be calculated from this blurred map and applied to the original density map in EMAN (*proc3d* or *e2proc3d.py* in EMAN2). A low-pass filter is then applied to the rescaled map at the calculated resolution and normalized so that the positive density values are between 0 and 1.

4.16. Cα to atomic model

An approximate atomic model can be reconstructed from the Cα model using a Cα-to-backbone builder such as the web-based SABBAC (Maupetit *et al.*, 2006). It is likely that the atomic model produced will have some missing or unassigned residues. If residues are missing, they can be manually added and fit to the density using X-ray crystallographic modeling packages like Coot.

It is important to note that building atomic models may not be possible even at near-atomic resolutions. Features in the map dictate the possibility and level of accuracy in model construction. Further model optimization and refinement requires a significant number of sidechain densities to be recognizable.

4.17. Model optimization

At near-atomic resolutions, all sidechains larger than Valine should be evident in the density map though this is generally not the case in practice. Therefore, precedence is typically given to mainchain atom positions within the density. However, sidechain density associated with positively charged and aromatic amino acids are discernable more frequently than other amino acids (Cong *et al.*, 2010; Kimura *et al.*, 1997; Zhang *et al.*, 2010) and thus provide landmarks for optimizing the model in the density map.

Conversely, Proline and Glycine are almost always marked by weak or broken density, though this can also be used to register the model to the density map as well.

In model optimization, small stretches of residues in the atomic model are adjusted to better fit the density map while maintaining good stereo-chemistry and enforcing sidechain and mainchain restraints. This process is generally interactive and begins in well-resolved regions such as α-helices, where the pitch of the helix and large, positively charged sidechains provide sufficient points to anchor the placement of atoms. Real-space refinement options found in computational modeling software, like Coot, allow the user to manually adjust atom positions while maintaining realistic biochemical properties.

Additionally, sidechain positions can be optimized using a rotamer search such that the corresponding atoms occupy any visible sidechain density. To achieve optimal rotamer assignment, mainchain atom position may need to be altered; optimization at this point should be done only over a small (3–5) number of amino acids. The entire model optimization procedure is relatively subjective and will likely be iterated until all the atoms are placed in the density map and registered with any visible features.

4.18. Monitoring model quality

Results of the previous optimization step can be monitored using a Rama-chandran plot (Fig. 1.5). When completed, all amino acids should fall in favorable or acceptable positions on the Ramachandran plot. Several itera-tions may be required to optimally assign all atoms. Once the final structure has been achieved, the model may be fit back into the entire assembly along with any other components.

4.19. Cα to atomic model optimization with Rosetta

The process of model optimization described above is relatively interactive. Alternatively, computational modeling is capable of semiautomatically opti-mizing an initial model, bypassing much of the user-intensive optimization steps. Rosetta (Bradley et al., 2005) can refine structures constrained by experimental electron density maps by optimizing an all-atom energy function that includes both statistical and chemical potential energy terms. When refining a model, a scoring term that assesses the fit of a model to the density map is simultaneously optimized with Rosetta's standard energy function (DiMaio et al., 2009).

From a Cα trace, Rosetta's *rebuilding-and-refinement* protocol is used to refine the structure. The Rosetta protocol *ca_to_allatom* infers an all-atom model and performs structure refinement. The protocol generates models by sampling different conformations of individual SSEs. The Cα positions in

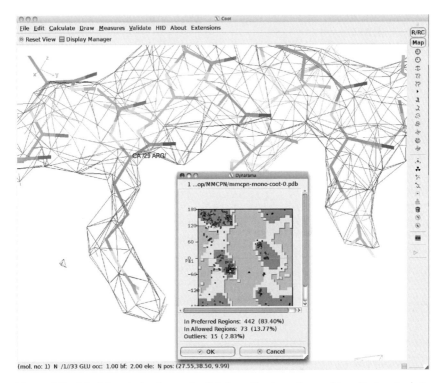

Figure 1.5   Model optimization and validation. Coot can be used to adjust mainchain and sidechain atom positions, optimizing the fit of the atomic model in the density map. A Ramachandran plot is shown overlaid with the map and model of Mm-cpn after optimization.

the starting model guide placement of the initial structure with a user-controllable parameter specifying how far Cαs are allowed to deviate from the starting model.

In the next stage, each atom is explicitly modeled and then evaluated using the complete all-atom energy function. Loops are rebuilt, sidechains are placed on the structure and the entire structure is relaxed with Rosetta's high-resolution energy function. Throughout the entire process, harmonic constraints keep Cα positions from deviating too far from their initial positions. This protocol generally requires significant sampling, on the order of thousands to tens of thousands of models. In addition, even the best models produced may still have loops and other features outside of the density contours. Thus, it is often necessary to follow this protocol with iterative rebuilding using Rosetta's *loopmodel*. A final all-atom optimization with a high-resolution energy function is performed to sample less-common sidechain rotamers and sidechain torsions.

4.20. Model optimization with Rosetta

Not limited to near-atomic resolutions, the aforementioned Rosetta protocols can also be adapted for use with lower resolution structures assuming α-helices can be identified within the density and a clear correspondence can be established with the predicted secondary structure. Similarly, Rosetta's *relax* and *rebuilding-and-refinement* protocols may also be applied to homology models to improve model accuracy and fit into a subnanometer resolution density map.

5. CASE STUDY: MM-CPN

To illustrate the above process, we have chosen to detail the process of constructing an atomic model directly from a cryo-EM density map, as done for the 4.3 Å resolution structure of Mm-cpn (Fig. 1.6A; Zhang *et al.*, 2010). Where appropriate, specific reference is made to the programs, parameters and results obtained for Mm-cpn using this protocol.

It should be noted that the level of detail found in the Mm-cpn density map is not required for analyzing and annotating subnanometer resolution protein structure. Rather, Mm-cpn simply provides a convenient and accessible vehicle to describe the variety of tools available for analyzing subnanometer resolution protein structure.

1. Sixteen subunits were isolated using *segment3d* from EMAN (*segment3d mmcpn.mrc segmented-mmcpn.mrc nseg = 16 split apix = 1.33 sym = d8*), which uses a K-means approach to identify subunits (Fig. 1.6A).

2. For Mm-cpn, a single subunit was padded to a 128^3 density map and the origin was reset to zero with EMAN using the following command: *proc3d mmcpn1.mrc mmcpn-monomer128.mrc clip = 128,128,128 origin = 0,0,0*. Note: mmcpn1.mrc is one the 16 segmented monomers from the prior step.

3. SSEHunter was used to identify SSEs using the following command in EMAN: *ssehunter3.py mmcpn-monomer128.mrc 1.33 4.5 0.4*. A single Mm-cpn subunit (543 amino acids per subunit) returned 201 pseudoatoms.

4. The results were loaded into Chimera along with the density map. Pseudoatoms were represented as spheres and bonds were hidden. The pseudoatoms were colored using Chimera's "Render by Attribute" option such that the most negative value in the B-factor column (− 3) was set to blue, the most positive value (3) was set to red, and zero was set to white.

5. For a single Mm-cpn subunit, five β-sheets and 17 α-helices were identified and built using SSEBuilder (Fig. 1.6B).

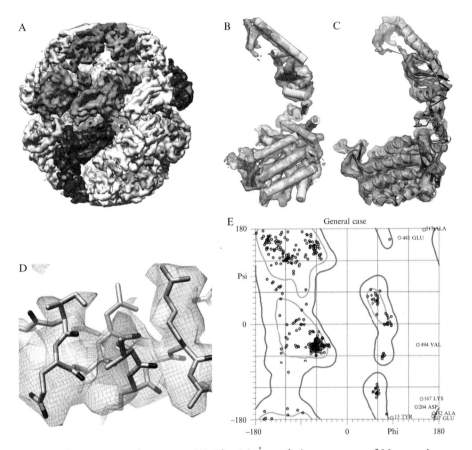

Figure 1.6 Structure of Mm-cpn. (A) The 4.3-Å resolution structure of Mm-cpn is shown (Zhang *et al.*, 2010). (B) Using SSEHunter, the secondary structure elements in the Mm-cpn subunit were identified: α-helices are shown as cylinders and β-strands are shown as planes. (C) Using the *de novo* modeling approach, an atomic model (residues 1–532) was constructed for one subunit of Mm-cpn. (D) Large, bulky sidechains in the model could be seen in the density. (E) The Ramachandran plot of an Mm-cpn monomer shows greater than 98% of all residues with allowable phi–psi angles. (See Color Insert.)

6. The structure of a related chaperone, the thermosome KS-1 (PDB ID: 1Q3Q; Shomura *et al.*, 2004), was used to construct a homology model for a single Mm-cpn subunit as done in a previous study (Booth *et al.*, 2008).

7. The homology model for the Mm-cpn subunit was fit to the segmented density map using Foldhunter (Jiang *et al.*, 2001) from EMAN with the following command: *foldhunter.py mmcpn-monomer128.mrc mmcpn-homology-model.pdb res = 4.3 apix = 1.33*. The resulting transformed PDB model was then loaded into Chimera to verify the fit to the density map and the agreement with the SSEs identified in the previous steps.

8. The fitted homology model was adjusted in Coot and Gorgon such that the model agreed with the α-helices identified by SSEHunter. The helices required only slight rotations and translations.

9. Once the helix positions were optimized strands and loops were adjusted to optimally fit the density while maintaining ∼3.8 Å Cα–Cα distances. The density skeleton was used to identify potential paths through the density.

10. In the Mm-cpn homology model, ∼30 amino acids were unresolved at the termini. In Coot, Cα atoms were added consecutively to the model starting at the ends of the model until the N-terminus was reached or until the density was not visible (C-terminus). The final model for the Mm-cpn monomer contained residues 1–532 (Fig. 1.6C).

11. After the addition of missing residues, optimization of the complete Cα model was performed to maximize the occupancy of the model in the subunit density map, while maintaining appropriate distance constraints.

12. Sixteen copies of the Mm-cpn Cα model were loaded into Chimera, manually moved into a subunit and fit to the density using the "Fit in map" option (Pettersen et al., 2004).

13. After eight iterations of refinement and fitting (steps 11 and 12) of the entire macromolecular assembly, all clashes were eliminated and model occupancy was optimized. A single PDB file was saved containing all 16 copies of the optimized Mm-cpn Cα model as separate chains.

14. From the full Mm-cpn Cα model, the structure factors for the model were calculated and applied to the original density map using EMAN (detailed description of the parameters can be found in the EMAN documentation).

   - pdb2mrc mmcpn.pdb mmcpn-simulated.mrc res = 4.3 apix = 1.33 box = 192
   - proc3d mmcpn-simulated.mrc junk.mrc calcsf = mmcpn-sf.txt apix = 1.33
   - proc3d mmcpn.mrc mmcpn-rescaled.mrc setsf = mmcpn-sf.txt apix = 1.33
   - proc3d mmcpn-rescaled.mrc mmcpn-lp.mrc lp = 4.3 apix = 1.33
   - proc3d mmcpn-lp.mrc mmcpn-final-map.mrc mult = 0.3

15. Using the SABBAC web server, the Cα model for one subunit was transformed into an atomic model. The resulting atomic model was loaded into Coot along with the rescaled density map.

16. Portions of the Mm–cpn atomic model created by SABBAC contained breaks in the polypeptide chain. The "Model/Fit/Refine" tools in Coot were used to move/add residues such that a complete polypeptide chain was constructed.

17. Once a complete all-atom model for an Mm-cpn subunit was produced, the "Model/Fit/Refine" tools in Coot were used to move the mainchain atoms in the density map. Small stretches of residues, from three amino acids to entire helices, were fit to the density such that (1) the mainchain atoms were encompassed by density, (2) secondary

structure constraints were maintained and registered with the density and (3) potential sidechain density for large, bulky sidechains were proximal to residues in the model containing corresponding sidechains. In this refinement round, torsion angles, planar peptide constraints, and Ramachandran constraints were enforced for the appropriate type of secondary structure.

18. For every amino acid, a rotamer search was performed allowing side-chains to be placed within corresponding sidechain density. In some instances, the "Model/Fit/Refine" tools in Coot were used to adjust the mainchain and sidechain positions to best fit the density (Fig. 1.6D).

19. The Ramachandran plot in Coot plots the phi–psi angles. Residues falling outside of the acceptable range are plotted in red. Clicking on these outliers, the map and model are recentered in the main Coot display. The steps 17 and 18 were performed over small stretches of residues (3–5 amino acids) containing the outliers until the residues were in acceptable or favorable conformations.

20. After eight iterations of steps 17–19, the Mm-cpn model had greater than 98% of all residues with acceptable Ramachandran angles (Fig. 1.6E).

21. Steps 11 and 12 were repeated except with a rescaled density map and the refined atomic model. A final model containing all 16 subunits was saved and deposited in the PDB.

▷ 6. Discussion

While the above protocol describes a complete approach to build and refine an atomic model from a cryo-EM density map at subnanometer resolution (Fig. 1.2), individual tools can be, and most often are, used independently. Thus, it is important to know when and what tools are most appropriate for a specific problem.

Limitations in the analysis of cryo-EM density maps, due in large part to resolution, are obvious, though not prohibitive, in describing salient structural features and functions in macromolecular assemblies. The ability to analyze a cryo-EM density map hinges on the map itself; size, complexity, and quality of the density map all play critical roles in annotating structure at any resolution. Even the most experienced scientists may not be able to reliably describe features in poorly resolved regions of density maps. One would not expect to see sidechain density at 9 Å resolution; conversely, the absence of β-strand separation at 3.5 Å resolution may indicate potential problems. As such, the analysis process requires a significant investment in time and understanding the quality of the original 2D data and the reconstructed volume.

Illustrating the dependance on the resolvability of density features, *de novo* model building is based on establishing a sequence-to-structure correspondence using SSEs, features unique to subnanometer resolutions. This necessitates the presence of clearly identifiable SSEs in the density map, though connecting loops may be ambiguous. Density maps vary in composition, quality, and resolution making it difficult to assign a clear resolution cut-off. Model building may be easier and more reliable at near-atomic resolutions (3.5–5 Å) but still possible at lower resolutions depending on the features that are resolved in the map, as in the case of Hepatitis B (Böttcher *et al.*, 1997; Conway *et al.*, 1997). At higher resolutions, sidechain density can aid in the placement of Cα atoms, thereby increasing the accuracy and reliability of models. Thus, in all cases of interpreting a subnanometer resolution density map, precedence must be given to the observable features in the density map and not the stated resolution.

ACKNOWLEDGMENTS

This work is supported by grants from NIH (P41RR02250, R01GM079429, R01AI0175208) and NSF (IIS-0705644, IIS-0705474). M. R. Baker is supported by a postdoctoral training fellowship from the National Library of Medicine Training Program in Computational Biology and Biomedical Informatics provided by the Keck Center and Gulf Coast Consortia (T15LM007093).

REFERENCES

Abeysinghe, S., Ju, T., Baker, M. L., and Chiu, W. (2008). Shape modeling and matching in identifying 3D protein structures. *Comput. Aided Des.* **40**, 708–720.

Altschul, S. F., Gish, W., Miller, W., Myers, E. W., and Lipman, D. J. (1990). Basic local alignment search tool. *J. Mol. Biol.* **215**, 403–410.

Arnold, K., Bordoli, L., Kopp, J., and Schwede, T. (2006). The SWISS-MODEL workspace: A web-based environment for protein structure homology modelling. *Bioinformatics* **22**, 195–201.

Baker, M. L., Jiang, W., Bowman, B. R., Zhou, Z. H., Quiocho, F. A., Rixon, F. J., and Chiu, W. (2003). Architecture of the herpes simplex virus major capsid protein derived from structural bioinformatics. *J. Mol. Biol.* **331**, 447–456.

Baker, M. L., Jiang, W., Rixon, F. J., and Chiu, W. (2005). Common ancestry of herpesviruses and tailed DNA bacteriophages. *J. Virol.* **79**, 14967–14970.

Baker, M. L., Jiang, W., Wedemeyer, W. J., Rixon, F. J., Baker, D., and Chiu, W. (2006a). Ab initio modeling of the herpesvirus VP26 core domain assessed by CryoEM density. *PLoS Comput. Biol.* **2**, e146.

Baker, M. L., Yu, Z., Chiu, W., and Bajaj, C. (2006b). Automated segmentation of molecular subunits in electron cryomicroscopy density maps. *J. Struct. Biol.* **156**, 432–441.

Baker, M. L., Ju, T., and Chiu, W. (2007). Identification of secondary structure elements in intermediate-resolution density maps. *Structure* **15**, 7–19.

Bates, P. A., Kelley, L. A., MacCallum, R. M., and Sternberg, M. J. (2001). Enhancement of protein modeling by human intervention in applying the automatic programs 3D-JIGSAW and 3D-PSSM. *Proteins* (Suppl. 5), 39–46.

Baumeister, W., and Steven, A. C. (2000). Macromolecular electron microscopy in the era of structural genomics. *Trends Biochem. Sci.* **25,** 624–631.

Booth, C. R., Meyer, A. S., Cong, Y., Topf, M., Sali, A., Ludtke, S. J., Chiu, W., and Frydman, J. (2008). Mechanism of lid closure in the eukaryotic chaperonin TRiC/CCT. *Nat. Struct. Mol. Biol.* **15,** 746–753.

Böttcher, B., Wynne, S. A., and Crowther, R. A. (1997). Determination of the fold of the core protein of hepatitis B virus by electron cryomicroscopy. *Nature* **386,** 88–91.

Bradley, P., Malmström, L., Qian, B., Schonbrun, J., Chivian, D., Kim, D. E., Meiler, J., Misura, K. M., and Baker, D. (2005). Free modeling with Rosetta in CASP6. *Proteins* **61** (Suppl 7), 128–134.

Chiu, W., Baker, M. L., Jiang, W., Dougherty, M., and Schmid, M. F. (2005). Electron cryomicroscopy of biological machines at subnanometer resolution. *Structure* **13,** 363–372.

Cole, C., Barber, J. D., and Barton, G. J. (2008). The Jpred 3 secondary structure prediction server. *Nucleic Acids Res.* **36,** W197–W201.

Cong, Y., and Ludtke, S. J. (2010). Single particle analysis at high resolution. *Methods Enzymol.* **482,** 211–236.

Cong, Y., Baker, M. L., Jakana, J., Woolford, D., Miller, E. J., Reissmann, S., Kumar, R. N., Redding-Johanson, A. M., Batth, T. S., Mukhopadhyay, A., *et al.* (2010). 4.0-Å resolution cryo-EM structure of the mammalian chaperonin TRiC/CCT reveals its unique subunit arrangement. *Proc. Natl. Acad. Sci. USA* **107,** 4967–4972.

Conway, J. F., Cheng, N., Zlotnick, A., Wingfield, P. T., Stahl, S. J., and Steven, A. C. (1997). Visualization of a 4-helix bundle in the hepatitis B virus capsid by cryo-electron microscopy. *Nature* **386,** 91–94.

DiMaio, F., Tyka, M. D., Baker, M. L., Chiu, W., and Baker, D. (2009). Refinement of protein structures into low-resolution density maps using rosetta. *J. Mol. Biol.* **392,** 181–190.

Emsley, P., and Cowtan, K. (2004). Coot: Model-building tools for molecular graphics. *Acta Crystallogr. D Biol. Crystallogr.* **60,** 2126–2132.

Fernández, J. J., Luque, D., Castón, J. R., and Carrascosa, J. L. (2008). Sharpening high resolution information in single particle electron cryomicroscopy. *J. Struct. Biol.* **164,** 170–175.

Frank, J. (2002). Single-particle imaging of macromolecules by cryo-electron microscopy. *Annu. Rev. Biophys. Biomol. Struct.* **31,** 303–319.

Ginalski, K., Elofsson, A., Fischer, D., and Rychlewski, L. (2003). 3D-Jury: A simple approach to improve protein structure predictions. *Bioinformatics* **19,** 1015–1018.

Gonen, T., Cheng, Y., Sliz, P., Hiroaki, Y., Fujiyoshi, Y., Harrison, S. C., and Walz, T. (2005). Lipid-protein interactions in double-layered two-dimensional AQP0 crystals. *Nature* **438,** 633–638.

Henderson, R., and Unwin, N. R. (1975). Three-dimensional model of purple membrane obtained by electron microscopy. *Nature* **257,** 28–32.

Henderson, R., Baldwin, J. M., Ceska, T. A., Zemlin, F., Beckmann, E., and Downing, K. H. (1990). Model for the structure of bacteriorhodopsin based on high-resolution electron cryo-microscopy. *J. Mol. Biol.* **213,** 899–929.

Humphrey, W., Dalke, A., and Schulten, K. (1996). VMD: Visual molecular dynamics. *J. Mol. Graph.* **14**(33–8), 27–28.

Jiang, W., Baker, M. L., Ludtke, S. J., and Chiu, W. (2001). Bridging the information gap: Computational tools for intermediate resolution structure interpretation. *J. Mol. Biol.* **308,** 1033–1044.

Jiang, W., Baker, M. L., Jakana, J., Weigele, P. R., King, J., and Chiu, W. (2008). Backbone structure of the infectious epsilon15 virus capsid revealed by electron cryomicroscopy. *Nature* **451**, 1130–1134.

Jones, T. A., Zou, J. Y., Cowan, S. W., and Kjeldgaard, M. (1991). Improved methods for building protein models in electron density maps and the location of errors in these models. *Acta Crystallogr. A* **47**(Pt 2), 110–119.

Ju, T., Baker, M. L., and Chiu, W. (2007). Computing a family of skeletons of volumetric models for shape description. *Comput. Aided Des.* **39**, 352–360.

Kelley, L. A., and Sternberg, M. J. (2009). Protein structure prediction on the Web: A case study using the Phyre server. *Nat. Protoc.* **4**, 363–371.

Kimura, Y., Vassylyev, D. G., Miyazawa, A., Kidera, A., Matsushima, M., Mitsuoka, K., Murata, K., Hirai, T., and Fujiyoshi, Y. (1997). Surface of bacteriorhodopsin revealed by high-resolution electron crystallography. *Nature* **389**, 206–211.

Kleywegt, G. J., and Jones, T. A. (1997). Detecting folding motifs and similarities in protein structures. *Methods Enzymol.* **277**, 525–545.

Kong, Y., and Ma, J. (2003). A structural-informatics approach for mining beta-sheets: Locating sheets in intermediate-resolution density maps. *J. Mol. Biol.* **332**, 399–413.

Kong, Y., Zhang, X., Baker, T. S., and Ma, J. (2004). A structural-informatics approach for tracing beta-sheets: Building pseudo-C(alpha) traces for beta-strands in intermediate-resolution density maps. *J. Mol. Biol.* **339**, 117–130.

Li, Z., Baker, M. L., Jiang, W., Estes, M. K., and Prasad, B. V. (2009). Rotavirus architecture at subnanometer resolution. *J. Virol.* **83**, 1754–1766.

Lipman, D. J., and Pearson, W. R. (1985). Rapid and sensitive protein similarity searches. *Science* **227**, 1435–1441.

Ludtke, S. J., Baldwin, P. R., and Chiu, W. (1999). EMAN: Semiautomated software for high-resolution single-particle reconstructions. *J. Struct. Biol.* **128**, 82–97.

Ludtke, S. J., Serysheva, I. I., Hamilton, S. L., and Chiu, W. (2005). The pore structure of the closed RyR1 channel. *Structure* **13**, 1203–1211.

Ludtke, S. J., Baker, M. L., Chen, D. H., Song, J. L., Chuang, D. T., and Chiu, W. (2008). De novo backbone trace of GroEL from single particle electron cryomicroscopy. *Structure* **16**, 441–448.

Maupetit, J., Gautier, R., and Tufféry, P. (2006). SABBAC: Online Structural Alphabet-based protein BackBone reconstruction from Alpha-Carbon trace. *Nucleic Acids Res.* **34**, W147–W151.

McGuffin, L. J., Bryson, K., and Jones, D. T. (2000). The PSIPRED protein structure prediction server. *Bioinformatics* **16**, 404–405.

Mizuguchi, K., and Go, N. (1995). Comparison of spatial arrangements of secondary structural elements in proteins. *Protein Eng.* **8**, 353–362.

Nakagawa, A., Miyazaki, N., Taka, J., Naitow, H., Ogawa, A., Fujimoto, Z., Mizuno, H., Higashi, T., Watanabe, Y., Omura, T., *et al.* (2003). The atomic structure of rice dwarf virus reveals the self-assembly mechanism of component proteins. *Structure* **11**, 1227–1238.

Pettersen, E. F., Goddard, T. D., Huang, C. C., Couch, G. S., Greenblatt, D. M., Meng, E. C., and Ferrin, T. E. (2004). UCSF Chimera–a visualization system for exploratory research and analysis. *J. Comput. Chem.* **25**, 1605–1612.

Pintilie, G. D., Zhang, J., Goddard, T. D., Chiu, W., and Gossard, D. C. (2010). Quantitative analysis of cryo-EM density map segmentation by watershed and scale-space filtering, and fitting of structures by alignment to regions. *J. Struct. Biol.* **170**, 427–438.

Pollastri, G., Przybylski, D., Rost, B., and Baldi, P. (2002). Improving the prediction of protein secondary structure in three and eight classes using recurrent neural networks and profiles. *Proteins* **47**, 228–235.

Roseman, A. M. (2000). Docking structures of domains into maps from cryo-electron microscopy using local correlation. *Acta Crystallogr. D Biol. Crystallogr.* **56**, 1332–1340.

Rossmann, M. G. (2000). Fitting atomic models into electron-microscopy maps. *Acta Crystallogr. D Biol. Crystallogr.* **56**, 1341–1349.

Rossmann, M. G., Morais, M. C., Leiman, P. G., and Zhang, W. (2005). Combining X-ray crystallography and electron microscopy. *Structure* **13**, 355–362.

Rusu, M., Birmanns, S., and Wriggers, W. (2008). Biomolecular pleiomorphism probed by spatial interpolation of coarse models. *Bioinformatics* **24**, 2460–2466.

Sali, A., Potterton, L., Yuan, F., van Vlijmen, H., and Karplus, M. (1995). Evaluation of comparative protein modeling by MODELLER. *Proteins* **23**, 318–326.

Schröder, G. F., Brunger, A. T., and Levitt, M. (2007). Combining efficient conformational sampling with a deformable elastic network model facilitates structure refinement at low resolution. *Structure* **15**, 1630–1641.

Serysheva, I. I., Hamilton, S. L., Chiu, W., and Ludtke, S. J. (2005). Structure of Ca^{2+} release channel at 14 Å resolution. *J. Mol. Biol.* **345**, 427–431.

Serysheva, I. I., Ludtke, S. J., Baker, M. L., Cong, Y., Topf, M., Eramian, D., Sali, A., Hamilton, S. L., and Chiu, W. (2008). Subnanometer-resolution electron cryomicroscopy-based domain models for the cytoplasmic region of skeletal muscle RyR channel. *Proc. Natl. Acad. Sci. USA* **105**, 9610–9615.

Shi, J., Blundell, T. L., and Mizuguchi, K. (2001). FUGUE: Sequence-structure homology recognition using environment-specific substitution tables and structure-dependent gap penalties. *J. Mol. Biol.* **310**, 243–257.

Shomura, Y., Yoshida, T., Iizuka, R., Maruyama, T., Yohda, M., and Miki, K. (2004). Crystal structures of the group II chaperonin from Thermococcus strain KS-1: Steric hindrance by the substituted amino acid, and inter-subunit rearrangement between two crystal forms. *J. Mol. Biol.* **335**, 1265–1278.

Siebert, X., and Navaza, J. (2009). UROX 2.0: An interactive tool for fitting atomic models into electron-microscopy reconstructions. *Acta Crystallogr. D Biol. Crystallogr.* **65**, 651–658.

Suhre, K., Navaza, J., and Sanejouand, Y. H. (2006). NORMA: A tool for flexible fitting of high-resolution protein structures into low-resolution electron-microscopy-derived density maps. *Acta Crystallogr. D Biol. Crystallogr.* **62**, 1098–1100.

Tama, F., Miyashita, O., and Brooks, C. L. (2004). Normal mode based flexible fitting of high-resolution structure into low-resolution experimental data from cryo-EM. *J. Struct. Biol.* **147**, 315–326.

Tan, R. K., Devkota, B., and Harvey, S. C. (2008). YUP.SCX: Coaxing atomic models into medium resolution electron density maps. *J. Struct. Biol.* **163**, 163–174.

Tang, G., Peng, L., Baldwin, P. R., Mann, D. S., Jiang, W., Rees, I., and Ludtke, S. J. (2007). EMAN2: An extensible image processing suite for electron microscopy. *J. Struct. Biol.* **157**, 38–46.

Topf, M., Baker, M. L., John, B., Chiu, W., and Sali, A. (2005). Structural characterization of components of protein assemblies by comparative modeling and electron cryo-microscopy. *J. Struct. Biol.* **149**, 191–203.

Topf, M., Baker, M. L., Marti-Renom, M. A., Chiu, W., and Sali, A. (2006). Refinement of protein structures by iterative comparative modeling and CryoEM density fitting. *J. Mol. Biol.* **357**, 1655–1668.

Topf, M., Lasker, K., Webb, B., Wolfson, H., Chiu, W., and Sali, A. (2008). Protein structure fitting and refinement guided by cryo-EM density. *Structure* **16**, 295–307.

Trabuco, L. G., Villa, E., Schreiner, E., Harrison, C. B., and Schulten, K. (2009). Molecular dynamics flexible fitting: A practical guide to combine cryo-electron microscopy and X-ray crystallography. *Methods* **49**, 174–180.

Volkmann, N. (2002). A novel three-dimensional variant of the watershed transform for segmentation of electron density maps. *J. Struct. Biol.* **138**, 123–129.

Volkmann, N. (2009). Confidence intervals for fitting of atomic models into low-resolution densities. *Acta Crystallogr. D Biol. Crystallogr.* **65**, 679–689.

Volkmann, N., and Hanein, D. (1999). Quantitative fitting of atomic models into observed densities derived by electron microscopy. *J. Struct. Biol.* **125**, 176–184.

Wriggers, W., Milligan, R. A., and McCammon, J. A. (1999). Situs: A package for docking crystal structures into low-resolution maps from electron microscopy. *J. Struct. Biol.* **125**, 185–195.

Wynne, S. A., Crowther, R. A., and Leslie, A. G. (1999). The crystal structure of the human hepatitis B virus capsid. *Mol. Cell* **3**, 771–780.

Yu, X., Jin, L., and Zhou, Z. H. (2008). 3.88 Å structure of cytoplasmic polyhedrosis virus by cryo-electron microscopy. *Nature* **453**, 415–419.

Zhang, X., Settembre, E., Xu, C., Dormitzer, P. R., Bellamy, R., Harrison, S. C., and Grigorieff, N. (2008). Near-atomic resolution using electron cryomicroscopy and single-particle reconstruction. *Proc. Natl. Acad. Sci. USA* **105**, 1867–1872.

Zhang, J., Baker, M. L., Schröder, G. F., Douglas, N. R., Reissmann, S., Jakana, J., Dougherty, M., Fu, C. J., Levitt, M., Ludtke, S. J., *et al.* (2010). Mechanism of folding chamber closure in a group II chaperonin. *Nature* **463**, 379–383.

Zhou, Z. H., Baker, M. L., Jiang, W., Dougherty, M., Jakana, J., Dong, G., Lu, G., and Chiu, W. (2001). Electron cryomicroscopy and bioinformatics suggest protein fold models for rice dwarf virus. *Nat. Struct. Biol.* **8**, 868–873.

Zhu, J., Cheng, L., Fang, Q., Zhou, Z. H., and Honig, B. (2010). Building and refining protein models within cryo-electron microscopy density maps based on homology modeling and multiscale structure refinement. *J. Mol. Biol.* **397**, 835–851.

CHAPTER TWO

METHODS FOR SEGMENTATION AND INTERPRETATION OF ELECTRON TOMOGRAPHIC RECONSTRUCTIONS

Niels Volkmann

Contents

Abstract

Electron tomography has become a powerful tool for revealing the molecular architecture of biological cells and tissues. In principle, electron tomography can provide high-resolution mapping of entire proteomes. The achievable resolution (3–8 nm) is capable of bridging the gap between live-cell imaging and atomic resolution structures. However, the relevant information is not readily accessible from the data and needs to be identified, extracted, and processed before it can be used. Because electron tomography imaging and image acquisition technologies have enjoyed major advances in the last few years and continue to increase data throughput, the need for approaches that allow automatic and objective interpretation of electron tomograms becomes more and more urgent. This chapter provides an overview of the state of the art in this field and attempts to identify the major bottlenecks that prevent approaches for interpreting electron tomography data to develop their full potential.

Sanford-Burnham Medical Research Institute, La Jolla, California, USA

Methods in Enzymology, Volume 483
ISSN 0076-6879, DOI: 10.1016/S0076-6879(10)83002-2

1. Introduction

Electron tomography is the most widely applicable method for obtaining 3D information by electron microscopy. In fact, it is the only method suitable for investigating polymorphic structures such as organelles, cells, and tissue at high resolution (current estimates 3–8 nm). Its principle is based on illuminating the sample from many different directions, usually tilt series around one or two axes, and to reconstruct it from those projection images. While the principles of electron tomography have been known for decades, its use has gathered momentum only in recent years. It has been realized that electron tomography, especially its cryovariant, is capable of providing a complete, molecular resolution 3D mapping of entire cellular proteomes including their detailed interactions (Leis *et al.*, 2009; Nickell *et al.*, 2006; Robinson *et al.*, 2007; Tocheva *et al.*, 2010; Volkmann and Hanein, 2009).

Electron tomography can depict unique structures and scenes but, due to the fact that all electron tomography preparations can only sustain a limited electron dose, the resulting reconstructions will inevitably suffer from low signal-to-noise ratios and relatively low resolution as compared to other electron microscopy techniques that can take advantage of some form of averaging. Consequently, maps obtained by electron tomography are difficult to interpret. This difficulty in interpretation is further aggravated in highly complex systems (Grünewald *et al.*, 2003).

Despite a recent surge in dedicated method development toward automatic interpretation of electron tomograms (reviewed in Best *et al.*, 2007; Frangakis and Förster, 2004; Sandberg, 2007), only relatively few algorithms for reliable detection and extraction of structural features from electron tomograms are available. Instead, the tasks of extracting and interpreting information from the highly complex, 3D scenes that make up cellular tomograms are, for the most part, painstakingly carried out manually. Apart from the subjectivity of the process, the time-consuming (and tiring) nature of this manual task all but precludes the prospects of the high throughput necessary to take full advantage of the method's potential. For example, it took over 9 months to manually segment and interpret roughly 1% of the volume of a pancreatic beta cell (Marsh, 2005; Marsh *et al.*, 2001). By combining sectioning with automatic data collection schemes (Chapter 12, Vol. 481) and large-scale montaging, it is now technically possible to reconstruct entire mammalian cells with high fidelity within a reasonable timeframe (Noske *et al.*, 2008). Extrapolating from the time required for segmenting 1% of a cell results in 75 man years for manually segmenting a single cell at similar detail. Conducting a meaningful study comparing a number of cells under disease conditions with a control set would literally take hundreds of man years. The need for computational tools to efficiently aid the process and automate the structure recognition, extraction, and interpretation process as much as possible is clearly vital for making these types of studies viable.

The quality of cryotomographic reconstructions can be correlated with the electron dose. A total dose of 50–300 $e^-/\text{Å}^2$ appears to give reasonable results with a "sweet spot" around 120 $e^-/\text{Å}^2$ (Iancu et al., 2006). Because this dose needs to be spread over the whole data set, the dose for each image needs to be kept low enough as to not exceed a total dose of 120 $e^-/\text{Å}^2$. For a $\pm 70°$ double tilt series with a $2°$ increment, the dose available for a single image is below 1 $e^-/\text{Å}^2$, which gives rise to extremely high noise levels in the individual images. The signal in the resulting 3D reconstructions is improved by the dose fractionation effect (McEwen et al., 1995) but the signal-to-noise ratio for these tomograms is still well below 1. Signal-to-noise ratios in cryo-images collected for single-particle reconstructions of ribosomes have been experimentally determined to be in the 0.05 range (Baxter et al., 2009). Given the fact that most contributions to the noise tend to be considerably worse in cryo-tomograms than in these relatively well behaved samples, signal-to-noise ratios in the neighborhood of 0.01 or less should be expected for cryo-tomograms. Together with complications from missing data and the electron microscope's contrast transfer, this makes noise from many other image-possessing disciplines look mundane.

Multiple scattering events in samples thicker than the mean free path length of the illuminating electrons dictate an upper limit of ~ 1 μm in thickness for high-resolution electron tomography, even if relatively high acceleration voltage and energy filtering are used (Grimm et al., 1996). Owing to the technically demanding nature of cryosample sectioning (Al-Amoudi et al., 2004), "conventional" electron tomography, which involves staining and plastic embedding, is often preferred in practice for samples that require sectioning (McEwen and Marko, 2001). In fact, cryosectioning is currently an art form practiced in only a handful of laboratories worldwide (see, e.g., Al-Amoudi et al., 2007; Gruska et al., 2008; Hsieh et al., 2006; Pierson et al., 2010; Salje et al., 2009). In embedded material, the main adverse effect of electron irradiation is not direct damage to the particles (which have been mostly exchanged for heavy metal stains anyway) it is damage to the embedding material. For example, the beam induces serious thinning (25–50%) perpendicular to the beam (Luther et al., 1988). The usual strategy around this is to preexpose the specimen to allow completion of a rapid initial thinning phase. The tilt series are then collected with low electron dose (up to about 100 $e^-/\text{Å}^2$ per image is deemed tolerable) to avoid beam-induced buckling of the sample or further thinning during data collection. While the signal-to-noise ratio is improved in these samples as compared to cryosamples, the resulting images and reconstructions still tend to be rather noisy. In addition, these samples can suffer additional problems such as uneven staining and other sample preparation artifacts that may complicate subsequent analysis.

While there have been substantial efforts during the last few years which specifically address data from electron tomography, progress has been much slower than in related imaging fields. As a consequence, the relative lack of adequate tools for automatic and/or objective interpretation in electron tomography has been recognized as a critical barrier to progress in the field (Ben-Harush *et al.*, 2010; Frey *et al.*, 2006; Koning and Koster, 2009; Leis *et al.*, 2006; Marsh, 2005).

Most development efforts in biological image processing are geared toward clinical medicine: Clinical data sets typically contain organs of well-defined shape with smooth and distinct boundaries and include only a few continuous objects. In contrast, electron tomograms contain amorphous structures, ragged contours, and numerous small objects which would make it extremely challenging to interpret electron tomograms even if noise would be absent. Unlike medical data sets, for which considerable knowledge is available to validate the results and tune parameters, the interpretation of electron tomograms often leave the user uncertain about the accuracy of the analysis: it is difficult to distinguish artifacts from real structures. Other reasons why image and signal processing methods developed for other domains are not straight forward to apply to electron tomography data include:

(i) As a consequence of the damaging effect of the electron beam which limits the amount of electrons available for image formation, electron tomograms tend to exhibit extremely high noise levels.

(ii) The geometry of the electron tomography sample holders and the shape of the electron microscopy chamber do not allow tilting more than $\sim 70°$ so a good third of data space is not accessible. In accordance with the projection theorem (Radon, 1917), the nonsampled region generates a wedge-shaped segment in Fourier space that contains no information (the "missing wedge"). This problem can be partially alleviated experimentally by taking a second data set after rotating the sample by 90° around the optical axis (Mastronarde, 1997; Nickell *et al.*, 2003; Penczek *et al.*, 1995), yet some of the data space is still not accessible so that some missing data artifacts will always remain.

(iii) Aberrations of the electron microscope optics give rise to a point-spread function that can be best described in Fourier space by its Fourier transform, the contrast transfer function. This function mainly depends on the amount of defocus used for imaging and changes the amplitudes and phases of the signal. In an electron tomography setting, the contrast transfer function is not well defined within the sample, especially for thick specimens where there can be a significant variance in focus. In addition, the tilting introduces a focus gradient, further obstructing the underlying signal.

(iv) The noise in the reconstruction results from a complex combination of different noise sources including signal-dependent shot noise due to the quantum nature of the electrons, digitization noise, and "structural noise" due to the support/embedding medium (Baxter *et al.*, 2009). In addition, the noise is highly correlated in space, and is possibly corrupted by the contrast transfer function (Scheres *et al.*, 2007) and the missing wedge. Thus, it is extremely hard to devise adequate noise models to take advantage of algorithms that explicitly take noise characteristics into account to improve performance.

The combination of these issues makes devising automatic and objective interpretation tools extremely challenging. While the best course of action will heavily depend on the actual data and the biological question to be addressed, the general workflow for interpreting electron tomograms can be divided into a number of successive stages. Each of these stages will be described in more detail below. Most of the approaches described have been implemented in publicly available computer programs (see Chapter 15, Vol. 482 for a comprehensive list).

2. Noise Reduction

Noise reduction has proven to be an indispensable tool for visualization and preprocessing of multidimensional images in the bioimaging field in general and in electron tomography in particular. Noise reduction is especially important for cryotomograms that suffer from the lowest contrast and highest noise levels. Noise-reduction schemes that were adapted for electron tomography relatively early on include wavelet transform techniques (Stoschek and Hegerl, 1997), and nonlinear anisotropic diffusion (Frangakis and Hegerl, 2001). A comparison between the two showed that nonlinear anisotropic diffusion appears to be preferable as a result of faster performance and the presence of better filtering properties (Frangakis *et al.*, 2001). The nonlinear anisotropic diffusion approach was later expanded by using more sophisticated local kernels that result in enhancement of curvilinear and planar structures (Fernandez and Li, 2003). In addition, the bilateral filter (Tomasi and Manduchi, 1998) which is mathematically essentially equivalent to nonlinear anisotropic diffusion (Barash and Comaniciu, 2004) was adapted for electron tomography (Jiang *et al.*, 2003), achieving—not entirely surprisingly—similar results. Some improvements in performance over the standard bilateral filter can be achieved by explicitly accounting for impulse noise during the filtering step (Pantelic *et al.*, 2006). The price to pay for the improvement is the need to tune one additional adjustable parameter.

In general, the noise-reduction performance of all approaches outlined above relies heavily on the choice of a set of user-specified parameters which are hard to predict for any given tomogram. In addition, these approaches can be very demanding in terms of computation time and memory requirements. These drawbacks make these algorithms cumbersome to use for nonexpert users. Approaches that are easier to use, less demanding on the computational side and that maintain fairly reasonable performance for typical electron tomograms include iterative median filtering (van der Heide *et al.*, 2007) and filtering based on a geometric diffusion flow called Beltrami flow (Kimmel *et al.*, 2000) that was recently adapted for electron tomography (Fernandez, 2009). For many practical purposes such as reducing oversegmentation in automatic segmentation procedures a simple low-pass filter, despite its tendency to blur edges and features, may also be adequate (Pintilie *et al.*, 2009).

3. SEGMENTATION

Interpreting an electron tomogram, even at the ultrastructural level, requires its decomposition into structural components such as membrane compartments, filamentous structures, or clusters of loosely associated macromolecules like polysomes. Various techniques have been proposed for automated or semiautomated segmentation in the field of image processing. Commonly used approaches include segmentation based on region growing, edge detection, active contours, and model-based segmentation. However, none of these proved to be straight-forward in use with electron tomography data and, until recently, most segmentations of electron tomographic volumes were carried out manually, using programs that allow tracing of volumes within slices to create an isocontour model of the structures of interest (Kremer *et al.*, 1996; Li *et al.*, 1997). This hand tracing tends to be tedious, time-consuming, and subjective (Frey *et al.*, 2006).

Many of the computational segmentation approaches developed for electron tomography attempt to improve upon manual segmentation using various types of surface fitting approaches. These range from simple spatial gradient optimization in two dimensions (Ress *et al.*, 2004) through the use of 3D geodesic active contours (Bartesaghi *et al.*, 2005), to implementations of static (fast marching method; Bajaj *et al.*, 2003), as well as full-fledged dynamic level-set based approaches (Osher and Sethian, 1988; Whitaker and Elangovan, 2002). Drawbacks for these energy-minimization based algorithms are their tendency to be subject to local optima and scalability issues resulting in the requirement of reasonably good initial surface models and careful fine-tuning and preconditioning of parameters to ensure correct convergence.

An alternative to these energy-based boundary detection algorithms is the use of region-based approaches where distinct regions are detected by some characteristic (intensity, texture) and their boundaries naturally become the output of the segmentation. In particular, the immersion-based watershed algorithm (Beucher and Meyer, 1993; Vincent and Soille, 1991) has been adapted specifically for use with electron tomography (Volkmann, 2002) and has been used successfully for various segmentation tasks in a wide variety of tomography projects (see, e.g., Auer *et al.*, 2008;

Janssen *et al.*, 2006; Marsh *et al.*, 2004; Rouiller *et al.*, 2008; Salvi *et al.*, 2008; Schietroma *et al.*, 2009). The algorithm is based on an analogy with a step-wise flooding of a topological relief by a fluid, with dams being built where independent flows meet. The method is in principle capable of fully automatic segmentation but, in practice, is more appropriately used in a semiautomatic fashion in order to optimize the few operating parameters. The speed of the method allows interactive refinement of these parameters (if the volume is not too large) and, consequently, has now been implemented in the popular graphics packages Amira (Pruggnaller *et al.*, 2008; Stalling *et al.*, 2004) and Chimera (Goddard *et al.*, 2007; Pintilie *et al.*, 2009). Other segmentation approaches that have shown promising potential with electron tomography data are based on normalized graph cut methods and eigenvector analysis (Frangakis and Hegerl, 2002), on orientation fields and line segment detection (Sandberg and Brega, 2007), and on fuzzy sets theory (Garduño *et al.*, 2008).

In principle, there are three possible sources of information that can guide segmentation algorithms in their task: (i) features that define an actual boundary point; (ii) features that define the inside or outside of a region; and (iii) shape information of the object to be segmented. All segmentation methods described above use only one of these information sources. A general strength of energy-based approaches is their ability to incorporate shape information. This is true for weak constraints such as boundary smoothness as well as strong constraints like adherence to an absolute shape. For many of the structures in electron tomograms, while the actual shape is not generally predictable, some distinctive geometric properties are known that can be exploited. For example, the tubular shape and distinct size of microtubules can be used in conjunction with active contours to extract them from tomograms of kinetochores (Jiang *et al.*, 2006). Similarly, template-based iterative boundary detection together with elliptic shape models can be used to generate high fidelity segmentations of *Caulobacter crescentus* cell membranes (Moussavi *et al.*, 2010). The downside of these approaches is that the models used are highly case specific and likely need non-trivial modifications and/or adjustments for each new application.

In contrast to energy-based algorithms, shape information is not easily incorporated into region-based approaches such as the watershed. While it is possible to impose weak constraints such as boundary smoothness *a posterioi*, the direct inclusion of more sophisticated shape models can not readily be done. A method called watersnakes (Nguyen *et al.*, 2003) combines the watershed transform and active contours (snakes) into a region growing technique with an energy function, allowing it to include shape information. This approach was used to generate high fidelity segmentation of membranes (Nguyen and Ji, 2008). However, the current implementation requires a rough manual segmentation in the form of a number of manually traced slices to define the shape model. Model-only approaches have also been used with encouraging results using patch templates for different types of membranes

(Coated vs. uncoated Lebbink *et al.*, 2007, 2010). However, these studies were limited to stained, plastic embedded sections.

In summary, there is now a fair number of segmentation approaches available that were tested with electron tomographic data. However, from a practical point of view, it is hard to predict which approach has the most promising prospects to yield the desired results. In addition, segmentation performance will likely also depend on which noise-reduction approach was used, making predictions even less certain.

4. DETECTION AND MAPPING OF MACROMOLECULAR ASSEMBLIES

While features like membranes, vesicles and, to some extent, filaments, can be detected, identified, and extracted with above-mentioned segmentation algorithms, macromolecular assemblies need to be addressed in a more direct fashion. One way is through explicit labeling of molecules of interest but this will only give access to a subset of assemblies. The more general and more attractive way of detection is through computational methods sometimes dubbed "visual proteomics" (Nickell *et al.*, 2006). The first feasibility test for detecting macromolecules in tomographic reconstructions using an algorithmic approach was done using correlation-based template matching on tomograms systems of purified thermosomes, 20S proteasomes, and GroEL, respectively (Böhm *et al.*, 2000). The results were quite encouraging with very high detection fidelity but the conditions of the specimen were far removed from the situation in cellular tomograms: cells are highly crowded with many different constituents and interacting partners. Follow-up tests were done with "phantom cells" (liposomes) filled with 20S proteasomes, thermosomes, or both (Frangakis *et al.*, 2002). In this environment, results are less convincing, but still encouraging. However, this constitutes still a best case scenario where crowding is absent and the templates are not only essentially perfect but also fairly dissimilar.

This type of template matching consists of using a "matched filter" which can be shown to be a Bayesian classifier (minimizing the probability of identification errors), as long as the template and the target are nearly identical and the noise is independent and identically distributed, Gaussian, and additive (Sigworth, 2004; van Trees, 1968). These conditions are not very well met for electron tomographic reconstructions: the noise is spatially correlated by the reconstruction process and the point-spread function; the tails of the noise distribution are often quite heavy, especially in stained samples (van der Heide *et al.*, 2007), making the noise distribution distinctly non-Gaussian; and the uncertainty in the magnification, the potential mix of conformations, and/or the presence of stain make it difficult to obtain sufficiently accurate templates. As a consequence, false hits tend to be generated by this method in areas of high density such as membranes or dense vesicles when used with cellular tomograms (Ortiz *et al.*, 2006; Rath *et al.*, 2003). Sensitivity of the detection performance to the template definition has also been observed (Rath *et al.*, 2003).

In summary, feasibility tests with simplified systems indicate that macromolecular assemblies in the size range of 0.5–1 MDa can be identified with satisfactory fidelity using correlation-based template matching in electron cryotomograms. However, these tests were done in the absence of molecular crowding and with essentially perfect templates. Applications to actual cellular tomograms indicate that there is still room for improvement, possibly through combination with other, alternative algorithms that are more robust to the noise features and to inaccuracies in the templates.

5. CLASSIFICATION AND AVERAGING

It has been realized that the quality and resolution of the raw densities of macromolecular assemblies extracted from electron tomograms are generally not good enough for direct structural interpretation or meaningful docking of atomic models. In order to boost the signal to make this feasible, the motifs must be aligned, classified, and averaged. The quality of the average depends critically on the accuracy of the 3D alignment. This alignment is not only hampered by the low signal-to-noise ratio in the tomograms but also by the missing data caused by the experimental setup. In addition, variations between the motifs (e.g., different conformations) may exist and must be sorted out before averaging to avoid blurring of details. These obstacles make the task of averaging motifs extracted from electron tomograms difficult. Several approaches have been developed recently that attempt to address these issues (Bartesaghi *et al.*, 2008; Förster *et al.*, 2007; Schmid and Booth, 2008; Winkler *et al.*, 2009).

An alternative and fairly general way of dealing with the missing data issue is by using weighting functions and Fourier space representations of the correlation coefficient which is then used as a scoring function for classification and alignment. The Pearson correlation coefficient can be expressed without loss of generality in Fourier space (Volkmann *et al.*, 1995). The missing data, which is expressed as an empty wedge or pyramid can easily accounted for during score calculation. We use the following mechanism for achieving this. The score is calculated according to the following formula:

$$C = \frac{\sum_{hkl} w_1 F_1 \cdot w_2 F_2 \cdot \cos(\phi_1 - \phi_2)}{\sqrt{\sum_{hkl}(w_1 F_1)^2 \cdot \sum_{hkl}(w_2 F_2)^2}}, \qquad (2.1)$$

where F_1 and F_2 are the amplitudes of the Fourier coefficients of the two densities, ϕ_i are the phases and w_i are weights. The sum is over all Fourier coefficients up to the resolution of the study. If all $w_i = 1$, this expression is exactly equivalent to the Pearson correlation coefficient in real space (Lunin and Woolfson, 1993). The weights can be exploited to account for the missing data. They can be set to 0 if a Fourier coefficient falls into the missing wedge and to 1 if it does not. Proper normalization is automatically taken care of. An additional advantage of this formulation is that arbitrary weights can be introduced without loss of generality. This will allow, for example, the use of fuzzy borders rather than sharp wedges, accounting for the uncertainty of the tilt angle determination or to weigh the Fourier terms in the average according to the number of contributing volumes. Application of a prototype version of this wedge-weighted correlation scheme has already proven useful in the determination of the Arp 2/3 branch junction structure, where it led to a resolution improvement from 32 to 26 Å (Rouiller et al., 2008). Another advantage of this scheme is that it can be equally well applied to account for missing data during correlation-based fitting of high-resolution structures into the averaged subvolumes (Volkmann, 2009; Volkmann and Hanein, 1999, 2003).

Classification or sorting of different conformations is an important step for improving the quality of the averages. This can be accomplished by pairwise scoring and hierarchical ascendant cluster analysis (Förster et al., 2007; Schmid and Booth, 2008) or alignment-through-classification techniques (Bartesaghi et al., 2008; Winkler et al., 2009), both of which have their strength and weaknesses. An alternative approach for sorting conformations is based on a locally focused classification procedure that uses the density distribution within selected variance peaks of the current average as a sorting criterion. This strategy enabled the separation of two distinctly different populations in the Bovine and Yeast Arp 2/3 mediated branch junction samples that were correlated with the orientation on the sample holder (Rouiller et al., 2008). Focused variance analysis was also used successfully in two dimensions for sorting two alternative positions of Arp2-attached GFP in actin branches (Egile et al., 2005). Through approximating the variance via bootstrapping (Penczek et al., 2006b), the same idea can be applied to single-particle reconstructions (Penczek et al., 2006a).

6. VALIDATION

Some of the major questions and reoccurring themes in the interpretation of electron tomograms are: Is this the best that can be done? Does the procedure used destroy information? Does it emphasize artifacts and leads to overinterpretation? Currently, there is no obvious answer to these questions. Other than the calculations for the template matching in the clean macromolecular and liposome systems, all evaluations are ultimately dependent on subjective human judgment. The associated difficulties are exemplified by the large discrepancies encountered by different human operators even for the relatively simple task of delineating membranes in relatively high contrast tomograms (Garduño et al., 2008; Nguyen and Ji, 2008). Furthermore, 3D scenes at the resolution of these tomograms are generally completely unfamiliar to the human eye and even in the absence of noise would likely appear rather chaotic due to the high degree of crowding in cells (Grünewald et al., 2003). As a consequence, the value of "ground truth" determination by human evaluation in this context is of questionable value. Humans are subject to all sorts of biases when evaluating visual clues (see, e.g., Harley et al., 2004; Maloney et al., 2005; Zhaoping and Jingling, 2008). While we could hope to average out individual subjectivity in scene evaluation tasks by averaging over many individuals, these species-dependent biases are not likely to be remedied by averaging.

The lack of adequate "ground truths" data dictates that the well-developed theories of signal detection (Green and Swets, 1989) and receiver operator characteristics (Hanley and McNeil, 1982) used in the medical imaging field are not readily applicable to the field of molecular resolution electron tomography. In medical imaging, the correctness of scene evaluation (diagnosis) can be readily evaluated (patient did/did not have cancer).

In summary, the perhaps most severe bottleneck for successful interpretation of molecular resolution electron tomograms is not necessarily the lack of tools; it is the lack of ability to evaluate the performance of available tools. Owing to general human biases and the alien nature of the molecular resolution scenes encountered in cellular environments, provision of hand-annotated data sets to establish "ground truth" for evaluation of algorithms is not likely to provide a satisfactory solution. Thus, the most viable path to remedy this situation may be the generation of realistic simulated data sets. However, this would not only require assembling realistic molecular scenes from atomic level structures, it would also require a push in developing appropriate imaging and noise models. Both are clearly difficult and challenging tasks. On the bright side, an added benefit of accurate noise models would be their potential usefulness in various image processing approaches that make explicit use of noise modeling such as maximum likelihood methods (Scheres et al., 2009, see also Chapter 10, Vol. 482).

ACKNOWLEDGMENTS

This work was supported by NIH grants GM076503, GM066311, and the NIGMS Cell Migration Consortium. I thank Dorit Hanein for critically reading the manuscript.

REFERENCES

Al-Amoudi, A., Chang, J. J., Leforestier, A., McDowall, A., Salamin, L. M., Norlen, L. P., Richter, K., Blanc, N. S., Studer, D., and Dubochet, J. (2004). Cryo-electron microscopy of vitreous sections. *EMBO J.* **23,** 3583–3588.

Al-Amoudi, A., Diez, D. C., Betts, M. J., and Frangakis, A. S. (2007). The molecular architecture of cadherins in native epidermal desmosomes. *Nature* **450,** 832–837.

Auer, M., Koster, A. J., Ziese, U., Bajaj, C., Volkmann, N., Wang, D. N., and Hudspeth, A. J. (2008). Three-dimensional architecture of hair-bundle linkages revealed by electron-microscopic tomography. *J. Assoc. Res. Otolaryngol.* (Epub ahead of print).

Bajaj, C., Yu, Z. Y., and Auer, M. (2003). Volumetric feature extraction and visualization of tomographic molecular imaging. *J. Struct. Biol.* **144,** 132–143.

Barash, D., and Comaniciu, D. (2004). A common framework for nonlinear diffusion, adaptive smoothing, bilateral filtering and mean shift. *Image Vis. Comput.* **22,** 73–81.

Bartesaghi, A., Sapiro, G., and Subramaniam, S. (2005). An energy-based three-dimensional segmentation approach for the quantitative interpretation of electron tomograms. *IEEE Trans. Image Process.* **14,** 1314–1323.

Bartesaghi, A., Sprechmann, P., Liu, J., Randall, G., Sapiro, G., and Subramaniam, S. (2008). Classification and 3D averaging with missing wedge correction in biological electron tomography. *J. Struct. Biol.* **162,** 436–450.

Baxter, W. T., Grassucci, R. A., Gao, H., and Frank, J. (2009). Determination of signal-to-noise ratios and spectral SNRs in cryo-EM low-dose imaging of molecules. *J. Struct. Biol.* **166,** 126–132.

Ben-Harush, K., Maimon, T., Patla, I., Villa, E., and Medalia, O. (2010). Visualizing cellular processes at the molecular level by cryo-electron tomography. *J. Cell Sci.* **123,** 7–12.

Best, C., Nickell, S., and Baumeister, W. (2007). Localization of protein complexes by pattern recognition. *Methods Cell Biol.* **79,** 615–638.

Beucher, S., and Meyer, F. (1993). The morphological approach to segmentation: The watershed transformation. *In* "Mathematical morphology in image processing," (E. Dougherty, ed.), pp. 433–481. Marcel Dekker, New York.

Böhm, J., Frangakis, A. S., Hegerl, R., Nickell, S., Typke, D., and Baumeister, W. (2000). From the cover: Toward detecting and identifying macromolecules in a cellular context: Template matching applied to electron tomograms. *Proc. Natl. Acad. Sci. USA* **97,** 14245–14250.

Egile, C., Rouiller, I., Xu, X. P., Volkmann, N., Li, R., and Hanein, D. (2005). Mechanism of filament nucleation and branch stability revealed by the structure of the Arp2/3 complex at actin branch junctions. *PLoS Biol.* **3,** e383.

Fernandez, J. J. (2009). TOMOBFLOW: Feature-preserving noise filtering for electron tomography. *BMC Bioinform.* **10,** 178.

Fernandez, J. J., and Li, S. (2003). An improved algorithm for anisotropic nonlinear diffusion for denoising cryo-tomograms. *J. Struct. Biol.* **144,** 152–161.

Förster, F., Pruggnaller, S., Seybert, A., and Frangakis, A. S. (2007). Classification of cryo-electron sub-tomograms using constrained correlation. *J. Struct. Biol.* **161,** 276–286.

Frangakis, A. S., and Förster, F. (2004). Computational exploration of structural information from cryo-electron tomograms. *Curr. Opin. Struct. Biol.* **14,** 325–331.

Frangakis, A. S., and Hegerl, R. (2001). Noise reduction in electron tomographic reconstructions using nonlinear anisotropic diffusion. *J. Struct. Biol.* **135,** 239–250.

Frangakis, A. S., and Hegerl, R. (2002). Segmentation of two- and three-dimensional data from electron microscopy using eigenvector analysis. *J. Struct. Biol.* **138,** 105–113.

Frangakis, A. S., Stoschek, A., and Hegerl, R. (2001). Wavelet transform filtering and nonlinear anisotropic diffusion assessed for signal reconstruction performance on multidimensional biomedical data. *IEEE Trans. Biomed. Eng.* **48,** 213–222.

Frangakis, A. S., Bohm, J., Forster, F., Nickell, S., Nicastro, D., Typke, D., Hegerl, R., and Baumeister, W. (2002). Identification of macromolecular complexes in cryoelectron tomograms of phantom cells. *Proc. Natl. Acad. Sci. USA* **99,** 14153–14158.

Frey, T. G., Perkins, G. A., and Ellisman, M. H. (2006). Electron tomography of membrane-bound cellular organelles. *Annu. Rev. Biophys. Biomol. Struct.* **35,** 199–224.

Garduño, E., Wong-Barnum, M., Volkmann, N., and Ellisman, M. H. (2008). Segmentation of electron tomographic data sets using fuzzy set theory principles. *J. Struct. Biol.* **162,** 368–379.

Goddard, T. D., Huang, C. C., and Ferrin, T. E. (2007). Visualizing density maps with UCSF Chimera. *J. Struct. Biol.* **157,** 281–287.

Green, D. M., and Swets, J. A. (1989). Signal Detection Theory and Psychophysics. Peninsula Publishing, Los Altos, CA.

Grimm, R., Typke, D., Barmann, M., and Baumeister, W. (1996). Determination of the inelastic mean free path in ice by examination of tilted vesicles and automated most probable loss imaging. *Ultramicroscopy* **63,** 169–179.

Grünewald, K., Medalia, O., Gross, A., Steven, A. C., and Baumeister, W. (2003). Prospects of electron cryotomography to visualize macromolecular complexes inside cellular compartments: Implications of crowding. *Biophys. Chem.* **100,** 577–591.

Gruska, M., Medalia, O., Baumeister, W., and Leis, A. (2008). Electron tomography of vitreous sections from cultured mammalian cells. *J. Struct. Biol.* **161,** 384–392.

Hanley, J. A., and McNeil, B. J. (1982). The meaning and use of the area under a receiver operating characteristic (ROC) curve. *Radiology* **143,** 29–36.

Harley, E. M., Carlsen, K. A., and Loftus, G. R. (2004). The "saw-it-all-along" effect: Demonstrations of visual hindsight bias. *J. Exp. Psychol. Learn. Mem. Cogn.* **30,** 960–968.

Hsieh, C. E., Leith, A., Mannella, C. A., Frank, J., and Marko, M. (2006). Towards high-resolution three-dimensional imaging of native mammalian tissue: Electron tomography of frozen-hydrated rat liver sections. *J. Struct. Biol.* **153,** 1–13.

Iancu, C. V., Wright, E. R., Heymann, J. B., and Jensen, G. J. (2006). A comparison of liquid nitrogen and liquid helium as cryogens for electron cryotomography. *J. Struct. Biol.* **153,** 231–240.

Janssen, M. E., Kim, E., Liu, H., Fujimoto, L. M., Bobkov, A., Volkmann, N., and Hanein, D. (2006). Three-dimensional structure of vinculin bound to actin filaments. *Mol. Cell* **21,** 271–281.

Jiang, W., Baker, M. L., Wu, Q., Bajaj, C., and Chiu, W. (2003). Applications of a bilateral denoising filter in biological electron microscopy. *J. Struct. Biol.* **144,** 114–122.

Jiang, M., Ji, Q., and McEwen, B. F. (2006). Model-based automated extraction of microtubules from electron tomography volume. *IEEE Trans. Inf. Technol. Biomed.* **10,** 608–617.

Kimmel, R., Malladi, R., and Sochen, N. A. (2000). Images as embedded maps and minimal surfaces: Movies, color, texture, and volumetric medical images. *Int. J. Comput. Vis.* **39,** 111–129.

Koning, R. I., and Koster, A. J. (2009). Cryo-electron tomography in biology and medicine. *Ann. Anat.* **191,** 427–445.

Kremer, J. R., Mastronarde, D. N., and McIntosh, J. R. (1996). Computer visualization of three-dimensional image data using IMOD. *J. Struct. Biol.* **116,** 71–76.

Lebbink, M. N., Geerts, W. J., van der Krift, T. P., Bouwhuis, M., Hertzberger, L. O., Verkleij, A. J., and Koster, A. J. (2007). Template matching as a tool for annotation of tomograms of stained biological structures. *J. Struct. Biol.* **158,** 327–335.

Lebbink, M. N., Jimenez, N., Vocking, K., Hekking, L. H., Verkleij, A. J., and Post, J. A. (2010). Spiral coating of the endothelial caveolar membranes as revealed by electron tomography and template matching. *Traffic* **11,** 138–150.

Leis, A. P., Beck, M., Gruska, M., Best, C., Hegerl, R., Baumeister, W., and Leis, J. W. (2006). Cryo-electron tomography of biological specimens. *IEEE Signal Process. Mag.* **23**, 95–103.

Leis, A., Rockel, B., Andrees, L., and Baumeister, W. (2009). Visualizing cells at the nanoscale. *Trends Biochem. Sci.* **34**, 60–70.

Li, Y., Leith, A., and Frank, J. (1997). Tinkerbell—A tool for interactive segmentation of 3D data. *J. Struct. Biol.* **120**, 266–275.

Lunin, V. Y., and Woolfson, M. M. (1993). Mean phase error and the map-correlation coefficient. *Acta Cryst.* **D49**, 530–535.

Luther, P. K., Lawrence, M. C., and Crowther, R. A. (1988). A method for monitoring the collapse of plastic sections as a function of electron dose. *Ultramicroscopy* **24**, 7–18.

Maloney, L. T., Dal Martello, M. F., Sahm, C., and Spillmann, L. (2005). Past trials influence perception of ambiguous motion quartets through pattern completion. *Proc. Natl. Acad. Sci. USA* **102**, 3164–3169.

Marsh, B. J. (2005). Lessons from tomographic studies of the mammalian Golgi. *Biochim. Biophys. Acta. Mol. Cell Res.* **1744**, 273–292.

Marsh, B. J., Mastronarde, D. N., Buttle, K. F., Howell, K. E., and McIntosh, J. R. (2001). Organellar relationships in the Golgi region of the pancreatic beta cell line, HIT-T15, visualized by high resolution electron tomography. *Proc. Natl. Acad. Sci. USA* **98**, 2399–2406.

Marsh, B. J., Volkmann, N., McIntosh, J. R., and Howell, K. E. (2004). Direct continuities between cisternae at different levels of the Golgi complex in glucose-stimulated mouse islet beta cells. *Proc. Natl. Acad. Sci. USA* **101**, 5565–5570.

Mastronarde, D. N. (1997). Dual-axis tomography: An approach with alignment methods that preserve resolution. *J. Struct. Biol.* **120**, 343–352.

McEwen, B. F., and Marko, M. (2001). The emergence of electron tomography as an important tool for investigating cellular ultrastructure. *J. Histochem. Cytochem.* **49**, 553–564.

McEwen, B. F., Downing, K. H., and Glaeser, R. M. (1995). The relevance of dose-fractionation in tomography of radiation-sensitive specimens. *Ultramicroscopy* **60**, 357–373.

Moussavi, F., Heitz, G., Amat, F., Comolli, L. R., Koller, D., and Horowitz, M. (2010). 3D segmentation of cell boundaries from whole cell cryogenic electron tomography volumes. *J. Struct. Biol.* **170**, 134–145.

Nguyen, H., and Ji, Q. (2008). Shape-driven three-dimensional watersnake segmentation of biological membranes in electron tomography. *IEEE Trans. Med. Imaging* **27**, 616–628.

Nguyen, H. T., Worring, M., and van den Boomgaard, R. (2003). Watersnakes: Energy-driven watershed segmentation. *IEEE Trans. Pattern Anal. Mach. Intell.* **25**, 330–342.

Nickell, S., Hegerl, R., Baumeister, W., and Rachel, R. (2003). Pyrodictium cannulae enter the periplasmic space but do not enter the cytoplasm, as revealed by cryo-electron tomography. *J. Struct. Biol.* **141**, 34–42.

Nickell, S., Kofler, C., Leis, A. P., and Baumeister, W. (2006). A visual approach to proteomics. *Nat. Rev. Mol. Cell Biol.* **7**, 225–230.

Noske, A. B., Costin, A. J., Morgan, G. P., and Marsh, B. J. (2008). Expedited approaches to whole cell electron tomography and organelle mark-up in situ in high-pressure frozen pancreatic islets. *J. Struct. Biol.* **161**, 298–313.

Ortiz, J. O., Forster, F., Kurner, J., Linaroudis, A. A., and Baumeister, W. (2006). Mapping 70S ribosomes in intact cells by cryoelectron tomography and pattern recognition. *J. Struct. Biol.* **156**, 334–341.

Osher, S., and Sethian, J. A. (1988). Fronts propagating with curvature-dependent speed—Algorithms based on Hamilton-Jacobi formulations. *J. Comput. Phys.* **79**, 12–29.

Pantelic, R. S., Rothnagel, R., Huang, C. Y., Muller, D., Woolford, D., Landsberg, M. J., McDowall, A., Pailthorpe, B., Young, P. R., Banks, J., Hankamer, B., and Ericksson, G. (2006). The discriminative bilateral filter: An enhanced denoising filter for electron microscopy data. *J. Struct. Biol.* **155**, 395–408.

Penczek, P., Marko, M., Buttle, K., and Frank, J. (1995). Double-tilt electron tomography. *Ultramicroscopy* **60**, 393–410.

Penczek, P. A., Frank, J., and Spahn, C. M. (2006a). A method of focused classification, based on the bootstrap 3D variance analysis, and its application to EF-G-dependent translocation. *J. Struct. Biol.* **154**, 184–194.

Penczek, P. A., Yang, C., Frank, J., and Spahn, C. M. (2006b). Estimation of variance in single-particle reconstruction using the bootstrap technique. *J. Struct. Biol.* **154**, 168–183.

Pierson, J., Fernandez, J. J., Bos, E., Amini, S., Gnaegi, H., Vos, M., Bel, B., Adolfsen, F., Carrascosa, J. L., and Peters, P. J. (2010). Improving the technique of vitreous cryosectioning for cryo-electron tomography: Electrostatic charging for section attachment and implementation of an anti-contamination glove box. *J. Struct. Biol.* **169**, 219–225.

Pintilie, G., Zhang, J., Chiu, W., and Gossard, D. (2009). Identifying components in 3D density maps of protein nanomachines by multi-scale segmentation. *IEEE NIH Life Sci. Syst. Appl. Workshop* **2009**, 44–47.

Pruggnaller, S., Mayr, M., and Frangakis, A. S. (2008). A visualization and segmentation toolbox for electron microscopy. *J. Struct. Biol.* **164**, 161–165.

Radon, J. (1917). Über die Bestimmung von Funktionen durch ihre Integralwerte längs gewisser Manningfaltigkeiten. *Math. Phys. Klasse* **69**, 262–277.

Rath, B. K., Hegerl, R., Leith, A., Shaikh, T. R., Wagenknecht, T., and Frank, J. (2003). Fast 3D motif search of EM density maps using a locally normalized cross-correlation function. *J. Struct. Biol.* **144**, 95–103.

Ress, D. B., Harlow, M. L., Marshall, R. M., and McMahan, U. J. (2004). Methods for generating high-resolution structural models from electron microscope tomography data. *Structure* **12**, 1763–1774.

Robinson, C. V., Sali, A., and Baumeister, W. (2007). The molecular sociology of the cell. *Nature* **450**, 973–982.

Rouiller, I., Xu, X. P., Amann, K. J., Egile, C., Nickell, S., Nicastro, D., Li, R., Pollard, T. D., Volkmann, N., and Hanein, D. (2008). The structural basis of actin filament branching by Arp2/3 complex. *J. Cell Biol.* **180**, 887–895.

Salje, J., Zuber, B., and Lowe, J. (2009). Electron cryomicroscopy of *E. coli* reveals filament bundles involved in plasmid DNA segregation. *Science* **323**, 509–512.

Salvi, E., Cantele, F., Zampighi, L., Fain, N., Pigino, G., Zampighi, G., and Lanzavecchia, S. (2008). JUST (Java User Segmentation Tool) for semi-automatic segmentation of tomographic maps. *J. Struct. Biol.* **161**, 287–297.

Sandberg, K. (2007). Methods for image segmentation in cellular tomography. *Methods Cell Biol.* **79**, 769–798.

Sandberg, K., and Brega, M. (2007). Segmentation of thin structures in electron micrographs using orientation fields. *J. Struct. Biol.* **157**, 403–415.

Scheres, S. H., Nunez-Ramirez, R., Gomez-Llorente, Y., San Martin, C., Eggermont, P. P., and Carazo, J. M. (2007). Modeling experimental image formation for likelihood-based classification of electron microscopy data. *Structure* **15**, 1167–1177.

Scheres, S. H., Melero, R., Valle, M., and Carazo, J. M. (2009). Averaging of electron subtomograms and random conical tilt reconstructions through likelihood optimization. *Structure* **17**, 1563–1572.

Schietroma, C., Fain, N., Zampighi, L. M., Lanzavecchia, S., and Zampighi, G. A. (2009). The structure of the cytoplasm of lens fibers as determined by conical tomography. *Exp. Eye Res.* **88**, 566–574.

Schmid, M. F., and Booth, C. R. (2008). Methods for aligning and for averaging 3D volumes with missing data. *J. Struct. Biol.* **161,** 243–248.

Sigworth, F. J. (2004). Classical detection theory and the cryo-EM particle selection problem. *J. Struct. Biol.* **145,** 111–122.

Stalling, D., Hege, H.-C., and Westerhoff, M. (2004). Amira—A highly interactive system for visual data analysis. *In* "Visualization Handbook," (C. R. Johnson and C. D. Hansen, eds.), Academic Press, Orlando, FL, USA.

Stoschek, A., and Hegerl, R. (1997). Denoising of electron tomographic reconstructions using multiscale transformations. *J. Struct. Biol.* **120,** 257–265.

Tocheva, E. I., Li, Z., and Jensen, G. J. (2010). Electron cryotomography. *Cold Spring Harb. Perspect. Biol.* **2,** a003442.

Tomasi, C., and Manduchi, R. (1998). Bilateral filtering for gray and color images. *Proc. ICCV '98* 839–846.

van der Heide, P., Xu, X. P., Marsh, B. J., Hanein, D., and Volkmann, N. (2007). Efficient automatic noise reduction of electron tomographic reconstructions based on iterative median filtering. *J. Struct. Biol.* **158,** 196–204.

van Trees, H. L. (1968). Detection, Estimation, and Modulation Theory. Wiley, New York.

Vincent, L., and Soille, P. (1991). Watersheds in digital space: An efficient algorithm based on immersion simulations. *IEEE Trans. Pattern Anal. Mach. Intell.* **13,** 583–598.

Volkmann, N. (2002). A novel three-dimensional variant of the watershed transform for segmentation of electron density maps. *J. Struct. Biol.* **138,** 123.

Volkmann, N. (2009). Confidence intervals for fitting of atomic models into low-resolution densities. *Acta Crystallogr. D. Biol. Crystallogr.* **65,** 679–689.

Volkmann, N., and Hanein, D. (2009). Electron microscopy in the context of systems biology. *In* "Structural Bioinformatics, 2nd edition," (J. Gu and P. Bourne, eds.), pp. 143–170. Wiley-Blackwell, New York.

Volkmann, N., and Hanein, D. (1999). Quantitative fitting of atomic models into observed densities derived by electron microscopy. *J. Struct. Biol.* **125,** 176–184.

Volkmann, N., and Hanein, D. (2003). Docking of atomic models into reconstructions from electron microscopy. *Methods Enzymol.* **374,** 204–225.

Volkmann, N., Schlünzen, F., Vernoslava, E. A., Urzhumtsev, A. G., Podjarny, A. D., Roth, M., Pebay-Peyroula, E., Berkovitch-Yellin, Z., Zaytzev-Bashan, A., and Yonath, A. (1995). On ab initio phasing of ribosomal particles at very low resolution. *Joint CCP4 ESF-EACBM Newslett.* **31.**

Whitaker, R. T., and Elangovan, V. (2002). A direct approach to estimating surfaces in tomographic data. *Med. Image Anal.* **6,** 235–249.

Winkler, H., Zhu, P., Liu, J., Ye, F., Roux, K. H., and Taylor, K. A. (2009). Tomographic subvolume alignment and subvolume classification applied to myosin V and SIV envelope spikes. *J. Struct. Biol.* **165,** 64–77.

Zhaoping, L., and Jingling, L. (2008). Filling-in and suppression of visual perception from context: A Bayesian account of perceptual biases by contextual influences. *PLoS Comput. Biol.* **4,** e14.

CHAPTER THREE

INTEGRATION OF CRYO-EM WITH ATOMIC AND PROTEIN–PROTEIN INTERACTION DATA

Friedrich Förster *and* Elizabeth Villa

Contents

Abstract

Cryoelectron microscopy (cryo-EM) is an increasingly popular method to elucidate the structures of macromolecular complexes. However, in many applications the resolution of cryo-EM densities is limited to the low or intermediate resolution regime, that is, $(10\,\text{Å})^{-1}$ or worse. Therefore, unambiguous molecular interpretation of cryo-EM densities requires efficient use of additional information, such as atomic structures of related subunits and protein–protein

Max-Planck Institute of Biochemistry, Department of Structural Biology, Martinsried, Germany

Methods in Enzymology, Volume 483
ISSN 0076-6879, DOI: 10.1016/S0076-6879(10)83003-4

interaction data. Here, we describe how information from different sources can be combined to determine the approximate molecular architecture of complexes. Molecular dynamics based flexible fitting protocols allow subsequent refinement of the atomistic models.

 1. INTRODUCTION

Cryoelectron microscopy (cryo-EM) is an increasingly popular method to elucidate the structures of macromolecular complexes. While X-ray crystallography and NMR spectroscopy are typically superior in resolution (both methods provide atomic models of the complex under scrutiny), the advantage of cryo-EM is its versatility. Compared to X-ray crystallography and NMR, cryo-EM is substantially less demanding in terms of sample amount, concentration, purity, and homogeneity. Thus, cryo-EM is particularly useful for obtaining structural insights into those complexes that are difficult to purify in high amounts, such as transient assemblies, membrane-associated protein complexes, and structurally heterogeneous macromolecules.

Single-particle analysis (SPA) and cryoelectron tomography (CET) are the cryo-EM methods of choice to study biochemically delicate macromolecular complexes. Both methods are restricted to relatively large assemblies; typically 250 kDa and larger, but continuous developments in hard- and software gradually reduce this limit (see also Vol. 482, Chapter 8). In many cases, structural heterogeneity of the assembly under scrutiny is nonnegligible. Then, the single-particle data need to be sorted into different bins according to the different conformers (see also Vol. 482, Chapters 11 and 12). Whereas in some cases sorting of single particles according to specific features yields high-resolution reconstructions, most notably for different states of the ribosome (Becker et al., 2009; Connell et al., 2007), structural heterogeneity limits the resolution to the medium or even low regime in most studies, that is, $(10–20 \text{ Å})^{-1}$ or worse than $(20 \text{ Å})^{-1}$, respectively.

CET is probably the most straightforward method to explore the structures of macromolecules associated to their native membranes. The method is capable of imaging pleiomorphic objects, such as virions or organelles, in 3D. When applying criteria from SPA, the resolution of cryoelectron tomograms does not exceed $(50 \text{ Å})^{-1}$ (Grünewald et al., 2003). However, the resolution can be increased by averaging subvolumes containing the same type of macromolecule, analogous to SPA (Förster et al., 2005). Using this approach, membrane-associated complexes can be resolved to $(20–30 \text{ Å})^{-1}$ (Förster and Hegerl, 2007).

Molecular interpretation of cryo-EM maps at low or medium resolutions is a challenge for a number of reasons: (i) Atomic models typically cannot be fitted into low or medium resolution maps precisely. As a rule of thumb, below a resolution of $(20 \text{ Å})^{-1}$ fits will be ambiguous. Nevertheless, this criterion depends ultimately on the size of the model: large models, for example whole subcomplexes, may often be fitted unambiguously at lower resolutions, while fitting of small fragments such as helices will require at least subnanometer-resolutions (Section 6). (ii) Accurate atomic models of many components may not be accessible. Many complexes studied by cryo-EM fail to crystallize due to structural flexibility of the subunits. For example, solenoid folds, which are present in many large eukaryotic complexes, have a substantial degree of structural heterogeneity to accommodate binding to different surfaces (Brohawn et al., 2009). (iii) Many complexes consist of evolutionary related subunits. Thus, the corresponding structures of these subunits are similar to each other, making it extremely hard to position these subunits based on geometric data only.

Here, we describe how to build models of assemblies using varied sources of data. The structures of many proteins or some of their domains can be predicted by comparative modeling based on known atomic structures. Using subunit models, EM maps, and protein–protein interaction data, the approximate quaternary structure of the subunits can be modeled. When the EM maps are of high resolution, the atomic models can be refined using flexible fitting methods.

2. The Problem of Placing Assembly Subunits into Cryo-EM Maps

As a prototypical application for interpretation of a cryo-EM map, we study the 26S proteasome. The 26S proteasome consists of the cylinder-shaped 20S core particle (CP), which is solved to atomic detail, and the regulatory particles (RPs), which associate to both cylinder ends (Förster et al., 2010). The RP possesses a high degree of inherent structural variability, which makes it hard to obtain high-resolution insights into the fully assembled 26S holocomplex. To date, the best-resolved structure $(20 \text{ Å})^{-1}$ of the 26S proteasome has been obtained using cryo-EM SPA (Nickell et al., 2009).

The RP contains six AAA-ATPases, Rpt1–6, which share a high degree of sequence similarity ($> 45\%$; Finley et al., 1998). AAA-ATPases typically assemble into hexameric rings (Vale, 2000). Thus, we expect that the six ATPase subunits assemble to a ring, but we do not know the topology of the six subunits within the ring. From the 26S proteasome EM map a density can be segmented consisting of two rings: the upper ring exhibits approximate threefold rotational symmetry, whereas the lower one is approximately sixfold symmetrical (Fig. 3.1A). The segmented density is placed at the cylinder ends, where numerous protein–protein interactions suggest that the AAA-ATPase is expected. Moreover, the estimated mass from

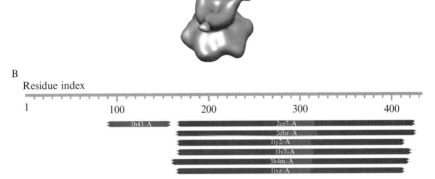

Figure 3.1 (A) Threefold rotationally symmetrized EM map of the proteasomal AAA-ATPases Rpt1–6. The upper ring is approximately threefold symmetrical, whereas the lower ring is approximately sixfold symmetrical. (B) Template search for *Drosophila melanogaster* Rpt1 using HHpred. The search results in one template covering residues 90–145 (3h43A) and several templates covering its AAA-ATPase domain (2ce7A, 2dhrA, 1iy2A, 1lv7A, 3h4mA, 1ixzA) ranging approximately from Rpt1 residue 150–420.

the extracted density (300 kDa) matches the mass of the Rpt1–6 (Nickell *et al.*, 2009). Thus, the segmented density almost certainly corresponds to the AAA-ATPase hexamer. While the resolution of 20 Å is sufficient to suggest quaternary structures, it is by no means sufficient to discern the different subunits, which are presumably highly similar to each other, as suggested by the high sequence identity. Thus, inferring the subunit topology can only be accomplished by a hybrid approach.

We describe how atomic models of the respective subunits are obtained and then fitted into the EM map in different topologies. These different candidate models are then assessed based on protein–protein interaction data. When high-resolution EM maps are available, the atomic coordinates can be refined further using molecular dynamics (MD) simulations.

3. Structure Prediction of Subunits

Structure prediction of proteins is the task of predicting a protein's 3D structure from its amino acid sequence. The most widely applicable approach to protein structure prediction is "comparative modeling": the structure of a protein is predicted based on features that are observed in other protein structures (templates; Marti-Renom *et al.*, 2000). This method is sometimes also referred to as "homology modeling" because the templates are usually derived from homologs. Alternative methods predict the protein structures without templates (*de novo*). While *de novo* structure prediction methods are increasingly successful, in particular the ROSETTA program (Schueler-Furman *et al.*, 2005), the methodology is still confined to relatively short proteins ($\sim$150 amino acids and less), which is prohibitive for many applications in cryo-EM modeling.

In comparative modeling, one structural template or multiple templates are chosen to build a model for the "target" sequence. Such features borrowed from the template are, for example, distances between corresponding atom pairs in the backbones of templates. Specifically, the distances between individual atoms in the template are determined and then imposed as "restraints" on the model. For a detailed description of comparative modeling, we refer to the more detailed book chapters, for example, Eswar and Sali (2007), Eswar *et al.* (2007).

3.1. Sequence alignment of target and template and finding appropriate structural templates

The accuracy of any comparative model will be determined by the correctness of the imposed restraints, and hence the similarity of the templates to the (unknown) structure of the target. Thus, the foremost requirement for building an accurate atomic model of a protein is the selection of appropriate templates. Moreover, the amino acid correspondences (sequence alignment) of target and template must be as precise as possible. Both problems are coupled and are therefore discussed in one section. Table 3.1 lists some of the respective web servers, which we find useful for our research.

3.1.1. Sequence based methods

The most straightforward way to compare the sequence of the target to another protein is a pairwise alignment of the target and template sequences. For example, the well-known Basic Local Alignment Search Tool (BLAST) is based on pairwise sequence–sequence comparison. Pairwise alignment methods produce largely correct sequence alignments when the sequence identity of the respective sequences exceeds 30–40%. Accordingly, these methods are successful in identifying (structural) homologs for these cases. However, below 30% sequence identity, these algorithms perform poorly in detecting homologs (Brenner *et al.*, 1998).

More sensitive methods use multiple sequences for sequence alignment. In sequence-profile alignment, the target sequence is compared to multiple prealigned sequences, that is, the "profile." Probably, the most popular variant of this class is Position-Specific Iterated (PSI)-BLAST. Another way to make use of multiple templates in the alignment is building a Hidden Markov Model (HMM) of the templates. The additional computational effort of sequence-profile and sequence-HMM alignment compared to

Table 3.1   Servers for sequence alignment and detection of structural templates

Server	Web address	Method
BLAST	http://blast.ncbi.nlm.nih.gov/Blast.cgi	Pairwise sequence alignment of target and proteins from PDB or other databases
CLUSTALW2	http://www.ebi.ac.uk/Tools/clustalw2/	Multiple-sequence alignment
MUSCLE	http://www.drive5.com/muscle/	Multiple-sequence alignment
T-COFFEE	http://tcoffee.vital-it.ch/	Multiple-sequence alignment; optionally builds a consensus alignment using different methods
PSI-BLAST	http://blast.ncbi.nlm.nih.gov/Blast.cgi http://www.ebi.ac.uk/Tools/psiblast/	Sequence-profile alignment of target and proteins from PDB or other databases
HHpred	http://toolkit.lmb.uni-muenchen.de/hhpred http://toolkit.tuebingen.mpg.de/hhpred	HMM–HMM alignment of target and proteins from PDB
FUGUE	http://tardis.nibio.go.jp/fugue/prfsearch.html	Threading-based homolog recognition
GenTHREADER	http://bioinf4.cs.ucl.ac.uk:3000/psipred/	Threading-based homolog recognition

The field is developing rapidly and the list is by no means comprehensive. For more extensive coverage of web servers we refer to Eswar and Sali (2007).

sequence–sequence alignment pays off in retrieving correct homologs, in particular, at low sequence identities.

The most sensitive methods employ multiple sequences instead of a single target sequence. Again, the methods can be categorized into profile–profile-based alignment and HMM–HMM-based alignment. For example, HHpred is a server for detecting appropriate structural templates from the PDB using HMM–HMM comparison (Soding, 2005; Soding *et al.*, 2005).

3.1.2. Threading

Another method for detecting remote homologs makes use of template structures. A coarse model of the target is built based on the target structure, typically by "threading" the target residues on the template structure, which is then assessed by a statistical potential. Popular softwares are FUGUE (Shi *et al.*, 2001) and GenTHREADER (Jones, 1999; Table 3.1).

3.2. Model building

In essence, the comparative target model is built using "features" observed in the template, for example, the distances between the Cα atoms of two residues, and generic features observed in protein structures, for example, bond angles, as restraints. Input for comparative modeling software, such as MODELLER (Sali and Blundell, 1993) are the structures of templates and the alignment of the target sequence and the templates. Optimization algorithms produce models, where the target features are as similar as possible to the corresponding template features. Thus, by definition, comparative modeling requires that the template(s) cover all parts of the modeled target. The generic restraints based on molecular force fields only allow modeling very short segments (< 6 amino acids) without a template when start and end positions are fixed ("loop modeling"). Since larger segments cannot be modeled precisely, larger loops typically need to be removed from the built model.

It may furthermore be necessary to dissect the target sequence into segments because longer proteins mostly consist of different domains, which often have different structural templates. In some cases, different templates may cover overlapping ranges of amino acids, which may then be used to approximate the quaternary structure of the respective domains. When templates do not overlap, the quaternary structure of the respective domains cannot be predicted by comparative modeling and the comparative models of the domains should be built separately.

The accuracy of comparative models is mostly limited by the structural deviations of the templates from the target and the goodness of the sequence alignments. Sequence identity and statistical measures may guide selection of appropriate templates, but biological knowledge on functional similarities may at least be equally important for selecting an appropriate template.

The accuracy of a comparative model typically increases as a function of sequence identity between target and template (Eswar et al., 2007). Whereas on average only about 80% of the Cα atoms are within 3.5 Å of their native positions when the sequence identity between target and template is approximately 30%, 95% of the Cα atoms are positioned largely correctly at 50% sequence identity (Eswar et al., 2007).

3.3. Building a structural model for the proteasomal AAA-ATPase hexamer

The six AAA-ATPases possess an AAA-fold that is preceded by an N-terminal fragment, which also contains its hallmark coiled–coiled domains (Frickey and Lupas, 2004). To select suitable structural templates, we used HHpred (Fig. 3.1B). We first searched structural templates for Rpt1. The search identified a plethora of different templates for the AAA-fold, and one template for the N-terminal fragment. From the different templates for the

AAA-fold, we chose the AAA-module of the proteasome-activating nucleotidase (PAN) because it is the prokaryotic homolog of the 26S proteasomal ATPases (AAA-PAN, PDB code: 3h4m). The template for the N-terminal segment is the N-terminal domain of PAN (N-PAN, PDB code: 3h43). The two templates do not overlap, that is, the quaternary structure of the two domains cannot be predicted. N-PAN is crystallized as a homohexamer (N-ring). AAA-PAN is not crystallized in the physiological homohexameric state; however, a homohexameric template (AAA-ring) can be obtained by superpositioning six AAA-PAN subunits on the HsLU hexamer (PDB code: 1DO2; Zhang *et al.*, 2009), for example, in UCSF Chimera (Pettersen *et al.*, 2004).

To build structural models of the AAA-ATPase hexamers, we save the alignments for the two templates from HHpred. We do not build models automatically in the *Bioinformatics Toolkit* (Biegert *et al.*, 2006) because this option is only designed for monomeric proteins. Analogously, we generate alignments for the remaining subunits Rpt2–Rpt6; in all cases, we select AAA-PAN and N-PAN as templates. To build an oligomer, the alignment of the different subunits is combined into a single file (Fig. 3.2A). Prior to model building, the PAN templates are positioned according to their probable quaternary structure (Section 5).

We then generate a comparative model using the oligomer alignment in MODELLER (Eswar *et al.*, 2007). The subunit topology can be varied by shuffling the sequence order in the alignment file (Fig. 3.2B). There are 5! different orders of subunits in the hexameric ring. Furthermore, the threefold symmetrical PAN structure has two structurally different monomers in the hexamer (*cis*- and *trans*-positions; Fig. 3.5B). Thus, 240 different quaternary structures can be generated. However, sequence conservation of a pivotal proline residue (corresponding to Pro-62 in PAN) suggests that the subunits Rpt2, Rpt3, and Rpt5 are placed at the *cis* positions of the template (Djuranovic *et al.*, 2009; Zhang *et al.*, 2009). Thus, the number of different models is reduced to 12. A script for systematic shuffling of the AAA-ATPase subunits in the alignment can be downloaded from www.biochem.mpg.de/foerster. The script is specific for the AAA-ATPase case, but can be easily adjusted for other applications. The remaining task is to combine the models for N- and AAA-ring into appropriate quaternary structures and to select the most probable configuration according to EM and protein–protein interaction data.

4. Protein–Protein Interaction Data

Here, we summarize different types of protein–protein interaction data. First, we discuss the most common protein–protein interaction assays, which are the basis of almost all to-date publications and databases. In addition,

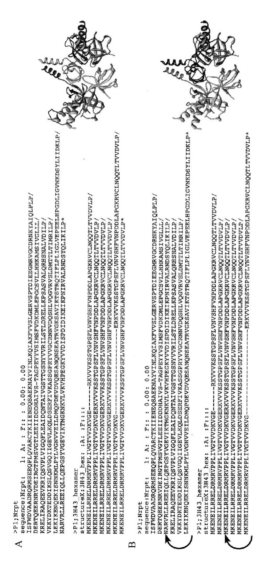

A

```
>P1;Nrpt
sequence:Nrpt:    1: A: : F:: : 0.00: 0.00
ISFWDVAADRQRMSEEQPLQVAARCTKIIENEQSAEKNAYVINLKQLAKFVVSLGERVSPTDIEEGEHRVGCDRNKYAIQLPLP/
DERTVQEERMRVDEIRGTPMSVGTLEEIIDDDRAIVS-TAGPEYYVSIMSFVDKDMLEPGCSVLLHHKAMSIVGLLL/
KRELIRAQEEVKRIQSVPLVIGQFLEAIDQNTAIVGSTTGSNYVVRILSTLDRELLKPSASVALQRHSNALVDIILP/
VKKYDKTEDDIKSLQSVGQIIGEVLKQLDSERFIVKASSGPRYVVGCRNNVDQSHLVQGVRVSLDMTTLAIMRILP/
LEKIKENQERKISNNKMLPYLVGNVVEILDMQPDEVDVQESAMQNSEATRVGKSAVIKTSTRQTIFLPLIGLVFPEELHFGDLIGVNKDSYLIIDKLP*
NARVRLLREEIQLLQEPGSYVGEVIKTMGRNKVLVKVHFPEGKYVVDISPDIDIKEIKPNIRVALRNDSYQLIKILP*
```

```
>P1;3H43_hexamer
structureX:3H43_hex: :A: :F:::
MKENEILRRELDRMRVPFPLIVGTVVDKVGE------------RKVVVKSSTGPSFLVNVSHFVNPDDLAPGKRVCLNQQTLTVVDVLP/
MKENEILRRELDRMRVPPLIVGTVVDKVGERKVVVKSSTGPSFLVNVSHFVNPDDLAPGKRVCLNQQTLTVVDVLP/
MKENEILRRELDRMRVPPLIVGTVVDKVGERKVVVKSSTGPSFLVNVSHFVNPDDLAPGKRVCLNQQTLTVVDVLP/
MKENEILRRELDRMRVPPLIVGTVVDKVGERKVVVKSSTGPSFLVNVSHFVNPDDLAPGKRVCLNQQTLTVVDVLP/
MKENEILRRELDRMRVPPLIVGTVVDKVG------------ERKVVVKSSTGPSFLVNVSHFVNPDDLAPGKRVCLNQQTLTVVDVLP*
MKENEILRRELDRMRVPPLIVGTVVDKVG*
```

B

```
>P1;Nrpt
sequence:Nrpt:    1: A: : F:: : 0.00: 0.00
ISFWDVAADRQRMSEEQPLQVAARCTKIIENEQSAEKNAYVINLKQLAKFVVSLGERVSPTDIEEGEHRVGCDRNKYAIQLPLP/
DERTVQEERMRVDEIRGTPMSVGTLEEIIDDDHAIVS-TAGPEYYVSIMSFVDKDMLEPGCSVLLHHKAMSIVGLLL/
NARVRLLREEIQLLQEPGSYVGEVIKTMGRNKVLVKVHFPEGKYVVDISPDIDIKEIKPNIRVALRNDSYQLIKILP/
KRELIRAQEEVKRIQSVPLVIGQFLEAIDQNTAIVGSTTGSNYVVRILSTLDRELLKPSASVALQRHSNALVDIILP/
VKKYDKTEDDIKSLQSVGQIIGEVLKQLDSERFIVKASSGPRYVVGCRNNVDQSHLVQGVRVSLDMTTLAIMRILP/
LEKIKENQERKISNNKMLPYLVGNVVEILDMQPDEVDVQESAMQNSEATRVGKSAVIKTSTRQTIFLPLIGLVFPEELHFGDLIGVNKDSYLIIDKLP*
```

```
>P1;3H43_hexamer
structureX:3H43_hex: :A: :F:::
MKENEILRRELDRMRVPPLIVGTVVDKVGERKVVVKSSTGPSFLVNVSHFVNPDDLAPGKRVCLNQQTLTVVDVLP/
MKENEILRRELDRMRVPPLIVGTVVDKVGERKVVVKSSTGPSFLVNVSHFVNPDDLAPGKRVCLNQQTLTVVDVLP/
MKENEILRRELDRMRVPPLIVGTVVDKVGERKVVVKSSTGPSFLVNVSHFVNPDDLAPGKRVCLNQQTLTVVDVLP/
MKENEILRRELDRMRVPPLIVGTVVDKVGERKVVVKSSTGPSFLVNVSHFVNPDDLAPGKRVCLNQQTLTVVDVLP/
MKENEILRRELDRMRVPPLIVGTVVDKVG------------ERKVVVKSSTGPSFLVNVSHFVNPDDLAPGKRVCLNQQTLTVVDVLP*
MKENEILRRELDRMRVPPLIVGTVVDKVG*
```

Figure 3.2 Sequence alignment of the six proteasome ATPases Rpt1–6 (*Nrpt*) and to the respective sequence of the N-PAN hexamer (*3H43_hexamer*) and the corresponding comparative models. (A) Alignment in subunit order Rpt1/Rpt2/Rpt3/Rpt4/Rpt5/Rpt6 (colors: red/orange/yellow/green/blue/purple). (B) The subunit order is varied in the alignment by placing Rpt6 at the 3rd position (Rpt1/Rpt2/Rpt6/Rpt3/Rpt4/Rpt5). Accordingly, the order of subunits is changed in the resulting comparative model. (For interpretation of the references to color in this figure legend, the reader is referred to the Web version of this chapter.)

we discuss emerging methods relying on mass spectrometry (MS). Finally, we briefly summarize computational means to predict protein–protein interactions.

4.1. State-of-the art protein–protein experimental methods

4.1.1. Two-hybrid assay

One candidate protein ("bait") is fused to a transcription factor DNA binding domain, and another to the activation domain ("prey"). Activation of a reporter gene implies a physical interaction between the two proteins. However, the assay is prone to false negatives as well as false-positives (Aloy and Russell, 2002b; Fields, 2005). One intuitive way to decrease the number of false-positive two-hybrid interactions is to consider only those interactions that were reported in two or more independent publications.

4.1.2. *In vivo* pulldown

Coimmunoprecipitation or tandem affinity purification (TAP) can be used to purify complexes that include a bait protein. Subsequent analysis, traditionally by Western blotting or more recently by MS, reveals the identity of affinity-purified complexes. Variation of elution buffers often allows purification of different subcomplexes for the same bait. The experiment indicates physical interactions of the detected subunits, but physical contacts between pairs of proteins typically cannot be deduced from the data because subcomplexes often comprise more than two proteins.

4.1.3. *In vitro* binding assays

A bait protein is recombinantly expressed and attached to glutathione *S*-transferase (GST) beads. The beads are incubated in cell lysate, again separated from the lysate, and analyzed, typically by Western blotting (Kaelin *et al.*, 1991). Alternatively, the complex may be dissociated using a 2D gel, transferred to a nitrocellulose matrix, and then exposed to *in vitro* and *in vivo* expressed protein, washed, and analyzed (Towbin *et al.*, 1979). *In vitro* binding assays reveal binary physical interactions.

4.1.4. Chemical cross-linking

The protein complex is chemically cross-linked, typically using relatively extended cross-linking agents such as bis-sulfosuccinimidylsuberate (BS^3) that cross-link specific residues (e.g., lysines) in spatial proximity (Sinz, 2003). The cross-linked samples are then typically separated in a denaturing 2D gel and subsequently analyzed by Western blotting or MS. The experiment reveals spatial proximity of two or more proteins. Direct physical interactions cannot be inferred with certainty, as typically used cross-linking agents are more than 12 Å long.

4.1.5. Coexpression

Expression of two or more proteins in a heterologous expression system or an *in vitro* translation system and subsequent purification of a complex of these proteins implies a stable subcomplex of the respective proteins.

4.2. Emerging experimental methods

4.2.1. Mass spectrometry of whole complexes

MS of whole, intact complexes isolated from cells reveals the mass of specific complexes (Sharon and Robinson, 2007). When the subunits are identified by conventional shotgun MS, the stoichiometry of the complex under scrutiny can be determined. In addition, subunits that are peripherally located in the complex can be identified by collision of the complex in the gas phase. Compared to TAP, the advantage is that the subunits of a subcomplex are comprehensive and the stoichiometry of the complexes can be revealed.

4.2.2. Residue-specific cross-linking

All of the aforementioned experimental methods have relatively low resolution: it is not resolved *where* proteins interact, that is, the interaction could, in principle, occur on arbitrary positions of the involved interfaces. Thus, the resolution of these techniques is on the order of the diameters of the involved proteins. For modeling, higher resolution information would be highly desirable. New developments in chemical cross-linking, MS, and bioinformatics allow the identification of residues that are connected in cross-linking experiments (Leitner *et al.*, 2010; Maiolica *et al.*, 2007). These approaches indicate that two cross-linked residues (lysines for most cross-linking agents) are in spatial proximity (distances smaller than the cross-link).

4.3. Computational interaction and interface prediction

4.3.1. Statistical potentials

It is tempting to exploit the structures of solved protein complexes in order to predict the quaternary structures of other proteins. Statistical potentials for complex modeling aim to devise statistical rules from the PDB, which can then be used to assess candidate models similarly to "threading" for fold prediction (Aloy and Russell, 2002a). However, in one instance the assembly order of the exosome core hexamer was predicted wrongly based on statistical potentials (Aloy *et al.*, 2002) while comprehensive two-hybrid screens (Raijmakers *et al.*, 2002), as well as *in vitro* pulldown assays (Lorentzen *et al.*, 2005), yielded the correct topology of the exosome core hexamer, as later determined by X-ray crystallographic analysis (Liu *et al.*, 2006).

4.3.2. Interface prediction

Various web servers predict the interfaces of a protein where a protein–protein interaction occurs (Zhou and Qin, 2007), which can then be used for subsequent modeling. Most of these servers focus on binary interactions. The different servers typically use sequence conservation and predicted physical properties such as solvent accessibility to identify patches on the protein where protein–protein interactions might occur. Satisfactory predictions typically require that the complex-forming proteins are well represented in the Protein Data Bank (Zhou and Qin, 2007).

4.4. Gathering interproteasomal interactions

To model the proteasomal ATPases, we compile all available experimental protein–protein interaction data from databases (Table 3.2; mostly The Biogrid) and publications (Förster *et al.*, 2009). The majority of experimental data stems from two-hybrid assays, which are particularly prone to false-positives. To improve the accuracy of the compiled data, we discard those two-hybrid interactions that were only reported in a single publication.

We store the interaction data in an *xml* file. The file contains information such as the interacting proteins, data source, specification of the interacting fragments if available, and length of cross-linker (Fig. 3.3). The *xml* format is readable for subsequent scoring (Section 5).

5. MODEL BUILDING OF A COMPLEX USING CRYO-EM AND ADDITIONAL DATA

The task of placing different "puzzle pieces" correctly is essentially approached in model building. Electron microscopy data have a pivotal role for assembling the pieces: a cryo-EM map provides a frame and

Table 3.2 Servers for protein–protein interaction data

Server	Web address	Method
Biogrid	http://www.thebiogrid.org/	Database of physical and genetic protein–protein interactions
iHop	http://www.ihop-net.org/UniPub/iHOP/	Text mining for associated proteins
Krogan Interactome	http://interactome-cmp.ucsf.edu/	Complexes identified by high-throughput TAP

```
<proteasome_restraints>
  <restraint_set name="xlink">
    <restraint span="11.">
      <protein name="Rpt1"/>
      <protein name="Rpt2"/>
      <src auth="Hartmann-Petersen" jour="Phys. Genomics" year="2001"/>
      <comment txt="cross linker BS3 (11.4 A)"/>
    </restraint>
  </restraint_set>
  <restraint_set name="pulldown">
    <restraint>
      <protein name="Rpn2" aa_start="797" aa_end="953"/>
      <protein name="Rpn13" aa_start="1" aa_end="124"/>
      <src auth="Yao" jour="NCB" year="2006"/>
      <src auth="Hamazaki" jour="EMBO" year="2006"/>
      <src auth="Gandhi" jour="Nature Genetics" year="2006"/>
      <src auth="Schreiner" jour="Nature" year="2008"/>
    </restraint>
  </restraint_set>
</proteasome_restraints>
```

Figure 3.3   File containing two example restraints from published data. The restraints are categorized into *restraint_sets* according to the experimental technique. The first restraint is based on chemical cross-linking. Rpt1 and Rpt2 are in spatial proximity, but not necessarily in direct contact; the spacer arm (*span*) sets an upper limit to the distance between the proteins. The second restraint stems from *in vivo pulldown*. In these experiments, the interaction could be narrowed down to two fragments in Rpn2 and Rpn13, which are specified by the first and last residues (*aa_start* and *aa_end*, respectively).

skeleton for the different subunits, which vastly reduces the assembly space. Different candidate models that are compatible with the EM data can then be ranked using protein–protein interactions.

5.1. Assembly representation

Prior to building a model, it must be decided how to represent a protein complex. Whereas, we represented proteins by their constituting atoms for comparative modeling, this choice may not necessarily be appropriate for assembly modeling: when the templates are of very low sequence identity (< 25%) the level of detail greatly exceeds the accuracy of the model. Moreover, the computational complexity, that is, the conformational space of subunits, may be too large to be sampled at the atomic level. Thus, coarser representations may be chosen for specific applications, which summarize several atoms or residues in one bead (Lasker *et al.*, 2010). While most modeling software is designed for atomic representation and few specific input data, the Integrative Modeling Platform (www. salilab.org/imp) offers flexibility to work at different granularity and to make use of self-defined restraints.

5.2. Scoring of assemblies

5.2.1. EM maps

EM maps provide the density of the protein assembly at a given resolution. Thus, a close-to-native assembly model should possess a density that closely matches the experimental data. The model density can be simulated by sampling the electron-optical density according to the resolution (see also Chapter 1). When coarse representations are chosen, it is essential to incorporate the diameter of the constituting beads.

The most straightforward measure of similarity between model density M and experimental density E is the standard error χ^2:

$$\chi^2 = \frac{1}{N} \sum_i^N \frac{(E_i - M_i)^2}{\sigma_i^2}. \tag{3.1}$$

Here, σ_i is the error of E at each voxel and N denotes the number of voxels N. It can be shown that χ^2 is proportional to the logarithm of the probability that the model matches the EM density (Jaynes, 2003). However, the error of the experimental densities is difficult to quantify and the gray value of EM data is arbitrary. Therefore, the error of the EM map is typically ignored (i.e., $\sigma = 1$) and the scaling problem is solved by either introducing a fitting parameter (Fabiola and Chapman, 2005), or by normalizing model, and EM density. The latter is more common: when E and M are mean-free and divided by their standard deviation the well-known cross-correlation coefficient is $CCC = 1 - \chi^2$ (see also Chapter 11). $CCC = 1$ indicates identity of model density and experimental density while $CCC = 0$ signifies that model and data are unrelated.

Alternative measures filter EM map and model map prior to correlation. For example, Laplacian filtering enhances high-frequency features and hence sharpens edges of molecules. While Laplacian filtering may increase accuracy when the EM map exhibits a large surface and SNR is high (Wriggers and Chácon, 2001), for most cases it holds a higher risk of obtaining entirely wrong subunit placements: when single subunits are fitted these will be preferably placed at the surface. Moreover, the Laplacian filtering greatly enhances noise because the upweighted high frequencies typically have a lower SNR, which may also decrease fitting accuracy.

Almost all software fits a model into the experimental density and does not incorporate other features into the scoring function (Table 3.3). Most available programs aim to find the position of a single rigid model in the EM density yielding the highest score. Some programs also retrieve multiple positions that are considered significant (Volkmann and Hanein, 1999). The sampling strategies vary: most methods perform an exhaustive search of the six-dimensional parameter space (three translations and three rotations)

Table 3.3   Selection of programs for fitting rigid models into EM maps

Program	Method	Reference
UCSF Chimera	Visualization program; interactive positioning followed by automated local CCC optimization	Goddard *et al.* (2007)
MOD-EM	Monte-Carlo optimization of CCC	Topf *et al.* (2005)
Foldhunter	Six-dimensional exhaustive search for maximum CCC and local optimization	Jiang *et al.* (2001)
SITUS	Six-dimensional exhaustive search for maximum CCC or Laplacian filtered CCC	Wriggers *et al.* (1999)
CoAn	Six-dimensional exhaustive search for significant $CCCs$	Volkmann and Hanein (1999)
NORMA	Six-dimensional search for max CCC using spherical harmonics	Suhre *et al.* (2006)

The programs use different methods to find the maximum correlation of a model and the EM data.

using different mathematical transformations to speed-up the process (Fourier transformations or spherical harmonics) while other methods perform rapid local optimization algorithms making use of the score gradient (e.g., steepest ascent). Since local optimization will result in local optima, it is essential for these methods that the intial position is approximately correct, which can be ensured by a coarse exhaustive sampling before or by approximate manual fitting.

The accuracy of the different correlation-based methods is largely comparable—but difficult to predict. In general, the accuracy will increase with the size of the fitted model and the resolution of the EM map because the number of compared voxels increases. One way to approximate the error bar of fits is to do independent reconstructions of the experimental data using random sets of particles, as done for Fourier ring correlation. The model can be fitted into these different densities, which will result in different $CCCs$ and peak positions. The corresponding standard deviations can be used to assess the significance of a fit; for example, a fit is significantly better compared to other fits if its CCC exceeds the CCC of competing fits by more than 3 standard deviations. We used this strategy to assess the fit of the 20S CP into the density of the 26S proteasome (Förster *et al.*, 2009). However, a disadvantage of this approach is that splitting the data into different bins will result in reconstructions with lower resolution compared to the reconstruction from the entire particle set. Hence, the estimate for the error bar is likely too big. The best way of validating a fit is certainly using orthogonal information, such as protein–protein interaction.

5.2.2. Protein–protein interactions

We suppose, we are interested in a complex containing 10 different subunits (subunits A–J). The knowledge, for example, from an *in vivo* pulldown experiment, that subunits A, B, and C form a ternary subcomplex would be valuable structural information: the subcomplex implies that the respective proteins are in physical contact, which greatly reduces the space of subunit configurations. However, almost all types of protein–protein interaction data do not specify *where* the subunits interact (Section 4). In our example, the data do not specify whether subunit A is in physical contact with subunit B or C (or both). Moreover, we do not know at which residues the subunits interact. Thus, the data are ambiguous in the sense that vastly different subunit conformations may be equally likely considering only this information.

For scoring a given model using protein–protein interaction data, we utilize the "conditional" or "ambiguous" restraint suggested by Alber *et al.* (2007, 2008). For our given ternary subcomplex, all interprotein distances are determined first (Fig. 3.4A). A ternary subcomplex requires at least two subunit–subunit contacts involving the three proteins (Fig. 3.4B): the so-called minimum spanning tree within the ternary subcomplex is calculated

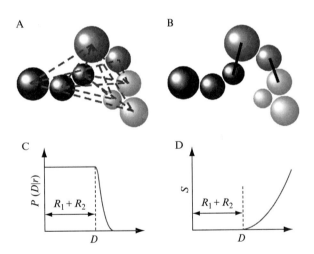

Figure 3.4 Scoring protein–protein interactions for a ternary subcomplex using a conditional restraint. (A) First, all interprotein distances are determined. (B) Then, only the closest two distances connecting the three proteins are retained (minimum spanning tree). (C) Only these two distances are restrained. For scoring, we approximate the probability distribution function $P(D|r)$ for each distance D given bead coordinates r as follows: P is constant if D is smaller than the sum of the bead radii of the involved particles (R_1 and R_2, respectively) and P decays as a Gaussian function for $D > R_1 + R_2$. The half width of the Gaussian reflects the accuracy of the experiment and the coordinates. (D) As a score S, we use the negative logarithm of P.

and only the two resulting distances are used for subsequent scoring. In essence, a model is penalized when the two distances exceed the sum of the radii of the interacting particles (Fig. 3.4C and D).

5.3. Building and assessment of AAA-ATPase models

For the example of the proteasomal AAA-ATPases, we use the EM data and the interaction restraints sequentially because they provide complementary information. From the EM data, we can deduce the possible positions of the subunits (Fig. 3.1A). Since the map and the template are threefold rotationally symmetrical, the EM data do not provide any information for inferring the subunit topology. In contrast, the interaction data contain information on subunit topology, but due to the low resolution it does provide information for subunit positioning.

First, we fit the PAN templates into the EM map using UCSF Chimera (Goddard *et al.*, 2007): the rings are first manually positioned and then refined using the Fitting tool (Fig. 3.5A). The threefold symmetrical N-ring fits neatly into the threefold symmetrical upper ring of the EM map. The sixfold symmetrical AAA-ring can be positioned in two different ways into the approximately sixfold symmetrical lower ring: the $C\alpha$–$C\alpha$ distances between the N-PAN C-terminus (aa 149) and the AAA-PAN N-terminus (aa 158) measure 25 and 26 Å, respectively, and they can be bridged by the eight residues in both arrangements, whereas the distances to the other four subunits are too large.

We then shuffle the order of the subunits by varying the alignment file for model building (see above). The subunits Rpt2, Rpt3, and Rpt5 are confined to the *cis* positions (Fig. 3.5B), according to the aforementioned structural argument. Thus, we have 60 different subunit topologies for the Rpt hexamer: 30 different subunit topologies for the N–ring, and for each

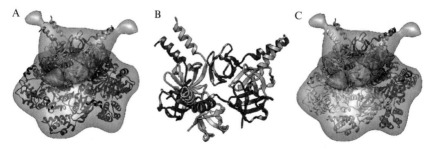

Figure 3.5 (A) N-PAN (green, cyan) and AAA-PAN (blue). (B) N-PAN hexamer. *cis-* (cyan) and *trans*-positions (green) of the N-PAN monomers in the N-ring. (C) Best-scoring configuration of comparative models (Rpt1/Rpt2/Rpt6/Rpt3/Rpt4/Rpt5) fitted into the EM map. (See Color Insert.)

N-ring topology two different options to position the AAA-ring (rotated by 60° with respect to each other).

We score the 60 different subunits based on the number of violated contacts using the subcomplex restraint. Specifically, we used subcomplex data of the Rpt subunits obtained by chemical cross-linking (8 restraints), two-hybrid assays (6), *in vitro* binding assays (4), and coexpression (1) (Table 1 in Förster *et al.*, 2009). We consider a restraint violated if the closest distance between two atoms from interacting proteins exceeds the standard deviation 2.5 Å, which we estimate to be the accuracy of the comparative models given the ∼40% sequence identity between targets and template (Eswar *et al.*, 2007). The scoring is performed in *IMP*, which we interface by our python library *sphIMP* (Förster and Lasker; www.biochem.mpg.de/foerster).

A single N-ring topology (Rpt1/Rpt2/Rpt6/Rpt3/Rpt4/Rpt5) violates a minimum of two restraints. The second-best scoring solution (Rpt1/Rpt2/Rpt4/Rpt5/Rpt6/Rpt3) violates three restraints. The protein–protein interaction data cannot discriminate between the two corresponding AAA-ring placements. Recently, the ring order of the proteasomal AAA-ATPases has been confirmed experimentally to be Rpt1/Rpt2/Rpt6/Rpt3/Rpt4/Rpt5 (Tomko *et al.*, 2010).

We can find the most probable position of the AAA-ring in the context of the 26S proteasome using the EM map of the entire 26S proteasome (Förster *et al.*, 2009): the 20S CP can be placed unambiguously into the map and the six different orientations of the AAA-ring (each rotated by 60°) can be scored using the protein–protein interactions of CP and AAA-ATPase, as described for the AAA-ATPase topology. Using the AAA-ring position, which results in minimum violations, we obtain a model for the quaternary structure of the whole AAA-ATPase (Fig. 3.5C). However, the quaternary structure of AAA-ring and N-ring is based on only three CP-AAA restraints and it will need to be solidified by further data.

6. REFINEMENT OF ATOMIC MODELS USING HIGH-RESOLUTION MAPS

When the macromolecule under scrutiny is structurally homogeneous, cryo-EM maps may reach subnanometer-resolution (see also Vol. 482, Chapter 8). At subnanometer-resolution, secondary structure elements of proteins are often discernible from the density. Atoms may be positioned into an EM map with an accuracy greatly exceeding the resolution of the EM map, similar to X-ray crystallography model building. In the latter, the amino acids are not necessarily discernible from the density, but their chemical structure and preferred isomers are inferred from *a priori* knowledge

(Brunger *et al.*, 1987). In flexible fitting, the atom positions of an input model (e.g., a crystal structure of the protein in a different context or a comparative model) are refined using the EM data. Flexible fitting is not only useful to study structural changes of proteins, but also permits the accurate segmentation of an EM map (Seidelt *et al.*, 2009).

Here, we describe the flexible fitting of models in EM maps using molecular dynamics flexible fitting (MDFF). Since the whole 26S proteasome is not available at high resolution yet, we illustrate the protocol for refining an atomic model of the 20S CP.

6.1. Molecular dynamics flexible fitting

MD simulations have been enormously successful in X-ray crystallographic refinement (Brunger *et al.*, 1987) and can likewise be employed for EM fitting. MD simulations are based on force fields that define the interactions between all atoms in a protein complex (Karplus and McCammon, 2002). Thus, MD preserves the correct stereochemistry of the protein complex, while morphing the protein according to its intrinsic flexibility. MD can be extended to incorporate other sources of data; notably, 3D EM density maps can be incorporated as an additional term in the force field that acts on the atomic model by driving atoms into high-density areas following the steepest descent of the 3D density (Fig. 3.6A; Trabuco *et al.*, 2008).

It is of utmost importance to note that flexibly fitting atomic models into EM maps bear the risk of overfitting; when the resolution of the EM data is too low, the ratio of unknown coordinates versus data points is poor. Moreover, EM maps, in particular, at moderate resolutions, may contain errors due to flexibility of the protein complex, but also artifacts introduced by imaging and processing. In general, the risk of overfitting is extremely high when the resolution is worse than $(10 \text{ Å})^{-1}$. It should also be kept in mind that resolution is a global measure (see also Vol. 482, Chapter 3) and that the quality of the map may be locally substantially worse due to structural heterogeneity.

To avoid overfitting, further restraints can be added to the force field in the form of springs to preserve the relative positions between pairs of atoms (distances or "bonds"), triads of atoms (angles), or between four atoms (dihedrals). MDFF imposes restraints to dihedral angles of secondary structure elements in proteins, that is, alpha helices and beta strands.

The MDFF simulations yield an ensemble of structures that conform to the EM data, which are representative of the uncertainty of the models. The quality of the fit is monitored by CCC along the simulation. When the value of this metric has equilibrated, all the conformations visited are likely to be present in the original vitrified sample. The deviation of the different conformations is a measure for the precision of the fitted model. The interpretation of flexibly fitted models must be done taking into account

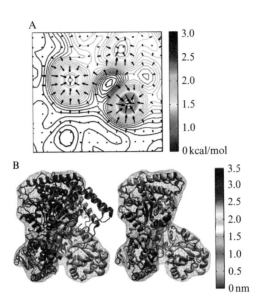

Figure 3.6   Molecular dynamics flexible fitting. (A) Cross-section of a 2D slice of the potential derived from the EM density of elongation factor Tu (EMD-5036) represented as a contour plot. Arrows represent forces driving the atomic structure toward high-density regions. The circular areas correspond to cross-sections of alpha helices. (B) We use carbon monoxide dehydrogenase to illustrate MDFF: carbon monoxide dehydrogenase adopts two different conformations in the same crystal (PDB code: 1OAO). The EM density of the first structure (gray ribbon) is simulated to $(10\ \text{Å})^{-1}$ resolution (gray mesh). We use the second crystal structure (left; color scheme corresponds to root-mean-squared deviation to the target structure) as a starting model for flexible fitting an atomic model to the EM density. The fitted model (right) exhibits dramatically reduced RMSD deviation. (See Color Insert.)

the quality of the initial atomic model and the EM data. Figure 3.6B and C shows an example of MDFF to model the conformational change of Acetyl-CoA synthase. A detailed tutorial of MDFF can be found on http://www. ks.uiuc.edu/Research/mdff/.

6.2. Refinement of a comparative model of the 20S core particle

Spanning the central region of the 26S proteasome, the 20S CP displays little structural variability compared to the RP. Therefore, the 20S CP can be resolved to subnanometer resolution comparably easily. Homologs of the eukaryotic 20S CP can be found in bacteria and archea, where they are activated by complexes different from the eukaryotic RP (Striebel et al., 2009). 20S CPs from all three domains of life have a common overall architecture (Voges *et al.*, 1999): they are built of 28 subunits arranged in a stack of four seven-membered rings. All CPs possess a twofold symmetry axis; the two polar rings and the two inner rings are identical, respectively. In its simplest form, for example, in the archeon *Thermoplasma acidophilum*, the seven CP subunits in the polar ring (α-subunits) and in the inner ring (β-subunits) are identical. In eukaryotes, both rings consist of seven different subunits each, which have typically $\sim 30\%$ identical sequences. The crystallographic structure of *T. acidophilum* CP revealed that the proteolytic sites are located in the inner cavity formed by the β-rings (Lowe *et al.*, 1995). Thus, the α-ring and the β-ring form two identical cavities above and below the central cavity, a phenomenon observed in proteases coined self-compartmentalization (Lupas *et al.*, 1997). In addition to *T. acidophilum*, CP crystal structures have been obtained for *Rhodococcus erythropolis*, *Saccharomyces cerevisiae*, and *Bos taurus*.

Here, we illustrate the use of flexible fitting for the following hypothetical scientific scenario: we assume the *T. acidophilum* crystal structure has not been solved, but the one of *S. cerevisiae* has been. We aim to obtain an atomic model of the *T. acidophilum* CP using the *S. cerevisiae* crystal structure and a *T. acidophilum* EM map at 7.5 $\mathring{A}^{-1}$ resolution (Yu *et al.*, 2010).

To build a starting model, we generated a sequence alignment of the CP subunits from *T. acidophilum* and *S. cerevisiae* using T-COFFEE. Using the sequence alignment and the crystal structure (PDB code: 1RYP) as input, we built a comparative model using MODELLER. Subsequently, this model was rigid-body fitted to the EM map (downloaded from the Electron Microscopy Database; EMDB-5130) using SITUS (Wriggers *et al.*, 1999). An MDFF simulation was set up with default parameters (see www.ks.uiuc. edu/Research/MDFF) and ran for 50,000 integration steps, taking a few hours on a typical desktop computer. For the MDFF simulation, the EM map was B-factor sharpened (Rosenthal and Henderson, 2003), which reverts the typical damping of high-frequencies in EM maps resulting from SPA (Saad *et al.*, 2001). In our experience, B-factor sharpening is generally beneficial for MDFF; however, it is only recommended if the resolution of the EM map approaches $(10–15 \mathring{A})^{-1}$ as it might be otherwise unreliable.

The initial model has a maximum CCC to the map of 0.74, which increases to 0.81 after the fit (Fig. 3.7A). The refinement process particularly changes the positions of the distal a subunits whereas the central β subunits change less because the initial model fits the EM map better (Fig. 3.7B). A comparison between the fitted model and the crystal structure (Yu *et al.*, 2010) can serve as a ground truth for the quality of the model. For example, the RMSD of an α-subunit model improves from 5.06 to 2.59 $\mathring{A}$ during the fit (Fig. 3.7D). The necessary files to reproduce this exercise can be downloaded from www.biochem.mpg.de/foerster.

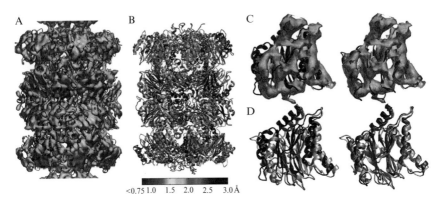

<0.75 1.0   1.5   2.0   2.5   3.0 Å

Figure 3.7   Flexible fitting of the 20S CP. (A) A comparative model is flexibly fitted to a 7.5 Å^{-1} resolution EM map of the *T. acidophilum* 20S proteasome (EMDB-5130). (B) The fitted model is colored by the root-mean-squared deviation (RMSD) of each residue, showing the changes undergone during the fit (initial vs. refined model). (C) One of the α-subunits is shown before (left, red) and after (right, blue) the fit. The segmented EM density for this region is shown in gray. The *CCC* improves from 0.75 to 0.84. (D) The same α-subunit compared to the crystal structure (PDB code: 3IPM). The RMSD of model and crystal structure improves from 5.06 to 2.59 Å for this subunit during the fit. (See Color Insert.)

One must note that MDFF does maintain the secondary structure of the initial model; thus, when the secondary structure of the initial model is erroneous MDFF will not fold the protein into the right conformation. Visual inspection of the fit after MDFF followed by manual reassignment of secondary structure to badly fitting segments can also help improve the fit; however, this approach is subjective and eludes automation. More objective would be methods combining *de novo* structure prediction and EM fitting and first attempts have been described recently (DiMaio *et al.*, 2009). However, these approaches are restricted to relatively small proteins (∼100 residues) and they require higher resolution maps.

7. Conclusion and Outlook

Molecular interpretation of low and intermediate resolution EM maps is a considerable challenge that can be aided through the incorporation of other sources of data. For the example of the proteasomal AAA-ATPases, we described how proteomics data can be pivotal for determining the molecular architecture of a given assembly. In this specific example, we used the EM data to generate different subunit permutations that were then assessed using protein–protein interaction data. This approach primarily applies to ring-like assemblies.

A generally applicable approach would score assemblies by a scoring function S that is a sum of different terms. Ideally, the score corresponds to the logarithm of a joint probability function (Alber *et al.*, 2008; Habeck *et al.*, 2006). For example, the different subscores reflect the agreement of the model with an EM map and protein–protein interaction data:

$$S = S_{EM} + S_{protein-protein}. \tag{3.2}$$

In this framework, the modeling task is to find the model, which minimizes the scoring function S. While calculating S for a given assembly using the above framework is straightforward, the problem of finding the model that minimizes S (sampling) is considerable. The configuration space of assemblies grows exponentially as a function of subunits. Thus, appropriate methods for assembly modeling will rely on appropriate simplified representations that reduce the sampling problem to an extent that is computationally feasible (Lasker *et al.*, 2010). We anticipate that the development of appropriate assembly representations and sampling algorithms will enable us to devise the molecular architecture of assemblies such as the entire 26S proteasome in the near future.

When the architecture of an assembly is established, methods for refining the positions of atoms in the context of the EM data can be used. A prerequisite for applying these methods is that the EM data are of high resolution, that is, $(10 \text{ Å})^{-1}$ or better. Here, we have presented how flexible fitting can be performed using MDFF, presently one of the most accurate protocols for flexible fitting. Depending on the starting model, accuracies of 2–4 Å RMSD are achieved using this method (Trabuco *et al.*, 2008). Probably, the most interesting future development will be to unify 3D reconstruction and modeling into a single optimization problem, as done in crystallography. Then, the accuracy of models will not be limited by the 3D reconstruction, which is, for example, a limitation when the single particles are not randomly oriented on an EM grid.

REFERENCES

Alber, F., *et al.* (2007). Determining the architectures of macromolecular assemblies. *Nature* **450**, 683–694.

Alber, F., *et al.* (2008). Integrating diverse data for structure determination of macromolecular assemblies. *Annu. Rev. Biochem.* **77**, 443–477.

Aloy, P., and Russell, R. B. (2002a). Interrogating protein interaction networks through structural biology. *Proc. Natl. Acad. Sci. USA* **99**, 5896–5901.

Aloy, P., and Russell, R. B. (2002b). Potential artefacts in protein-interaction networks. *FEBS Lett.* **530**, 253–254.

Aloy, P., *et al.* (2002). A complex prediction: Three-dimensional model of the yeast exosome. *EMBO Rep.* **3**, 628–635.

Becker, T., *et al.* (2009). Structure of monomeric yeast and mammalian Sec61 complexes interacting with the translating ribosome. *Science* **326**, 1369–1373.

Biegert, A., *et al.* (2006). The MPI Bioinformatics Toolkit for protein sequence analysis. *Nucleic Acids Res.* **34**, W335–W339.

Brenner, S. E., *et al.* (1998). Assessing sequence comparison methods with reliable structurally identified distant evolutionary relationships. *Proc. Natl. Acad. Sci. USA* **95**, 6073–6078.

Brohawn, S. G., *et al.* (2009). The nuclear pore complex has entered the atomic age. *Structure* **17**, 1156–1168.

Brunger, A. T., *et al.* (1987). Crystallographic R factor refinement by molecular dynamics. *Science* **235**, 458–460.

Connell, S. R., *et al.* (2007). Structural basis for interaction of the ribosome with the switch regions of GTP-bound elongation factors. *Mol. Cell* **25**, 751–764.

DiMaio, F., Tyka, M. D., Baker, M. L., Chiu, W., and Baker, D. (2009). Refinement of protein structures into low-resolution density maps using rosetta. *J. Mol. Biol.* **392**, 181–190.

Djuranovic, S., *et al.* (2009). Structure and activity of the N-terminal substrate recognition domains in proteasomal ATPases. *Mol. Cell* **34**, 580–590.

Eswar, N., and Sali, A. (2007). Comparative modeling of drug target proteins. *In* "Computer-Assisted Drug Design," (J. Taylor and D. Triggle, eds.), pp. 215–236. Elsevier, Oxford.

Eswar, N., *et al.* (2007). Comparative protein structure modeling using MODELLER. *Curr. Protoc. Protein Sci.* Chapter 2, Unit 2 9.

Fabiola, F., and Chapman, M. S. (2005). Fitting of high-resolution structures into electron microscopy reconstruction images. *Structure* **13**, 389–400.

Fields, S. (2005). High-throughput two-hybrid analysis. The promise and the peril. *FEBS J.* **272**, 5391–5399.

Finley, D., *et al.* (1998). Unified nomenclature for subunits of the *Saccharomyces cerevisiae* proteasome regulatory particle. *Trends Biochem. Sci.* **23**, 244–245.

Förster, F., and Hegerl, R. (2007). Structure determination in situ by averaging of tomograms. *Methods Cell Biol.* **79**, 741–767.

Förster, F., *et al.* (2005). Retrovirus envelope protein complex structure in situ determined by cryo-electron tomography. *Proc. Natl. Acad. Sci. USA* **102**, 4729–4734.

Förster, F., *et al.* (2009). An atomic model AAA-ATPase/20S core particle sub-complex of the 26S proteasome. *Biochem. Biophys. Res. Commun.* **388**, 228–233.

Förster, F., Lasker, K., Nickell, S., Sali, A., and Baumeister, W. (2010). Towards an integrated structural model of the 26S proteasome. *Mol. Cell. Proteomics* **9**, 1666–1677.

Frickey, T., and Lupas, A. N. (2004). Phylogenetic analysis of AAA proteins. *J. Struct. Biol.* **146**, 2–10.

Goddard, T. D., *et al.* (2007). Visualizing density maps with UCSF Chimera. *J. Struct. Biol.* **157**, 281–287.

Grünewald, K., *et al.* (2003). Three-dimensional structure of herpes simplex virus from cryo-electron tomography. *Science* **302**, 1396–1398.

Habeck, M., *et al.* (2006). Weighting of experimental evidence in macromolecular structure determination. *Proc. Natl. Acad. Sci. USA* **103**, 1756–1761.

Jaynes, E. T. (2003). *Probability theory: The logic of science* Cambridge University Press, Cambridge (UK).

Jiang, W., *et al.* (2001). Bridging the information gap: Computational tools for intermediate resolution structure interpretation. *J. Mol. Biol.* **308**, 1033–1044.

Jones, D. T. (1999). GenTHREADER: An efficient and reliable protein fold recognition method for genomic sequences. *J. Mol. Biol.* **287**, 797–815.

Kaelin, W. G., Jr., *et al.* (1991). Identification of cellular proteins that can interact specifically with the T/E1A-binding region of the retinoblastoma gene product. *Cell* **64**, 521–532.

Karplus, M., and McCammon, J. A. (2002). Molecular dynamics simulations of biomolecules. *Nat. Struct. Biol.* **9**, 646–652.

Lasker, K., Phillips, J. L., Russel, D., Velázquez-Muriel, J., Schneidman-Duhovny, D., Tjioe, E., Webb, B., Schlessinger, A., and Sali, A. (2010). Integrative Structure Modeling of Macromolecular Assemblies from Proteomics Data. *Mol. Cell. Proteomics* **9**, 1689–1702.

Leitner, A., Walzthoeni, T., Kahraman, A., Herzog, F., Rinner, O., Beck, M., and Aebersold, R. (2010). Probing native protein structures by chemical cross-linking, mass spectrometry and bioinformatics. *Mol. Cell. Proteomics* **9**, 1634–1649.

Liu, Q., *et al.* (2006). Reconstitution, activities, and structure of the eukaryotic RNA exosome. *Cell* **127**, 1223–1237.

Lorentzen, E., *et al.* (2005). The archaeal exosome core is a hexameric ring structure with three catalytic subunits. *Nat. Struct. Mol. Biol.* **12**, 575–581.

Lowe, J., *et al.* (1995). Crystal structure of the 20S proteasome from the archaeon *T. acidophilum* at 3.4 A resolution. *Science* **268**, 533–539.

Lupas, A., *et al.* (1997). Self-compartmentalizing proteases. *Trends Biochem. Sci.* **22**, 399–404.

Maiolica, A., *et al.* (2007). Structural analysis of multiprotein complexes by cross-linking, mass spectrometry, and database searching. *Mol. Cell. Proteomics* **6**, 2200–2211.

Marti-Renom, M. A., *et al.* (2000). Comparative protein structure modeling of genes and genomes. *Annu. Rev. Biophys. Biomol. Struct.* **29**, 291–325.

Nickell, S., *et al.* (2009). Insights into the molecular architecture of the 26S proteasome. *Proc. Natl. Acad. Sci. USA* **106**, 11943–11947.

Pettersen, E. F., *et al.* (2004). UCSF Chimera—A visualization system for exploratory research and analysis. *J. Comput. Chem.* **25**, 1605–1612.

Raijmakers, R., *et al.* (2002). Protein–protein interactions between human exosome components support the assembly of RNase PH-type subunits into a six-membered PNPase-like ring. *J. Mol. Biol.* **323**, 653–663.

Rosenthal, P. B., and Henderson, R. (2003). Optimal determination of particle orientation, absolute hand, and contrast loss in single-particle electron cryomicroscopy. *J. Mol. Biol.* **333**, 721–745.

Saad, A., *et al.* (2001). Fourier amplitude decay of electron cryomicroscopic images of single particles and effects on structure determination. *J. Struct. Biol.* **133**, 32–42.

Sali, A., and Blundell, T. L. (1993). Comparative protein modelling by satisfaction of spatial restraints. *J. Mol. Biol.* **234**, 779–815.

Schueler-Furman, O., *et al.* (2005). Progress in modeling of protein structures and interactions. *Science* **310**, 638–642.

Seidelt, B., *et al.* (2009). Structural insight into nascent polypeptide chain-mediated translational stalling. *Science* **326**, 1412–1415.

Sharon, M., and Robinson, C. V. (2007). The role of mass spectrometry in structure elucidation of dynamic protein complexes. *Annu. Rev. Biochem.* **76**, 167–193.

Shi, J., *et al.* (2001). FUGUE: Sequence–structure homology recognition using environment-specific substitution tables and structure-dependent gap penalties. *J. Mol. Biol.* **310**, 243–257.

Sinz, A. (2003). Chemical cross-linking and mass spectrometry for mapping three-dimensional structures of proteins and protein complexes. *J. Mass Spectrom.* **38**, 1225–1237.

Soding, J. (2005). Protein homology detection by HMM–HMM comparison. *Bioinformatics* **21**, 951–960.

Soding, J., *et al.* (2005). The HHpred interactive server for protein homology detection and structure prediction. *Nucleic Acids Res.* **33**, W244–W248.

Striebel, F., *et al.* (2009). Controlled destruction: AAA+ ATPases in protein degradation from bacteria to eukaryotes. *Curr. Opin. Struct. Biol.* **19**, 209–217.

Suhre, K., *et al.* (2006). NORMA: A tool for flexible fitting of high-resolution protein structures into low-resolution electron-microscopy-derived density maps. *Acta Crystallogr. D Biol. Crystallogr.* **62**, 1098–1100.

Tomko, R. J., Jr., *et al.* (2010). Heterohexameric ring arrangement of the eukaryotic proteasomal ATPases: Implications for proteasome structure and assembly. *Mol. Cell* **38**, 393–403.

Topf, M., *et al.* (2005). Structural characterization of components of protein assemblies by comparative modeling and electron cryo-microscopy. *J. Struct. Biol.* **149**, 191–203.

Towbin, H., *et al.* (1979). Electrophoretic transfer of proteins from polyacrylamide gels to nitrocellulose sheets: Procedure and some applications. *Proc. Natl. Acad. Sci. USA* **76**, 4350–4354.

Trabuco, L. G., *et al.* (2008). Flexible fitting of atomic structures into electron microscopy maps using molecular dynamics. *Structure* **16**, 673–683.

Vale, R. D. (2000). AAA proteins. Lords of the ring. *J. Cell Biol.* **150**, F13–F19.

Voges, D., *et al.* (1999). The 26S proteasome: A molecular machine designed for controlled proteolysis. *Annu. Rev. Biochem.* **68**, 1015–1068.

Volkmann, N., and Hanein, D. (1999). Quantitative fitting of atomic models into observed densities derived by electron microscopy. *J. Struct. Biol.* **125**, 176–184.

Wriggers, W., and Chácon, P. (2001). Modeling tricks and fitting techniques for multi-resolution structures. *Structure* **9**, 779–788.

Wriggers, W., *et al.* (1999). Situs: A package for docking crystal structures into low-resolution maps from electron microscopy. *J. Struct. Biol.* **125**, 185–195.

Yu, W., *et al.* (2010). Interactions of PAN's C-termini with archaeal 20S proteasome and implications for the eukaryotic proteasome-ATPase interactions. *EMBO J.* **29**, 692–702.

Zhang, F., *et al.* (2009). Structural insights into the regulatory particle of the proteasome from *Methanocaldococcus jannaschii*. *Mol. Cell* **34**, 473–484.

Zhou, H. X., and Qin, S. (2007). Interaction-site prediction for protein complexes: A critical assessment. *Bioinformatics* **23**, 2203–2209.

CHAPTER FOUR

UNIFIED DATA RESOURCE FOR CRYO-EM

Catherine L. Lawson

Contents

Abstract

Three-dimensional (3D) cryoelectron microscopy reconstruction methods are uniquely able to reveal structures of many important macromolecules and macromolecular complexes. EMDataBank.org, a joint effort of the Protein Databank in Europe (PDBe), the Research Collaboratory for Structural Bioinformatics (RCSB), and the National Center for Macromolecular Imaging (NCMI), is a "one-stop shop" resource for global deposition and retrieval of cryo-EM map, model, and associated metadata. The resource unifies public access to the two

Department of Chemistry and Chemical Biology and Research Collaboratory for Structural Bioinformatics, Rutgers, The State University of New Jersey, USA

Methods in Enzymology, Volume 483
ISSN 0076-6879, DOI: 10.1016/S0076-6879(10)83004-6

major EM Structural Data archives: EM Data Bank (EMDB) and Protein Data
Bank (PDB), and facilitates use of EM structural data of macromolecules and
macromolecular complexes by the wider scientific community.

1. INTRODUCTION

Structural biology of macromolecules has become an indispensable
branch of molecular biology. Researchers use the results from structural
studies to explain the functions and mechanisms of biological processes at
the molecular level, leading to more targeted experiments to explore
structure and function. Many key biological processes are carried out by
large macromolecular complexes, including signal transduction, genome
replication, transcription, translation, chaperonin-assisted protein folding,
viral infection, and motility. It is becoming increasingly feasible to determine three-dimensional structures of these complexes in different functional or chemical states using cryoelectron microscopy (cryo-EM).

Specimens for cryo-EM studies come in many forms and shapes, for
example, two- or three-dimensional crystals (Gonen *et al.*, 2005;
Henderson *et al.*, 1990; Schmid *et al.*, 2004), one-dimensional filaments or
tubular crystals possessing helical symmetry (Unwin, 2005; Wang *et al.*,
2006), and individual particles with or without symmetry (Gabashvili *et al.*,
2000; Olson *et al.*, 1990; Zhou *et al.*, 2001). Cryo-EM is also being applied to
large samples consisting of irregular ensembles of complexes and cells using
tomographic reconstruction methods (Baumeister, 2004; Murphy and
Jensen, 2007). Other chapters in this volume describe preparation and
image reconstruction methods for the different specimen types including
2D crystals (Chapters 1, 4, and 11, Vol. 481; Chapter 4, Vol. 482), helical
arrays (Chapter 2, Vol. 481; Chapters 5 and 6, Vol. 482), single particles
(Chapters 1, 8, and 14, Vol. 482), and unique structures (Chapter 12, Vol.
481; Chapters 13, Vol. 482).

At present, cryo-EM researchers are rapidly producing a large body of
knowledge regarding the 3D structural arrangements of components within
large macromolecular complexes, within subcellular assemblies, and even
within whole cells, based on map volumes with resolution limits ranging
from 80 to 2 Å. Interpretation varies according to the map resolution,
available tools, and additional knowledge of the system and/or its components and may involve either segmentation, rigid-body fitting of atomic
coordinates determined using X-ray crystallography or NMR, or *ab initio*
model building.

Public access to cryo-EM map volumes and their associated fitted model
interpretations permits independent assessment and interpretation of structural
results and stimulates development of new tools for visualization, fitting, and

validation. The EM Data Bank (EMDB) is the major repository for 3D map volumes solved using electron microscopy (EM) (Tagari et al., 2002), while the Protein Data Bank (PDB) collects atomic coordinates fitted into EM map volumes (Dutta et al., 2009). The Unified Data Resource for cryo-EM (http://www.emdatabank.org) was created in order to unify data deposition, processing, and retrieval of maps and fitted models. This chapter provides an overview of the EM structural data archives and the unified resource, including historical context, current content and use, and future prospects.

2. EM Structural Data Archives

2.1. Maps

The EMDB was established at the European Bioinformatics Institute (EBI) in Hinxton, United Kingdom, and began operations in 2002. It was initially supported by two European Union-funded projects, the Integration of Information about Macromolecular Structure project (IIMS) and the 3DEM Network of Excellence (3DEM NoE). An IIMS-sponsored workshop was held in November 2002 that focused on data exchange, harvesting, deposition issues, and presentation of EM data to nonspecialists. Guidelines and release policies were set for the newly founded EMDB, and the workshop established the database as a resource for the international community, with an announcement published in Structure (Fuller, 2003), followed by an editorial in Nature Structural Biology (2003). The workshop concluded with a strong endorsement of EM map volume deposition and linkage of EMDB with other archival databases in biomedical research.

Working closely with IIMS project partners, leading European EM laboratories and PDB partners, an initial data model was produced for EM-derived maps. A web-based deposition system, EMDEP, was developed to handle data capture (Henrick et al., 2003). EMDEP validates data via an interactive depositor-driven operation, and it relies on the knowledge and expertise of the experimenters for the complete and accurate description of the structural experiment and its results. The captured metadata–for example, sample description, specimen preparation, imaging, reconstruction, fitting details–are stored in a "header" file, and the deposited map is converted to a common format for redistribution. A database query tool, EMSEARCH, was also designed and implemented to enable web-based searches.

By the time the IIMS project was completed in December 2003, the EMDB had become an operational public database with 65 map volumes deposited by major EM laboratories in Europe and the United States. At this time, the PDB began to see a significant increase in EM-related coordinate depositions; in many cases, models that were fitted into maps deposited to EMDB.

2.2. Models

The PDB archive was established in 1971 as a public repository for X-ray crystal structures of biological macromolecules (Bernstein *et al.*, 1977), and is presently maintained by the global organization worldwide PDB (wwPDB; Berman *et al.*, 2003), a consortium consisting of the RCSB–PDB (www.pdb. org), the Protein Data Bank in Europe (PDBe) at EBI (www.pdbe.org), the Protein Data Bank Japan (PDBj; www.pdbj.org), and the Biological Magnetic Resonance Bank (http://www.bmrb.wisc.edu). The number of structures in PDB has grown from the initial seven to over 65,000 entries. Over time, the PDB began to collect coordinates of structures determined by methods other than X-ray crystallography. In the 1980s, coordinates and restraint data determined from NMR methods began to be included in the PDB and these now represent about 15% of the archive. In the 1990s, model coordinates for structures determined using EM began to be archived, beginning with models for bacteriorhodopsin (Henderson *et al.*, 1990), and the RecA hexamer (Yu and Egelman, 1997). EM structures currently account for less than 0.5% of all PDB entries, but the rate of deposition is increasing more rapidly than for any other experimental method.

2.3. EM dictionary development

Two workshops held in 2004 invited the EM community to participate in development of an improved data model for describing cryo-EM experiments, and also set in motion efforts to unify deposition and access to EM maps and models.

The 2004 3DEM NoE workshop at EBI reviewed tools and software practices used in the field of cryo-EM, examined data items and data models required to fully describe EM experiments, and allocated tasks to different groups to develop the required standards. The workshop also defined goals for further development of the database including providing archiving capabilities for cryoelectron tomography, providing cross-referencing between EMDB maps and PDB coordinate models, and converting common map formats in a lossless manner.

The second 2004 workshop held at the Research Collaboratory for Structural Bioinformatics (RCSB) at Rutgers, and cosponsored by the National Center for Macromolecular Imaging (NCMI) aimed to develop a global community consensus on data items needed for deposition of 3D map volumes and fitted atomic models derived from cryo-EM studies. In addition to discussion of desired improvements in the areas of visualization, data mining and data integration, a unanimous recommendation of workshop attendees was the need to develop a "one-stop shop" for deposition of EM map and model data in order to eliminate the duplication of effort involved in creating separate depositions to EMDB and PDB.

Based on recommendations gathered at the 2004 workshops, a revised and expanded EM dictionary was created in a three-way collaboration between EBI, RCSB, and NCMI with broad community input and was presented in 2005 at the 3DEM Gordon Research Conference in New Hampshire as well as a 3DEM Developers workshop held in the United Kingdom. Recommended conventions for exchange of cryo-EM data were published (Heymann *et al.*, 2005), and a standards task force was created to gather information on the different cryo-EM map and image conventions and formats to facilitate conversion. In June 2005, a notice posted to the 3DEM community e-mail bulletin board (http://3dem.ucsd.edu) announced the intention of EBI and RCSB to jointly collaborate toward further development of 3DEM database services.

2.4. Unified resource

Following the recommendations of the EM community to create a one-stop shop for EM, the Unified Data Resource for Cryo-EM (EMDataBank. org) was established in 2007 with funding from the National Institutes of Health/NIGMS as a joint effort of EBI, RCSB, and NCMI. The resource is creating a global deposition and retrieval network for cryo-EM map, model, and associated metadata, as well as a portal for software tools for standardized map format conversion, map, segmentation and model assessment, visualization, and data integration. The first goal of this three-way collaboration was completed in early 2008 when RCSB joined EBI as a second EMDB deposition and retrieval site. Joint EMDB (map) and PDB (model) deposition systems were developed and put into operation at both EBI and RCSB in early 2009, and web-based 3D visualization tools have been integrated into EMDB atlas pages. Efforts to improve uniformity and usability of the EM structural data in both EMDB and PDB databases are ongoing, and additional services for data harvesting and evaluation are planned.

3. DEPOSITION AND CONTENT

The EMDB currently holds more than 800 map volume entries while PDB holds more than 300 entries of coordinates fitted into EM map volumes (Fig.4.1). Map volume and fitted model deposition rates are on the rise, currently ~ 150 and ~ 40 per year, respectively, with roughly 40% of all published EM structures being captured in the databases. As the importance of EM-derived structural information continues to increase, it is anticipated that more journals and funding agencies will require deposition.

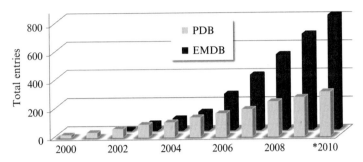

Figure 4.1   EM entries in EMDB and PDB, cumulative by year (*2010 data is through May).

Each EMDB entry holds a single map volume plus associated experimental metadata; each associated PDB entry holds the fitted coordinate models and associated experimental metadata plus primary sequence information for each polymer. The metadata information that is shared by both databases is automatically transferred during the joint deposition process and includes:

- Sample description, including components and stoichiometry
- Sample and specimen preparation, including buffer, grid, vitrification, staining
- Imaging experiment, including microscope type, microscope settings, detector
- Particle selection, image correction, reconstruction method, resolution, resolution method
- Coordinate fitting description

The correspondences between maps and associated fitted coordinate models are maintained in both archives. EMDB entries can optionally hold associated masks, structure factors, and/or layerline data; PDB entries can also hold structure factors. The underlying dictionaries for the two databases have direct translations and are regularly being updated to reflect changes in experimental apparatus and methods. For readers interested in depositing EM structural data, some guidelines are provided in the last section of this chapter.

The map archive includes several different types of maps generated by EM imaging (Fig. 4.2). The majority of entries are single particle reconstructions (81%), which represent ensemble averages of thousands of individual imaged particles, often with additional symmetry averaging. The largest class of single particle specimens represented are the viruses (20% of all holdings), the majority of which are icosahedrally averaged (Chiu and Rixon, 2002; Huiskonen and Butcher, 2007; Lee and Johnson, 2003). Virus entries typically represent distinct states of maturation, or complexes with antibodies or receptors.

The second largest class of single particle specimens represented are the ribosomes (15%). Ribosome entries define key structural conformations

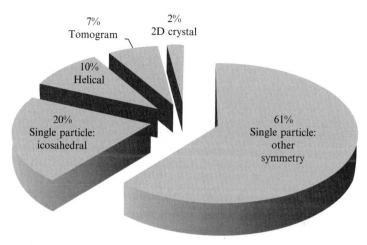

Figure 4.2   Current distribution of EM map entry types.

encountered in translating messenger RNA into protein, or elucidate structural variations across diverse species (Frank, 2009; Mitra and Frank, 2006). Other single particle specimens represented include macromolecular machines involved in protein folding, protein degradation, energy metabolism, cell cycle processes, DNA replication, DNA repair, RNA transcription, and RNA splicing.

The map archive also holds densities for 2D crystals and for helical arrays, including intracellular filaments and microtubules, flagella, and helical crystals. There are also several tomographic maps of unique structures as well as maps that represent 3D averages of aligned tomograms. Diverse specimens currently held of this type include flagellar motors, insect flight muscle tissue, and desmosomes. A gallery of representative map volumes is presented in Fig. 4.3.

Coordinates of EM entries are obtained using a variety of modeling methods including manual docking, rigid-body fitting, homology modeling, and computational refinement algorithms. EM entries in the PDB are classified either under EM or electron crystallography as the experimental method. For structures with regular point or helical symmetry, coordinates are given for the asymmetric unit along with a set of transformation matrices to build the biological assembly (Lawson *et al.*, 2008).

4. ACCESS

Access to EM structural data and related services is through the http://www.emdatabank.org web site. The EMDB and PDB archives are updated weekly on Wednesdays at 00:00 GMT. EMDB is distributed on two ftp

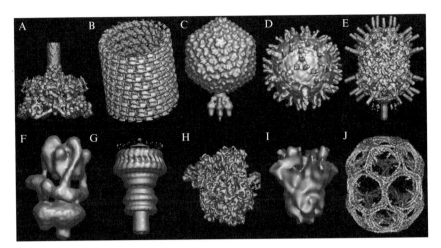

Figure 4.3 Image gallery of representative EM map entries (arbitrary scales). *Top row from left to right*: (A) EMD-1048: bacteriophage T4 baseplate (Kostyuchenko *et al.*, 2003); (B) EMD-1129: GDP-tubulin (Wang and Nogales, 2005); (C) EMD-1222: bacteriophage P22 (Chang *et al.*, 2006); (D) EMD-1234: West Nile Virus decorated with neutralizing antibody Fab (Kaufmann *et al.*, 2006); (E) EMD-1265: bacteriophage phi29 (Xiang *et al.*, 2006). *Bottom row from left to right*: (F) EMD-1590: vacuolar ATPase motor (Muench *et al.*, 2009); (G) EMD-1617: *Shigella flexneri* T3SS needle complex (Hodgkinson *et al.*, 2009); (H) EMD-5036: *E. coli* 70S ribosome complex (Villa *et al.*, 2009); (I) EMD-5114: BK potassium channel (Wang and Sigworth, 2009); (J) EMD-5119: clathrin lattice (Fotin *et al.*, 2004). Images were created using AstexViewer.

mirrors supported by the two EMDataBank.org distribution partners in the United Kingdom and the United States, while PDB is distributed on ftp mirrors supported by each of the wwPDB partners.

The EMSEARCH web service is also maintained and updated weekly at both distribution sites. EMSEARCH enables browsing and searching of EMDB metadata uploaded into a relational database. Simple searches can be performed based on author name, title, sample name, citation abstract word, aggregation type, resolution, and release date range. Search summaries link to atlas pages for each entry, which include summary, visualization, sample, experiment, processing, map information, and download pages (Fig. 4.4).

The atlas visualization page provides several ways to view EM maps. In addition to a static 2D image provided by the deposition author, it is also possible to launch two different Java-based 3D viewers (Fig. 4.5, left and right top panels). EMViewer, developed by Powei Feng and Joe Warren at Rice University in collaboration with NCMI, provides a simple, single isosurface representation of a map at a predetermined contour level. Clicking on the "Launch EMViewer" button will bring up a simple 3D representation of the map that can be manipulated by mouse drag and click actions. AstexViewer, a molecular graphics program originally developed to display crystallographic

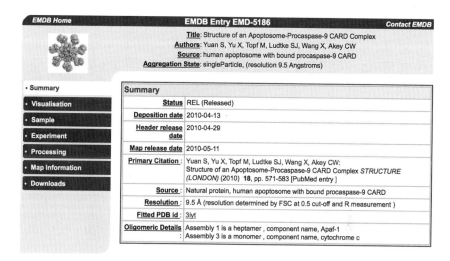

Figure 4.4   Example EMDB atlas page.

data (Hartshorn, 2002) was recently re-released under an open source license and has been adapted by EMDataBank.org for display of EM maps and associated PDB coordinate models. To improve web download speed and minimize memory requirements, a compact map format is used (BRIX), and larger maps are also down-sampled by a factor of 2–5. Current capabilities include ability to control map contour level, opacity, color, solid versus mesh surface rendering, and concurrent display of a PDB coordinate entry. Viewing large maps may require increasing Java Applet Runtime memory allocation.

Map volumes are distributed in CCP4 format and can be viewed with locally installed software such as UCSF Chimera (Pettersen *et al.*, 2004), Pymol (Delano, 2002), VMD (Hsin et al., 2008), Coot (Emsley and Cowtan, 2004), or other graphics programs enabling investigation with a more extensive set of tools. Links for map download are available on atlas "Download" and "Map Information" pages. Recent distributions of UCSF Chimera enable direct downloads of EMDB maps *plus* associated fitted PDB models via simple queries (e.g., author name, title) to the EMSEARCH relational database through the EMDataBank.org beta-web service, which is a self-contained programming interface based on the SOAP protocol (Fig. 4.5, bottom panel).

EM Navigator at PDBj (http://emnavi.protein.osaka-u.ac.jp) is an additional resource for browsing, viewing, and downloading maps and fitted coordinate models from the EMDB and PDB databases. Each of the wwPDB partners (wwpdb.org) has web interfaces to database representations of the PDB archive with advanced searching and browsing capabilities. ViperDB (Natarajan *et al.*, 2005; http://viperdb.scripps.edu), a database specifically for icosahedral viruses, also holds some EM maps and related coordinates.

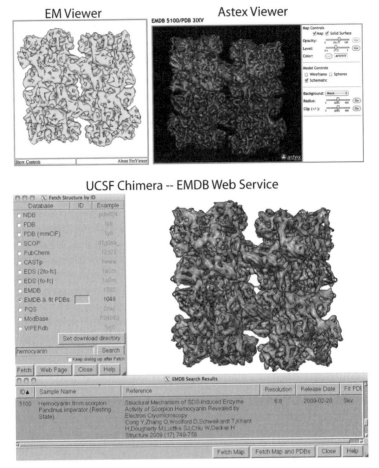

Figure 4.5 Tools for 3D visualization of EM structural data. EM Viewer (top left) and Astex Viewer (top right) are Java-based 3D viewers that can be launched from the EMDB atlas "Visualization" pages. EM Viewer employs a compact single contour-level mesh representation for lightweight map viewing; Astex Viewer enables exploration of maps and their fitted models with adjustable map contour level. The EMDB web service implemented within UCSF Chimera (bottom panel) expedites search of EMDB and subsequent map + model download. Scorpion hemocyanin (Cong *et al.*, 2009, EMD-5100 and PDB id 3ixv) is the example shown in each of the panels. (See Color Insert.)

5. EXAMPLES OF USE

A major goal of the EMDataBank.org unified data resource is to archive EM-derived structural information in a way that will enable further research. The impact of availability of structural data for smaller proteins has already been amply demonstrated by the success of the PDB, which is accessed globally by thousands of individuals every day. Listed here are a few examples of how archived EM data facilitates subsequent scientific exploration by other investigators.

5.1. Building up molecular pictures of large macromolecular assemblies

The availability of a coordinate model representation of a macromolecular complex leads directly to a dramatic increase in our fundamental understanding of biological function. Atomic coordinates permit 3D mapping of a wide variety of data including sites of mutations leading to disease, modification sites, antibody recognition sites, amino acid sequence variability, and electrostatic properties. But in many cases, high-resolution structural models are not available for every component within a cryo-EM map at the time it is first interpreted. By preserving map and model information together in a freely accessible database, new structural information can be incorporated as it becomes available. Examples include: (1) reinterpretation of ribosome stalk regions in multiple EMDB-archived maps after the crystal structure of the L7/L12 complex was determined (Diaconu *et al.*, 2005), and (2) progression toward a complete fitted coordinate model for bacteriophage T4 EM reconstructions based on fitting of components as they have become available (Aksyuk *et al.*, 2009a,b).

5.2. Structural basis for interpreting data

Structural knowledge can be crucial for interpreting biochemical and biophysical data. For instance, dengue and West Nile virus cryo-EM structures (Kuhn *et al.*, 2002; Mukhopadhyay *et al.*, 2003) have led the way to mapping sites of glycosylation (Hanna *et al.*, 2005), amino acid sequence variability (Modis *et al.*, 2005), neutralizing antibody binding (Nybakken *et al.*, 2005), design of vaccines (Ledizet *et al.*, 2005), and a structural basis for understanding membrane fusion (Modis *et al.*, 2004). Recently, multiple EMDB map entries were examined in light of biochemical data designed to distinguish between two distinct models for the hexameric subunit arrangement of two closely related disaggregating proteins, ClpB and Hsp10 (Wendler and Saibil, 2010).

5.3. Crystal structure phasing

Cryo-EM maps can be used to initiate crystallographic phasing. Isomorphous replacement phasing methods that are routinely used in X-ray crystallography are technically difficult to apply to large macromolecular complexes. A low-resolution cryo-EM map of the complex under study thus provides a valuable complementary source of phase information. For high-symmetry structures, such as icosahedral viruses, application of robust computational averaging and extension algorithms to initial cryo-EM-derived phases is often sufficient to complete the structure determination without additional phase information. Examples of crystal structures phased with cryo-EM envelopes include proteases (Bosch *et al.*, 2001; Wang *et al.*, 1998), ribosomes (Ban *et al.*, 1998; Cate *et al.*, 1999; Thygesen *et al.*, 1996), and icosahedral viruses (Dokland *et al.*, 1997, 1998; Grimes *et al.*, 1998; Helgstrand *et al.*, 2003; Prasad *et al.*, 1999; Reinisch *et al.*, 2000; Wynne *et al.*, 1999). Additional case studies are described by Xiong (2008), and use of a stain EM reconstruction for phasing is described by Trapani *et al.* (2010). Routine archiving of EM-derived maps facilitates the use of this method by the crystallography community.

5.4. Comparative studies

By capturing the coordinates, maps, and metadata and making them available through searchable databases, it is possible to easily compare structures, select structures for further analysis using a variety of criteria, perform experimental design for new analyses, and design new algorithms to improve the state-of-the-art methodology. For example, observation of publicly available maps and coordinates led to the conclusion that the double-stranded DNA tailed phages and herpesvirus have a similar fold in their major capsid proteins though there is little sequence similarity among them (Baker *et al.*, 2005). In a second example, a flexible fitting approach applied to 43 EMDB map volumes of bacterial 70S ribosome in various functional states revealed global conformational differences between the EM structures involving large-scale ratchet-like deformations (Matsumoto and Ishida, 2009).

5.5. Software development

Availability of EM map volumes in the EMDB facilitates development and validation of software for map viewing, analysis, manipulation, coordinate model fitting, and validation. Development of algorithms for fitting of EM maps with atomic coordinates is a particularly active area. For low- to medium-resolution studies in which the source coordinates are likely to be derived from crystallographic or NMR studies, algorithms being explored include rigid-body fitting (Wriggers, 2010), normal mode analysis (Tama *et al.*, 2004), spatial interpolation (Rusu *et al.*, 2008), conformational sampling under low-resolution restraints (Schroder *et al.*, 2007), molecular dynamics flexible fitting (Trabuco *et al.*, 2009), and simulated annealing approaches (Tan *et al.*, 2008; Topf *et al.*, 2008). For medium- to high-resolution studies where *de novo* model building becomes possible, methods are being developed for skeletonization and secondary structure element detection (Baker *et al.*, 2007) and incorporation of structure prediction from primary sequence (DiMaio *et al.*, 2009). Additional examples of algorithm development facilitated by public availability of EMDB maps include investigations of map denoising (Jiang *et al.*, 2003), B-factor sharpening (Fernandez *et al.*, 2008), map resolution determination (Sousa and Grigorieff, 2007), and automated segmentation (Baker *et al.*, 2006; Pintilie *et al.*, 2010). The UCSF Chimera team has made particularly effective use of the EMDB resource for development and testing of a large, versatile set of tools for manipulating volume data (Goddard *et al.*, 2007; Pettersen *et al.*, 2004).

6. FUTURE PROSPECTS

3D cryo-EM reconstruction methods are uniquely able to reveal structural aspects of many important macromolecules and macromolecular complexes, and for this reason the field is in a period of rapid expansion and development. Based on current growth of EM entries and publication of EM structures, the total number of structures of large biological assemblies contributed by EM is anticipated to approach 10,000 by the year 2020.

In the near future, deposition and archiving of EM structural data will be integrated in a common tool that is being developed by the wwPDB partners to handle depositions from all structural biology methods. In addition, as the field matures, validation tools and criteria for assessment of map and fitted coordinate models will play an important role in providing guidance to users of cryo-EM-derived structural data. To this end, a validation task force is being assembled by the Unified Data Resource partners (NCMI, RCSB, and EBI) to develop recommendations as to how best to assess the quality of both maps and models that have been obtained from cryo-EM data. The recommendations will form the basis for a validation suite that will be used by EMDB and PDB.

7. GUIDE TO DEPOSITION OF EM STRUCTURAL DATA

This section provides an overview of joint map + fitted coordinate model deposition to the EM DataBank (EMDB) and PDB. To prepare for deposition, the following items should be available:

- *Map File*: One map file may be uploaded per EMDB entry. If an experiment yields more than one 3D map volume, the maps must be deposited in separate entries. Map formats accepted for upload include SPIDER, MRC, and CCP4; maps can be compressed (tar, zip, gzip). The uploaded map will be converted to EMDB redistribution format (currently CCP4).

- *Model File*: Atomic models in PDB or mmCIF format must be in the same coordinate frame as the map. Models may include species homologs or homology models, and/or may contain either full-residues or alpha-carbon/phosphate backbone traces. For structures with regular point or helical symmetry, coordinates for one asymmetric unit may be deposited along with transformation matrices to build the assembly (Lawson *et al.*, 2008). Detailed information about coordinate formats, annotation, and processing policies can be found at wwpdb.org/docs.html.
- *Experiment Info*: Basic information about the experiment, including a description of the assembly, specimen preparation, microscope imaging, image processing and reconstruction, and coordinate fitting will be requested during deposition.
- *External Database IDs*: In addition, external database IDs such as GO (geneontology.org), INTERPRO (www.ebi.ac.uk/interpro), Virus ICTVdB (www.ncbi.nlm.nih.gov/ICTVdb), PDB (coordinates used for fitting), and PUBMED (if published) will be requested.

7.1. Map deposition using EMDEP

EM structural data can be submitted to deposition sites at PDBe (United Kingdom) or RCSB-PDB (United States). Go to emdatabank.org/deposit. html to select the deposition site. To initiate an EMDEP session, click on "Start Session." Select "new deposition" or select "based on previous submission" if the experiment is related to a prior deposition.

The map deposition is created page by page. When all of the relevant information on the page is entered and saved, the symbol for the page on the left hand menu changes from a red arrow to a green circle. Before submission, it is possible to go back and change answers on completed pages. Help text is available for every data item by clicking on the item.

Many pages contain sections for entering metadata information about the experiment that can be duplicated. For instance, two "sample component" sections should be completed for each unique component in an Fab: virus complex assembly, one section for the Fab, and a second section for the virus. Multiple imaging sessions/microscopes can be defined.

Atomic coordinates from existing PDB entries (e.g., X-ray or NMR structures) used in fitting of the map are specified by the depositor on the "fitting" page, which can also be duplicated as needed. In contrast, fitted atomic coordinates representing the depositor's molecular interpretation of the EM map are deposited to the PDB following the map deposition (see below), and the resulting PDB id is associated with the map entry by the database curator.

7.2. Fitted model deposition

After completing map deposition using EMDEP at either site, a link will appear on the EMDEP left hand menu to initiate deposition of one or more models to PDB with automatic transfer of experimental metadata (sample description, microscope type, etc.). At the PDBe site, the link opens an AutoDep session; at the RCSB–PDB site, the link opens an EM-Adit session. Sequence information must be provided for each protein or nucleic acid entity should include the entire sequence of the imaged material, including any mutations or expression tags.

7.3. Accession IDs

After completion of map and model depositions, accession IDs are assigned by the EMDB and PDB, respectively. The accession IDs are associated with each other by the two databases. All provided IDs should be included in the primary publication describing the EM structure.

ACKNOWLEDGMENTS

Many current and past EMDataBank.org team members have made significant contributions to the development of the Unified Data Resource for Cryo-EM including Kim Henrick, Wah Chiu, Helen Berman, Gerard Kleywegt, Richard Newman, John Westbrook, Glen van Ginkel, Batsal Devkota, Matt Baker, Tom Oldfield, Christoph Best, Gaurav Sahni, Raul Sala, Chunxiao Bi, Powei Feng, Joe Warren, Matt Dougherty, Steve Ludtke, and Ian Rees. The resource is funded by National Institutes of Health GM079429 to Baylor College of Medicine, Rutgers University, and the European Bioinformatics Institute.

REFERENCES

Aksyuk, A. A., *et al.* (2009a). The tail sheath structure of bacteriophage T4: A molecular machine for infecting bacteria. *EMBO J.* **28,** 821–829.

Aksyuk, A. A., *et al.* (2009b). The structure of gene product 6 of bacteriophage T4, the hinge-pin of the baseplate. *Structure* **17,** 800–808.

Baker, M. L., *et al.* (2005). Common ancestry of herpesviruses and tailed DNA bacterio-phages. *J. Virol.* **79,** 14967–14970.

Baker, M. L., *et al.* (2006). Automated segmentation of molecular subunits in electron cryomicroscopy density maps. *J. Struct. Biol.* **156,** 432–441.

Baker, M. L., *et al.* (2007). Identification of secondary structure elements in intermediate-resolution density maps. *Structure* **15,** 7–19.

Ban, N., *et al.* (1998). A 9 A resolution X-ray crystallographic map of the large ribosomal subunit. *Cell* **93,** 1105–1115.

Baumeister, W. (2004). Mapping molecular landscapes inside cells. *Biol. Chem.* **385,** 865–872.

Berman, H. M., *et al.* (2003). Announcing the worldwide Protein Data Bank. *Nat. Struct. Biol.* **10,** 980.

Bernstein, F. C., et al. (1977). Protein Data Bank: A computer-based archival file for macromolecular structures. *J. Mol. Biol.* **112,** 535–542.

Bosch, J., et al. (2001). Purification, crystallization, and preliminary X-ray diffraction analysis of the Tricorn protease hexamer from Thermoplasma acidophilum. *J. Struct. Biol.* **134,** 83–87.

Cate, J. H., et al. (1999). X-ray crystal structures of 70S ribosome functional complexes. *Science* **285,** 2095–2104.

Chang, J., et al. (2006). Cryo-EM asymmetric reconstruction of bacteriophage P22 reveals organization of its DNA packaging and infecting machinery. *Structure* **14,** 1073–1082.

Chiu, W., and Rixon, F. J. (2002). High resolution structural studies of complex icosahedral viruses: A brief overview. *Virus Res.* **82,** 9–17.

Cong, Y., et al. (2009). Structural mechanism of SDS-induced enzyme activity of scorpion hemocyanin revealed by electron cryomicroscopy. *Structure* **17,** 749–758.

Delano, W. L. (2002). The Pymol Molecular Graphics System. Delano Scientific, Palo Alto.

Diaconu, M., et al. (2005). Structural basis for the function of the ribosomal L7/12 stalk in factor binding and GTPase activation. *Cell* **121,** 991–1004.

DiMaio, F., et al. (2009). Refinement of protein structures into low-resolution density maps using Rosetta. *J. Mol. Biol.* **392,** 181–190.

Dokland, T., et al. (1997). Structure of a viral procapsid with molecular scaffolding. *Nature* **389,** 308–313.

Dokland, T., et al. (1998). Structure determination of the phiX174 closed procapsid. *Acta Crystallogr. D Biol. Crystallogr.* **54**(Pt 5), 878–890.

Dutta, S., et al. (2009). Data deposition and annotation at the worldwide protein data bank. *Mol. Biotechnol.* **42,** 1–13.

Nature Structural Biology Editorial (2003). A database for 'em. *Nat. Struct. Biol.* **10,** 313.

Emsley, P., and Cowtan, K. (2004). Coot: Model-building tools for molecular graphics. *Acta Crystallogr. D Biol. Crystallogr.* **60,** 2126–2132.

Fernandez, J. J., et al. (2008). Sharpening high resolution information in single particle electron cryomicroscopy. *J. Struct. Biol.* **164,** 170–175.

Fotin, A., et al. (2004). Molecular model for a complete clathrin lattice from electron cryomicroscopy. *Nature* **432,** 573–579.

Frank, J. (2009). Single-particle reconstruction of biological macromolecules in electron microscopy—30 years. *Q. Rev. Biophys.* **42,** 139–158.

Fuller, S. D. (2003). Depositing electron microscopy maps. *Structure (Cambridge)* **11,** 11–12.

Gabashvili, I. S., et al. (2000). Solution structure of the E. coli 70S ribosome at 11.5 A resolution. *Cell* **100,** 537–549.

Goddard, T. D., et al. (2007). Visualizing density maps with UCSF Chimera. *J. Struct. Biol.* **157,** 281–287.

Gonen, T., et al. (2005). Lipid-protein interactions in double-layered two-dimensional AQP0 crystals. *Nature* **438,** 633–638.

Grimes, J. M., et al. (1998). The atomic structure of the bluetongue virus core. *Nature* **395,** 470–478.

Hanna, S. L., et al. (2005). N-linked glycosylation of west nile virus envelope proteins influences particle assembly and infectivity. *J. Virol.* **79,** 13262–13274.

Hartshorn, M. J. (2002). AstexViewer: A visualisation aid for structure-based drug design. *J. Comput. Aided Mol. Des.* **16,** 871–881.

Helgstrand, C., et al. (2003). The refined structure of a protein catenane: The HK97 bacteriophage capsid at 3.44 Å resolution. *J. Mol. Biol.* **334,** 885–899.

Henderson, R., et al. (1990). Model for the structure of bacteriorhodopsin based on high-resolution electron cryo-microscopy. *J. Mol. Biol.* **213,** 899–929.

Henrick, K., et al. (2003). EMDep: A web-based system for the deposition and validation of high-resolution electron microscopy macromolecular structural information. *J. Struct. Biol.* **144,** 228–237.

Heymann, J. B., et al. (2005). Common conventions for interchange and archiving of three-dimensional electron microscopy information in structural biology. J. Struct. Biol. **151**, 196–207.

Hodgkinson, J. L., et al. (2009). Three-dimensional reconstruction of the Shigella T3SS transmembrane regions reveals 12-fold symmetry and novel features throughout. Nat. Struct. Mol. Biol. **16**, 477–485.

Hsin, J., et al. (2008). Using VMD: An introductory tutorial. Curr. Protoc. Bioinformatics Chapter 5, Unit 5 7.

Huiskonen, J. T., and Butcher, S. J. (2007). Membrane-containing viruses with icosahedrally symmetric capsids. Curr. Opin. Struct. Biol. **17**, 229–236.

Jiang, W., et al. (2003). Applications of bilateral denoising filter in biological electron microscopy. J. Struct. Biol. **144**, 114–122.

Kaufmann, B., et al. (2006). West Nile virus in complex with the Fab fragment of a neutralizing monoclonal antibody. Proc. Natl Acad. Sci. USA **103**, 12400–12404.

Kostyuchenko, V. A., et al. (2003). Three-dimensional structure of bacteriophage T4 baseplate. Nat. Struct. Biol. **10**, 688–693.

Kuhn, R. J., et al. (2002). Structure of dengue virus: Implications for flavivirus organization, maturation, and fusion. Cell **108**, 717–725.

Lawson, C. L., et al. (2008). Representation of viruses in the remediated PDB archive. Acta Crystallogr. D Biol. Crystallogr. **64**, 874–882.

Ledizet, M., et al. (2005). A recombinant envelope protein vaccine against West Nile virus. Vaccine **23**, 3915–3924.

Lee, K. K., and Johnson, J. E. (2003). Complementary approaches to structure determination of icosahedral viruses. Curr. Opin. Struct. Biol. **13**, 558–569.

Matsumoto, A., and Ishida, H. (2009). Global conformational changes of ribosome observed by normal mode fitting for 3D Cryo-EM structures. Structure **17**, 1605–1613.

Mitra, K., and Frank, J. (2006). Ribosome dynamics: Insights from atomic structure modeling into cryo-electron microscopy maps. Annu. Rev. Biophys. Biomol. Struct. **35**, 299–317.

Modis, Y., et al. (2004). Structure of the dengue virus envelope protein after membrane fusion. Nature **427**, 313–319.

Modis, Y., et al. (2005). Variable surface epitopes in the crystal structure of dengue virus type 3 envelope glycoprotein. J. Virol. **79**, 1223–1231.

Muench, S. P., et al. (2009). Cryo-electron microscopy of the vacuolar ATPase motor reveals its mechanical and regulatory complexity. J. Mol. Biol. **386**, 989–999.

Mukhopadhyay, S., et al. (2003). Structure of West Nile virus. Science **302**, 248.

Murphy, G. E., and Jensen, G. J. (2007). Electron cryotomography. Biotechniques **43**, 413, 415, 417 passim.

Natarajan, P., et al. (2005). Exploring icosahedral virus structures with VIPER. Nat. Rev. Microbiol. **3**, 809–817.

Nybakken, G. E., et al. (2005). Structural basis of West Nile virus neutralization by a therapeutic antibody. Nature **437**, 764–768.

Olson, N. H., et al. (1990). The three-dimensional structure of frozen–hydrated Nudaurelia capensis beta virus, a T = 4 insect virus. J. Struct. Biol. **105**, 111–122.

Pettersen, E. F., et al. (2004). UCSF Chimera—A visualization system for exploratory research and analysis. J. Comput. Chem. **25**, 1605–1612.

Pintilie, G. D., et al. (2010). Quantitative analysis of cryo-EM density map segmentation by watershed and scale-space filtering, and fitting of structures by alignment to regions. J. Struct. Biol. **170**, 427–438.

Prasad, B. V., et al. (1999). X-ray crystallographic structure of the Norwalk virus capsid. Science **286**, 287–290.

Reinisch, K. M., et al. (2000). Structure of the reovirus core at 3.6 Å resolution. Nature **404**, 960–967.

Rusu, M., *et al.* (2008). Biomolecular pleiomorphism probed by spatial interpolation of coarse models. *Bioinformatics* **24**, 2460–2466.

Schmid, M. F., *et al.* (2004). Structure of the acrosomal bundle. *Nature* **431**, 104–107.

Schroder, G. F., *et al.* (2007). Combining efficient conformational sampling with a deformable elastic network model facilitates structure refinement at low resolution. *Structure* **15**, 1630–1641.

Sousa, D., and Grigorieff, N. (2007). Ab initio resolution measurement for single particle structures. *J. Struct. Biol.* **157**, 201–210.

Tagari, M., *et al.* (2002). New electron microscopy database and deposition system. *Trends Biochem. Sci.* **27**, 589.

Tama, F., *et al.* (2004). Normal mode based flexible fitting of high-resolution structure into low-resolution experimental data from cryo-EM. *J. Struct. Biol.* **147**, 315–326.

Tan, R. K., *et al.* (2008). YUP.SCX: Coaxing atomic models into medium resolution electron density maps. *J. Struct. Biol.* **163**, 163–174.

Thygesen, J., *et al.* (1996). The suitability of multi-metal clusters for phasing in crystallography of large macromolecular assemblies. *Structure* **4**, 513–518.

Topf, M., *et al.* (2008). Protein structure fitting and refinement guided by cryo-EM density. *Structure* **16**, 295–307.

Trabuco, L. G., *et al.* (2009). Molecular dynamics flexible fitting: A practical guide to combine cryo-electron microscopy and X-ray crystallography. *Methods* **49**, 174–180.

Trapani, S., *et al.* (2010). Macromolecular crystal data phased by negative-stained electron-microscopy reconstructions. *Acta Crystallogr. D* **66**, 514–521.

Unwin, N. (2005). Refined structure of the nicotinic acetylcholine receptor at 4A resolution. *J. Mol. Biol.* **346**, 967–989.

Villa, E., *et al.* (2009). Ribosome-induced changes in elongation factor Tu conformation control GTP hydrolysis. *Proc. Natl. Acad. Sci. USA* **106**, 1063–1068.

Wang, H. W., and Nogales, E. (2005). Nucleotide-dependent bending flexibility of tubulin regulates microtubule assembly. *Nature* **435**, 911–915.

Wang, L., and Sigworth, F. J. (2009). Structure of the BK potassium channel in a lipid membrane from electron cryomicroscopy. *Nature* **461**, 292–295.

Wang, J., *et al.* (1998). Crystal structure determination of Escherichia coli ClpP starting from an EM-derived mask. *J. Struct. Biol.* **124**, 151–163.

Wang, Y. A., *et al.* (2006). The structure of a filamentous bacteriophage. *J. Mol. Biol.* **361**, 209–215.

Wendler, P., and Saibil, H. R. (2010). Cryo electron microscopy structures of Hsp100 proteins: Crowbars in or out? *Biochem. Cell Biol.* **88**, 89–96.

Wriggers, W. (2010). Using Situs for the integration of multi-resolution structures. *Biophys. Rev.* **2**, 21–27.

Wynne, S. A., *et al.* (1999). The crystal structure of the human hepatitis B virus capsid. *Mol. Cell* **3**, 771–780.

Xiang, Y., *et al.* (2006). Structural changes of bacteriophage phi29 upon DNA packaging and release. *EMBO J.* **25**, 5229–5239.

Xiong, Y. (2008). From electron microscopy to X-ray crystallography: Molecular-replacement case studies. *Acta Crystallogr. D Biol. Crystallogr.* **64**, 76–82.

Yu, X., and Egelman, E. H. (1997). The RecA hexamer is a structural homologue of ring helicases. *Nat. Struct. Biol.* **4**, 101–104.

Zhou, Z. H., *et al.* (2001). Electron cryomicroscopy and bioinformatics suggest protein fold models for rice dwarf virus. *Nat. Struct. Biol.* **8**, 868–873.

CHAPTER FIVE

Electron Crystallography and Aquaporins

Andreas D. Schenk,* Richard K. Hite,* Andreas Engel,†
Yoshinori Fujiyoshi,‡ and Thomas Walz*,§

Contents

Abstract

Electron crystallography of two-dimensional (2D) crystals can provide information on the structure of membrane proteins at near-atomic resolution. Originally developed and used to determine the structure of bacteriorhodopsin (bR), electron crystallography has recently been applied to elucidate the structure of aquaporins (AQPs), a family of membrane proteins that form pores mostly for water but also other solutes. While electron crystallography has made major contributions to our understanding of the structure and function of AQPs, structural studies on AQPs, in turn, have fostered a number of technical

* Department of Cell Biology, Harvard Medical School, Boston, Massachusetts, USA
† Department of Pharmacology, Case Western Reserve University, Cleveland, Ohio, USA
‡ Department of Biophysics, Kyoto University, Oiwake, Kitashirakawa, Sakyo-ku, Kyoto, Japan
§ Howard Hughes Medical Institute, Harvard Medical School, Boston, Massachusetts, USA

Methods in Enzymology, Volume 483
ISSN 0076-6879, DOI: 10.1016/S0076-6879(10)83005-8

developments in electron crystallography. In this contribution, we summarize the insights electron crystallography has provided into the biology of AQPs, and describe technical advancements in electron crystallography that were driven by structural studies on AQP 2D crystals. In addition, we discuss some of the lessons that were learned from electron crystallographic work on AQPs.

ABBREVIATIONS

EM	electron microscopy
bR	bacteriorhodopsin
2D	two-dimensional
3D	three-dimensional
AQP	aquaporin
DMPC	dimyristoyl phosphatidylcholine
DOPC	dioleyl phosphatidylcholine
cmc	critical micellar concentration
MBCD	methyl-β-cyclodextrin
MIP	major intrinsic protein
SNR	signal-to-noise ratio

1. ELECTRON CRYSTALLOGRAPHY

Until recently, when the first density maps at near-atomic resolution were determined by single-particle electron microscopy (EM) (reviewed in Cheng and Walz, 2009; Zhou, 2008), electron crystallography was the only EM-based technique that allowed the determination of density maps at a resolution suitable for building atomic models. Electron crystallography was originally developed to determine the structure of bacteriorhodopsin (bR) (Henderson *et al.*, 1990; Mitsuoka *et al.*, 1999a) (Fig. 5.1A), and although it eventually also yielded the structure of the $\alpha\beta$ tubulin dimer (Nogales *et al.*, 1998), electron crystallography remains particularly powerful for the determination of membrane protein structures. Essentially, membrane proteins are reconstituted into lipid bilayers at a very low lipid-to-protein ratio, with the goal of inducing the protein to form regular arrays in the membrane. Such two-dimensional (2D) crystals can then be used to collect EM images and electron diffraction patterns, which are analyzed and combined to produce a three-dimensional (3D) density map. If the resolution reaches about 3.5 Å, an atomic model can be built into the density map. Although X-ray crystallography now produces the majority of membrane protein

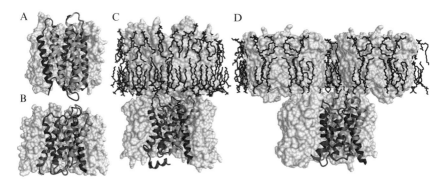

Figure 5.1 Atomic models of bacteriorhodopsin and aquaporins determined by electron crystallography. (A) Atomic model of the bacteriorhodopsin trimer (PDB entry 2AT9). Two subunits are shown as molecular surfaces and the front most subunit is shown in ribbon representation. The bound retinal is represented by an orange stick model. (B) Atomic model of the AQP1 tetramer (PDB entry 1FQY). Three subunits are shown as molecular surfaces and the front most subunit is shown in ribbon representation. (C, D) Atomic models of the membrane junctions formed by AQP0 (PDB entry 2B60) (C) and AQP4 (PDB entry 2ZZ9) (D). Three subunits of the bottom tetramers are shown as white molecular surfaces and the front most subunits are shown in ribbon representation. The top tetramers are shown as yellow molecular surfaces, together with ordered lipids represented by red stick models. Figures were created using *OpenStructure* (www.openstructure.org). The surfaces were calculated using *MSMS* (Sanner *et al.*, 1996). (See Color Insert.)

structures, electron crystallography can still make important contributions to the structural biology of membrane proteins (reviewed in Hite *et al.*, 2007). The greatest advantage of electron crystallography, however, is that it allows the structure of a membrane protein to be determined in its native environment, the lipid bilayer (reviewed in Fujiyoshi and Unwin, 2008; Raunser and Walz, 2009).

2. CONTRIBUTIONS OF ELECTRON CRYSTALLOGRAPHY TO THE STRUCTURAL BIOLOGY OF AQUAPORINS

The MIP family, named for its founding member, the major intrinsic protein, initially contained a small number of homologous membrane proteins that were suspected to form membrane channels (Pao *et al.*, 1991). In 1991, Peter Agre and coworkers established that one member of the MIP family, CHIP28, functions as a water channel (Preston *et al.*, 1992). Soon after, additional members of the MIP family were shown to conduct water, and water channels were identified in a wide variety of cells and organisms, which prompted the MIP family to be renamed aquaporins (AQPs) (Agre *et al.*, 1993). AQPs are now recognized as a family of ubiquitous membrane proteins that not only conduct water very efficiently but also conduct a number of other, mostly small, neutral solutes (recently reviewed in Gomes *et al.*, 2009). While AQPs conduct a variety of solutes across biological membranes, a hallmark of all AQPs is that they are completely impermeable to protons, a characteristic that is important for the maintenance of transmembrane proton gradients. Electron crystallography has made many contributions to our understanding of the structure and function of AQPs (reviewed in Andrews *et al.*, 2008; Gonen and Walz, 2006; Walz *et al.*, 2009), ranging from the first atomic structure of an AQP (Murata *et al.*, 2000), providing first insights into water selectivity and proton exclusion, to defining the interactions with annular lipids (Gonen *et al.*, 2005; Hite *et al.*, 2010a,b).

2.1. Aquaporin-1

In 1991, Peter Agre and coworkers identified a 28-kDa membrane protein in red blood cells that they named channel-forming integral protein of 28 kDa (CHIP28) (Preston and Agre, 1991). Functional assays with mRNA expressed in *Xenopus* oocytes and purified protein reconstituted into proteoliposomes identified CHIP28 as the long-sought red blood cell water channel (Preston *et al.*, 1992; Zeidel *et al.*, 1992), and the protein was subsequently renamed AQP1.

Electron crystallographic studies on AQP1 started soon after its discovery. The first structural data were obtained with negatively stained 2D crystals that were produced with AQP1 purified from red blood cells, providing first low-resolution 2D (Mitra *et al.*, 1994; Walz *et al.*, 1994a) and 3D maps of the tetramer (Walz *et al.*, 1994b). Vesicular 2D crystals were also used to perform functional assays, which established that the crystallized protein retained full water channel activity (Walz *et al.*, 1994a). Higher resolution projection maps were obtained when three groups analyzed AQP1 2D crystals by cryo-EM. Although three different lipids were used to prepare the 2D crystals, *E. coli* polar lipids, dimyristoyl phosphatidylcholine (DMPC) and dioleyl phosphatidylcholine (DOPC), the crystals had very similar unit cell dimensions and the tetramers displayed identical shapes. The projection maps, ranging in resolution from $\sim$6 Å (Mitra *et al.*, 1995; Walz *et al.*, 1995) to 3.5 Å (Jap and Li, 1995), clearly resolved the wedge-shaped subunits in the AQP1 tetramer and the water channel in the center of each monomer. The positions of the six predicted transmembrane helices were less clear in the projection maps. The arrangement of the helices was eventually seen in 3D density maps of AQP1 at a resolution of approximately 6 Å, which showed the helices to be tilted and to form a right-handed bundle (Cheng *et al.*, 1997; Walz *et al.*, 1997). The two density maps did not suffice, however, to assign the helices and to fully resolve a density in the center of the monomer, which, when the resolution increased to 4.5 Å, was seen to be formed in part by two short, pore-lining α-helices (Mitsuoka *et al.*, 1999b).

In 2000, an electron crystallographic density map was published at a resolution of about 3.8 Å, which allowed the first atomic model to be built for AQP1 (Murata *et al.*, 2000) (Fig. 5.1B). The novel AQP fold was seen to consist of two pseudo-symmetrical halves, each consisting of three trans-membrane α-helices and a reentrant loop, inserted into the membrane in opposite orientations and connected by the extended extracellular loop C. The six transmembrane helices H1–H6 form the periphery of the monomer and surround the two reentrant loops. Reentrant loops B (between helices H2 and H3) and E (between helices H5 and H6) enter from opposite sides of the membrane and connect with each other in the center of the membrane through the proline residues of the two highly conserved NPA motifs. After interacting with each other, the reentrant loops turn back and form the two short pore helices HB and HE. In addition to defining the highly conserved AQP fold, the structure also provided first insights into the water specificity and proton exclusion mechanisms. In particular, the structure revealed the importance of the two NPA motifs, both for stabilizing the AQP fold as well as for proton exclusion (Murata *et al.*, 2000).

Shortly after publication of the AQP1 structure, the structure of the *E. coli* glycerol facilitator GlpF was determined by X-ray crystallography (Fu *et al.*, 2000), and the next year brought the publication of the X-ray structure of bovine AQP1 (Sui *et al.*, 2001) and of another EM structure of human AQP1 (Ren *et al.*, 2001). It is interesting to note that the X-ray structure confirmed to a large extent the refined AQP1 structure derived from electron crystallography (de Groot *et al.*, 2001, 2003). The increasing number of AQP structures paved the way for many molecular dynamics studies that provided much insight into the mechanisms governing water permeation, proton exclusion, and solute selectivity (recently reviewed in Hub *et al.*, 2009).

2.2. Aquaporin-0

AQP0 is the most abundant membrane protein in lens fiber cells and was thus initially named MIP (Gorin *et al.*, 1984). Like AQP1, it forms a channel that is highly specific for water (Mulders *et al.*, 1995), but it also mediates the formation of membrane junctions and was initially even thought to be part of gap junctions (Bok *et al.*, 1982). The adhesive properties of AQP0 were first demonstrated *in vitro* by showing that reconstitution of AQP0 into proteoliposomes induced the vesicles to cluster (Dunia *et al.*, 1987). Later studies established that proteolytic cleavage of its cytoplasmic termini increases the adhesiveness of the extracellular surface of AQP0 (Gonen *et al.*, 2004a).

First 2D crystals of AQP0 were reported in 1998 (Hasler *et al.*, 1998a), and further studies revealed that the 2D crystals consist of two layers that mimic membrane junctions (Fotiadis *et al.*, 2000). Electron crystallography of such double-layered 2D crystals, grown with a mixture of full-length and cleaved AQP0 purified from the core of sheep lenses, eventually yielded a density map at 3 Å resolution, which made it possible to build an atomic model for the AQP0-mediated membrane junction (Gonen *et al.*, 2004b) (Fig. 5.1C). The atomic model not only identified the junction-forming interactions but also suggested that the water channel in junctional AQP0 may be in a closed or low-conductivity state. When the resolution of the density map was increased to 1.9 Å, only three water molecules were visible in the water channel, which were also too far apart to form hydrogen bonds (Gonen *et al.*, 2005). While this finding supported the notion that the water channel in junctional AQP0 is in a closed conformation, molecular dynamics calculations suggested that AQP0 in junctions still functions as an open water channel (Han *et al.*, 2006; Jensen *et al.*, 2008). When the structure of full-length AQP0 was determined by X-ray crystallography (Harries *et al.*, 2004), comparison with the EM structure of the truncated protein provided potential mechanisms for how cleavage induces junction formation and how junction formation may close the water pore (Gonen *et al.*, 2005).

In addition to visualizing water molecules, the 1.9 Å structure also resolved nine lipid molecules surrounding each AQP0 subunit, making it possible to describe the interaction of AQP0 with its annular lipids (Gonen *et al.*, 2005) (Fig. 5.1C). Comparison with the X-ray structure of AQP0 in a detergent micelle demonstrated the stabilizing effect of lipids on the residues they contact (Hite *et al.*, 2008). Most recently, a 2.5 Å resolution electron crystallographic structure of AQP0 in a bilayer formed by *E. coli* polar lipids provided the first view of a membrane protein in two different lipid environments and allowed the deduction of first principles that govern the interaction of annular lipids with membrane proteins (Hite *et al.*, 2010a).

2.3. Aquaporin-4

AQP4 is the predominant water channel in the mammalian brain, where it is expressed in glial cells (Jung *et al.*, 1994; Nielsen *et al.*, 1997). AQP4 forms orthogonal arrays in the plasma membrane of astrocytes and ependymocytes (Rash *et al.*, 1998; Verbavatz *et al.*, 1997), which vary in size depending on the physiological conditions of the water homeostasis in the brain (Landis and Reese, 1981). AQP4 is expressed in two splicing variants, AQP4M1 and AQP4M23 (Jung *et al.*, 1994), of which only the shorter splicing variant AQP4M23 promotes the formation of large square arrays (Furman *et al.*, 2003; Silberstein *et al.*, 2004). The longer splicing variant AQP4M1 appears to restrict orthogonal array formation due to palmitoylation of N-terminal cysteine residues that are missing in AQP4M23 (Suzuki *et al.*, 2008).

Electron crystallography produced a density map of rat AQP4M23 at 3.2 Å resolution, which made it possible to build an atomic model (Hiroaki *et al.*, 2006) (Fig. 5.1D) that was later confirmed by an X-ray crystal structure of the human homolog (Ho *et al.*, 2009). AQP4M23 formed double-layered 2D crystals, and the interactions between AQP4 molecules in the adjoining membranes were mediated by a short 3_{10} helix in extracellular loop C, suggesting that AQP4 may be involved in the formation of membrane junctions *in vivo*. The notion that AQP4 makes membrane junctions was further supported by expression of AQP4 in L-cells, which resulted in cell clustering (Hiroaki *et al.*, 2006), and a recent higher resolution electron crystallographic structure of AQP4M23, which showed that AQP4 in one membrane interacts with lipids in the adjoining membrane (Tani *et al.*, 2009) (Fig. 5.1D). AQP4 thus became the second member of an AQP subfamily whose members are involved in the formation of membrane junctions (reviewed in Engel *et al.*, 2008).

2.4. Electron crystallographic studies of other AQPs

In addition to producing atomic models for the three mammalian water channels AQP1, AQP0, and AQP4, electron crystallography has provided lower resolution information on the structure of several other AQPs.

AQPZ: AQPZ is the exclusively water-permeable AQP expressed in *E. coli* (Calamita *et al.*, 1995). Before any atomic models were available for AQPs, a projection map of AQPZ at 8 Å resolution revealed its pronounced structural similarity with the red blood cell water channel AQP1, thus demonstrating that the high sequence homology between members of the AQP family is reflected in their structure (Ringler *et al.*, 1999). The structural similarity of AQPZ with AQP1 was later confirmed by X-ray structures of AQPZ (Jiang *et al.*, 2006; Savage *et al.*, 2003).

GlpF: The *E. coli* glycerol facilitator GlpF was the first structure of a member of the AQP family that was determined by X-ray crystallography (Fu *et al.*, 2000). In the same year, electron crystallography also produced a 3.7 Å projection map (Braun *et al.*, 2000) and a 6.9 Å 3D density map of GlpF (Stahlberg *et al.*, 2000). Based on the 6.9 Å 3D density map and the atomic model of AQP1, a homology model was created for GlpF to explore the nature of the GlpF channel. The homology model and the X-ray structure both independently showed the channel to be amphipathic, allowing glycerol molecules to pass through the channel in a single file arrangement by forming hydrogen bonds with the hydrophilic side of the channel. Furthermore, the hydrophobic wall of the channel allows permeating water molecules to interact only with the two neighboring waters, whereas water in bulk solution typically interacts with 4–5 waters. This limited coordination of water in the channel provided an attractive explanation for GlpF conducting water at a lower rate than glycerol.

α-*TIP*: AQPs play many important roles in plants (reviewed in Maurel *et al.*, 2008). α-TIP, highly expressed in vacuolar membranes of cotyledons during seed maturation, was the first plant AQP whose structure was investigated. 2D crystals of bean α-TIP allowed calculation of a projection map at 7.7 Å resolution, which revealed the same structural architecture previously seen with mammalian and bacterial AQPs, providing first evidence that the structure of AQPs is conserved in all kingdoms (Daniels *et al.*, 1999).

SoPIP2;1: The plant aquaporin SoPIP2;1 is one of the major integral proteins in spinach leaf plasma membranes (Fraysse *et al.*, 2005). The first structural information on SoPIP2;1 came in the form of a 5 Å resolution 3D density map determined by electron crystallography (Kukulski *et al.*, 2005). Comparison with AQP1 suggests that a cysteine at the C terminus of loop A, a residue that is highly conserved in plasma membrane intrinsic protein (PIP)-type AQPs, is located near the fourfold axis of the tetramer, but the physiological significance of this conserved residue remains unclear. 2D crystals formed by SoPIP2;1 are double-layered, with the extracellular side of helix 1 protruding into the adjoining layer. This interaction between the extracellular surfaces of tetramers in the double-layered 2D crystals might be an indication that SoPIP2;1, like AQP0 and AQP4, has the potential to form membrane junctions *in vivo* (Engel *et al.*, 2008).

X-ray structures of SoPIP2;1 at 2.1 and 3.9 Å resolution showing the channel in a closed and open conformation, respectively, suggested potential mechanisms underlying AQP channel gating (Törnroth-Horsefield *et al.*, 2006). It was proposed that drought stress would cause channel closure by dephosphorylation of two highly conserved serine residues, Ser 115 and Ser 274, whereas flooding would cause channel closure by protonation of His 193. Later studies showed, however, that the two single mutants S115E and S274E and the corresponding double mutant, mimicking serines in a phosporylated state, did not result in the opening of the water pore (Nyblom *et al.*, 2009). Recent molecular dynamics simulations on SoPIP2;1 now suggest that residues Ser 36, Arg 190, and Asp 191 may play important roles in channel gating (Khandelia *et al.*, 2009).

AQP2: AQP2, expressed in the principal cells of the collecting duct, is responsible for the vasopressin-regulated water re-absorption in the kidney (Deen *et al.*, 1994; Fushimi *et al.*, 1993). The 4.5 Å resolution 3D density map of human AQP2 determined by electron crystallography represents the first structure of a recombinantly expressed human membrane channel (Schenk *et al.*, 2005). AQP2 also forms double-layered 2D crystals, but in contrast to AQP0, AQP4, and SoPIP2;1, AQP2 tetramers in the two layers of the crystals interact with each other through their cytoplasmic surfaces. While this interaction cannot be physiologically relevant, it has the advantage that it likely immobilizes the C terminus of AQP2, which may therefore be resolved in future higher resolution density maps. The C terminus of AQP2 is of particular interest, because phosphorylation of Ser 256 by cAMP-activated protein kinase A is involved in the vasopressin-dependent trafficking of AQP2 from cytoplasmic vesicles to the apical membrane (reviewed in Nedvetsky *et al.*, 2009).

AQP9: AQP9, expressed in liver and other tissues (Elkjaer *et al.*, 2000), displays an unusually broad substrate selectivity and allows permeation of molecules much larger than water while remaining completely impermeable to protons (Carbrey *et al.*, 2003; Tsukaguchi *et al.*, 1998). The only structural information on AQP9 available to date is a projection map at 7 Å resolution (Viadiu *et al.*, 2007). The projection map, together with a comparison of a homology model of AQP9 with the GlpF structure, suggests that substitution in the pore-lining residues of AQP9 occur predominantly at the hydrophobic edge of the tripathic GlpF pore, potentially explaining the broader substrate specificity of AQP9.

3. CONTRIBUTIONS OF STRUCTURAL STUDIES ON AQUAPORINS TO ADVANCES IN ELECTRON CRYSTALLOGRAPHY

Methodological developments are usually driven by the desire to answer a question that cannot be addressed with current technology. The driving force for the conception of electron crystallography and most of the early developments was the lack of methods that allowed determination of membrane protein structures. Because bR forms highly ordered 2D crystals *in vivo*, it was the ideal test specimen for developing the methodology of electron crystallography (Amos *et al.*, 1982; Unwin and Henderson, 1975), which soon produced a density map of bR at 7 Å resolution (Henderson and Unwin, 1975). The density map revealed that the membrane-spanning regions of bR form an α-helical bundle and thus provided the first information of the transmembrane organization of a membrane protein. While initial progress was fast, it took another 15 years to develop the technology to a point that made it possible to produce a density map at 3.5 Å resolution, which was sufficient to build the first atomic model for bR (Henderson *et al.*, 1990) (Fig. 5.1A). During this time, important advances were made in every step involved in electron crystallographic structure determination.

First data of bR were collected at room temperature with membranes embedded in glucose (Unwin and Henderson, 1975), but soon data collection was performed at cryogenic temperatures, as cooling the specimen to liquid nitrogen temperature was found to greatly reduce the visible effects of beam damage (Hayward and Glaeser, 1979). The benefits of data collection at low temperature led to the development of the first cryo-specimen holders (Hayward and Glaeser, 1980; Henderson *et al.*, 1991). These developments culminated in the design of an electron microscope equipped with a very stable top-entry specimen stage cooled to liquid helium temperature (Fujiyoshi, 1989). Data collected with this instrument of purple membranes embedded and frozen in a layer of trehalose (Hirai *et al.*, 1999) produced a density map at 3 Å resolution (Kimura *et al.*, 1997). With steadily improving instrumentation, more attention had to be paid to the preparation of flat specimens (Vonck, 2000). As a result, copper grids were replaced by molybdenum grids, which minimized the incidence of cryo-crinkling, the formation of wrinkles in the carbon film when a carbon-coated grid is cooled down to low temperatures (Booy and Pawley, 1993; Fujiyoshi, 1998). Much effort was also invested in finding ways to prepare atomically flat carbon support films (e.g., Fujiyoshi, 1998; Glaeser, 1992; Han *et al.*, 1994; Williams and Glaeser, 1972). In stark contrast, the software used to process electron crystallographic data has advanced surprisingly little over the years. With the exception of small improvements, the MRC software (Crowther *et al.*, 1996) used to process electron crystallographic data has remained almost unchanged since it was developed for determining the first bR structure.

Determining the bR structure was clearly the main motivation for advancing the methods used in electron crystallography, but work on other proteins also prompted methodological advances. Work on the plant light-harvesting complex II introduced tannic acid as an embedding medium (Wang and Kühlbrandt, 1991), and when its structure was determined by electron crystallography in 1994 (Kühlbrandt *et al.*, 1994), it provided proof that *in vitro* assembled 2D crystals can be sufficiently well ordered to yield atomic models. Work on $\alpha\beta$ tubulin led to the development of spot scanning (Downing, 1991), and the structure published in 1998 (Nogales *et al.*, 1998) demonstrated that electron crystallography can also be used to determine the structure of soluble proteins. Finally, the recent structure of microsomal prostaglandin E synthase 1 (Jegerschöld *et al.*, 2008) illustrated that omit maps can be used to pinpoint the location of substrates in membrane transporters.

While work on bR dominated the early developments in electron crystallography, AQPs provided most of the motivation for more recent methodological developments. Several reasons contribute to the role AQPs play in electron crystallography. Many AQPs are expressed at sufficiently high levels that they can be purified from natural sources, and many others can be produced in large amounts in heterologous expression systems. Several AQPs also form 2D arrays already in their native membranes, but because all AQPs are tetrameric, even AQPs that do not form regular arrays *in vivo* constitute ideal building blocks to form 2D crystals *in vitro*. Finally, AQPs are expressed in a large variety of cells and organisms, and different AQPs have different characteristics in terms of substrate specificity, permeation efficiency, and regulation. As the AQP fold is highly conserved, much can thus be learned from analyzing how small structural differences cause changes in function, providing strong motivation to determine the structure of more AQPs. The following sections describe how structural studies on AQPs led to advances in the different steps in electron crystallographic structure determination.

3.1. 2D Crystallization

Many AQPs form 2D crystals quite readily and are not as sensitive to reconstitution parameters as other membrane proteins (e.g., Zhao *et al.*, 2010). Although the quality and morphology of the crystals varied, given an appropriate lipid-to-protein ratio, AQP1 crystallized under almost any buffer condition. Indeed, highly ordered 2D crystals could be produced with three different lipids, DMPC (Jap and Li, 1995), *E. coli* polar lipids (Murata *et al.*, 2000), and DOPC (Ren *et al.*, 2001). One parameter that did have a remarkable effect on the 2D crystallization of AQP1 was the addition of Mg^{2+} to the dialysis buffer, which favored the growth of sheet-like crystals over the formation of vesicular crystals (Fig. 5.2). In the case of AQP9, growth of large 2D crystals depended on raising the protein concentration from 1 to 5 mg/ml. Specific to AQP0 was the finding that full-length protein isolated from the lens cortex formed single-layered crystals, whereas a mixture of full-length and truncated protein isolated from the lens core promoted the formation of double-layered crystals, presumably reflecting its function *in vivo* as a junction-forming membrane protein (Gonen *et al.*, 2004a). In the case of AQP4, only the shorter AQP4M23 isoform formed large 2D crystals, reflecting the situation *in vivo*, in which the longer AQP4M1 isoform restricts the growth of orthogonal arrays. Like AQP0, AQP4M23 formed double-layered 2D crystals, which suggests that it can also form membrane junctions *in vivo*. Non-tagged SoPIP2;1 expressed in *P. pastoris* could be reconstituted into two crystal forms (Kukulski *et al.*, 2005). Tubular vesicles that formed exhibited a specific surface texture, resulting from up–down oriented tetramers that were packed into alternating rows and were anisotropically ordered. In addition, highly ordered double-layered membrane sheets formed that exhibited p4 symmetry and a unit cell size of 65 Å. In contrast to the coaxially packed double-layered AQP0 crystals, the two crystalline layers of SoPIP2;1 were shifted against each other by half a unit cell in the x and y directions and were thus packed

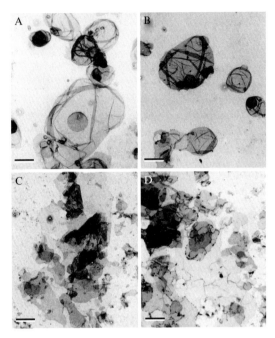

Figure 5.2   Effect of Mg^{2+} on the morphology of 2D crystals formed by AQP1. Reconstitution of AQP1 with *E. coli* lipids at a lipid-to-protein ratio of 0.5 yielded highly folded crystalline vesicles at a Mg^{2+} concentration of 10 mM (A), vesicles and a few small sheets at 20 mM Mg^{2+} (B), mostly sheets at 40 mM Mg^{2+} (C), and large 2D crystalline sheets at 50 mM Mg^{2+} (D). Scale bars: 1 μm. Figure adapted from Hasler *et al.* (1998b).

in precise register as well. Few of the membrane sheets were single-layered crystals, which were generally less well ordered, suggesting that a stabilizing crystal contact exists between the two layers.

Dialysis has been most effective in reconstituting membrane proteins into large, well-ordered 2D crystals (Jap *et al.*, 1992) and it is the method that yielded the 2D crystals that produced atomic models of AQP1 (Murata *et al.*, 2000), AQP0 (Gonen *et al.*, 2004b), and AQP4 (Hiroaki *et al.*, 2006). The reproducibility of 2D crystals produced by dialysis, however, has proven to be quite poor. Part of the problem may be that it is difficult to accurately control the kinetics of detergent removal, which depends on many factors, such as the critical micellar concentration (cmc) of the detergent used, the initial concentration of the detergent, the detergent gradient across the dialysis membrane, the pore size of the dialysis membrane, and the temperature. Dialysis efficiency also depends on the area of the dialysis membrane, which can inadvertently be reduced by the formation of air bubbles and aggregates that block the pores in the dialysis membrane. AQPs were thus used as test specimens to evaluate methods designed to allow for a better control of detergent removal.

In one approach, a device was designed that allowed controlled dilution of a mixture of membrane protein and lipid in detergent solution (Rémigy *et al.*, 2003). By adding sub-microliter volumes of detergent-free buffer to the sample, the detergent concentration was gradually lowered below its cmc, and dynamic light scattering was used to follow crystal formation. The dilution device was tested with the bacterial porin OmpF, but also with AQP2 (unpublished experiments) and AQP1 (Rémigy *et al.*, 2003). While controlled dilution did produce 2D crystals, the method has disadvantages. The detergent is not completely removed from the sample, which may make the crystals more fragile and complicate subsequent specimen preparation. Furthermore, dilution not only reduces the detergent concentration but also the protein and lipid concentrations, which can potentially interfere with the growth of 2D crystals, especially when the detergent has a low cmc and requires substantial dilution.

Another way to lower the detergent concentration below the cmc is detergent removal by adsorption, for which Bio-Beads have traditionally been used in 2D crystallization experiments (Rigaud *et al.*, 1997). Bio-Beads are granules of variable sizes, and so it is difficult to control and fractionate the amount of Bio-Beads to be added to a sample. To more finely tune the detergent removal, methyl-β-cyclodextrin (MBCD), a water-soluble compound that chelates detergents, was introduced to 2D crystallization experiments to substitute for the granular Bio-Beads (Signorell *et al.*, 2007). By gradually adding small volumes of an MBCD solution to the reconstitution mixture, detergent removal can be very finely controlled. An additional advantage of MBCD is that all components of the reconstitution experiments are soluble, which allows crystal growth to be still monitored by dynamic light scattering. MBCD-based detergent chelation also provided the basis for automation of 2D crystallization, and recently the first 2D crystallization robot was introduced (Iacovache *et al.*, 2010) (Fig. 5.3). In addition to facilitating the long and tedious process of finding and optimizing crystallization conditions, the robot also requires much smaller protein quantities for each sample. In addition to OmpF and aerolysin, the aquaporins SoPIP2;1 and AQP8 were extensively used as model proteins to establish 2D crystallization by detergent chelation and to test the crystallization robot.

3.2. Specimen preparation

AQPs also played a role in fully appreciating the importance of specimen preparation and in the development of the carbon sandwich technique (Gyobu *et al.*, 2004). Collecting the data used to build the first atomic model for AQP1 took many years (Murata *et al.*, 2000). The main problem was that AQP1 2D crystals were very sensitive to dehydration. It was thus very difficult to produce a sugar layer with the perfect thickness: if the layer

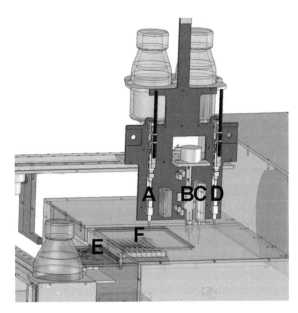

Figure 5.3   Schematic drawing of the 2D crystallization robot. (A) Contact-less cyclodextrin dispenser; (B) contact-less volume measurement sensor; (C) light scattering detector; (D) contact-less water dispenser maintaining the targeted sample volume; (E) orbital microplate shaker for homogenization of the crystallization mixture, allowing for simultaneous light scattering measurement; (F) commercially available, transparent 96-well microplate. The temperature is regulated by a Peltier device. Figure adapted from Iacovache *et al.* (2010).

was too thick the 2D crystals were not flat, and if it was too thin the 2D crystals became dehydrated and lost their order.

AQP4 forms double-layered 2D crystals, which initially caused a problem with data processing. While the protein order in the individual membrane layers was very high, there was significant variation in the distance and register between the two layers. These differences cannot be detected in diffraction patterns, but diffraction patterns from crystals with different arrangements of the two layers cannot be merged. It was thus necessary to collect images of 2D crystal that produced good diffraction patterns, which could then be used to classify the 2D crystals based on their arrangement of the two layers. This task would have been virtually impossible with the low yield of useable images collected from highly tilted specimens prepared with conventional specimen preparation methods. Overcoming this problem was the motivation for developing the carbon sandwich technique, which increased the yield of good images from highly tilted specimens to over 90% (Gyobu *et al.*, 2004). The carbon sandwich technique was thus the key to determining the structure of AQP4 (Hiroaki *et al.*, 2006).

AQP0 2D crystals were initially prepared with the well-established method of glucose embedding, and data were collected at liquid nitrogen temperature, resulting in a density map at 3 Å resolution (Gonen et al., 2004b). Crystals prepared in exactly the same way but cooled to liquid helium temperature produced diffraction patterns that had much poorer quality than those collected at liquid nitrogen temperature. Amazingly, when the same crystals were prepared with the carbon sandwich technique, they produced diffraction patterns showing reflections to better than 1.9 Å resolution (Gonen et al., 2005). The carbon sandwich technique was thus essential for data collection on AQP0 2D crystals at liquid helium temperature. To obtain high-resolution diffraction patterns, it was important that the carbon films were not only very flat but also had the correct thickness. When the carbon films were too thin, large areas of the specimen were unusable due to breakage. When the carbon films were too thick, the top carbon film was too rigid, preventing the two carbon films from coming close together during blotting. This resulted either in a trehalose layer that was too thick and deteriorated the quality and resolution of the diffraction patterns or in the creation of an air pocket between the crystals and the top carbon layer that led to dehydration of the crystals. Although specimen preparation depended on several factors, once a good preparation could be obtained, a single specimen allowed collection of a large number of high-quality diffraction patterns.

3.3. Single-layered versus double-layered 2D crystals

In most cases, membrane proteins form single-layered 2D crystals. Some membrane proteins form double-layered crystals, and one protein, connexin-26, even forms triple-layered crystal (Oshima et al., 2007). Multilayered crystals can be the result of the biological function of the protein, namely junction formation, as in the case of connexins (Oshima et al., 2007; Unger et al., 1999), AQP0 (Gonen et al., 2004b), AQP4 (Hiroaki et al., 2006), and potentially SoPIP2;1 (Kukulski et al., 2005), which all make contacts between the layers through their extracellular surfaces, or are merely caused by the crystallization conditions, as in the case of AQP2, which makes contacts between the layers through its cytoplasmic surface (Schenk et al., 2005). In all double-layered AQP 2D crystals, the individual layers have p4 symmetry due to the tetrameric organization of AQPs, but the symmetry of the double-layered crystal depends on the register between the two crystalline membranes. In the case of AQP0, the two layers were exactly in register with a tetramer in one membrane interacting with a tetramer in the other membrane, thus leading to an overall p422 symmetry. In double-layered AQP4 crystals, the tetramers were embedded in the membrane at a 30° angle. Thus, although the size of the AQP0 and AQP4 tetramers are virtually identical, the unit cell size in the AQP4 crystals (69 Å) was larger than in the AQP0 crystals (65.5 Å). Furthermore, the AQP4 tetramers in the two membranes were shifted with respect to each other by half a unit cell in the x and y directions, resulting in a tight tongue-into-grooves interaction between the two crystalline layers and an overall $p42_12$ symmetry. A similar packing arrangement was observed for SoPIP2;1 2D crystals, although the tetramers were not tilted, yielding a unit cell size of 65 Å (Kukulski *et al.*, 2005). Double-layered crystals of AQP2 showed $p22_12$ symmetry (see below). Irrespective of the symmetry, 2D crystals with multiple layers are advantageous for data collection, because they have a higher mass-per-area, and thus diffract electrons more strongly and produce data with a better signal-to-noise ratio (SNR) (Fig. 5.4). Problems arise, however, when the number or arrangement of the layers differ from crystal to crystal, which was the case with AQP4 and made merging data from different crystals very challenging. As described above, the problem was overcome by using images to classify the crystals and to obtain a homogeneous set of electron diffraction patterns (Hiroaki *et al.*, 2006). The double-layered crystals formed by AQP2 were suspected to suffer from merohedral twinning. In addition, the two layers were shifted with respect to each other in one direction by half a unit cell ($p22_12$ symmetry), which made it impossible to discern individual tetramers in

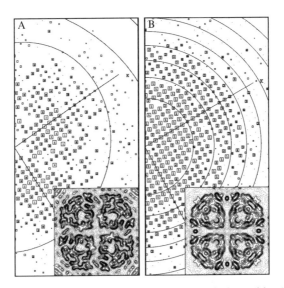

Figure 5.4   Single-layered and double-layered 2D crystals formed by AQP0. (A) IQ plot calculated from a cryoelectron micrograph of a glucose-embedded, single-layered 2D crystal. (B) IQ plot calculated from a cryoelectron micrograph of a glucose-embedded, double-layered 2D crystal. Both samples allowed calculation of a 4 Å projection map (insets), but the double-layered crystals produced reflections with higher signal-to-noise ratios due to their higher mass-per-area. Figures adapted from Gonen *et al.* (2004a,b).

projection maps. To resolve the organization of the AQP2 tetramers, the two layers were computationally separated using the image processing library and toolbox (*IPLT*) (Philippsen *et al.*, 2003, 2007) by an iterative process that exploited the p4 symmetry present in each individual layer but absent in the $p22_12$-symmetric double-layered crystal (Schenk *et al.*, 2005) (Fig. 5.5).

Membrane proteins usually form either single- or multi-layered crystals, but usually not both (although it is not uncommon that proteins that form preferentially multi-layered crystals also form some poorly ordered single-layered crystals, as was the case for AQP2 and SoPIP2;1). AQP0 is a special case, because full-length protein forms well-ordered single-layered crystals (Fig. 5.4A), whereas the presence of truncated protein promotes the formation of well-ordered double-layered crystals (Gonen *et al.*, 2004a) (Fig. 5.4B). Electron crystallographic data collection of the two crystal types showed that the double-layered crystals, as expected, produced diffraction patterns with much stronger reflections that extended to higher resolution. The higher resolution reflections may be visible simply because of the better SNR, but there may be additional reasons. The interactions between AQP0 tetramers in the adjoining layers are very strong and may help to increase the crystalline order in the two layers. The presence of two membranes may also increase the rigidity of the 2D crystals, so that the flatness of the crystals may be less affected by small defects in the surface of the carbon support film. This notion is supported by the fact that it was significantly easier to prepare double-layered crystals of AQP0 that showed high-resolution diffraction spots than single-layered crystals.

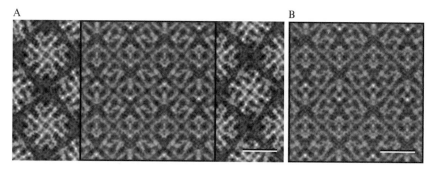

Figure 5.5　Symmetry of double-layered 2D crystals formed by AQP2. (A) Projection map of a double-layered AQP2 crystal with $p22_12$ symmetry produced by overlaying two simulated projection maps of single-layered crystals. (B) Experimental $p22_12$-symmetrized projection map of AQP2. The black boxes show the corresponding regions in the overlay of two simulated projections of single-layered crystals and the experimental projection of a double-layered crystal. Scale bars: 50 Å. Figure adapted from Schenk *et al.* (2005).

3.4. Data processing

The software for processing electron crystallographic data (Crowther *et al.*, 1996) that was developed at the Medical Research Council in Cambridge (United Kingdom) has changed very little over the years. Work on AQPs, in particular AQP2 and SoPIP2;1, made it clear, however, that data processing was becoming a bottleneck and that efficient data processing would require automation and would benefit from a user-friendly graphical interface. This realization motivated the development of two new software packages, *2dx* (Gipson *et al.*, 2007a,b) and *IPLT* (Philippsen *et al.*, 2003, 2007). *2dx* started out as a graphical user interface that provided mostly simplified access to the *MRC* programs, but it now also includes new developments, such as the use of the maximum likelihood approach for processing images of 2D crystals (Zeng *et al.*, 2007) (Fig. 5.6A–C). *IPLT*, which was designed to be highly modular and adaptable, represents a complete overhaul of data processing in electron crystallography and features many newly developed algorithms for processing images and diffraction patterns. The flexibility built into *IPLT* already proved very valuable for the processing of images of 2D crystals formed by the L-arginine/agmatine antiporter AdiC and DtpD, a member of the peptide transporter family, which produced much better projection maps after modifying the generic processing procedure (Casagrande *et al.*, 2008, 2009) (Fig. 5.6D, E). So far, *2dx* has focused on the processing of images, whereas *IPLT* improved the processing of diffraction patterns. Efforts are now underway to integrate the two software packages to allow processing of all electron crystallographic data using a common graphical user interface.

Images contain phase information, and the possibility to collect images is thus an advantage of electron microscopy over X-ray diffraction, which only allows the measurement of diffraction intensities. Collecting high-resolution images, however, is substantially more difficult and time-consuming than recording high-resolution diffraction patterns (Hite *et al.*, 2010b). Obtaining phase information by indirect methods thus would have the potential to significantly speed up electron crystallographic structure determination. It was unclear, however, whether indirect phasing methods would work in electron crystallography, because reflection intensities measured by electron diffraction have a much poorer quality (higher R_{sym} and R_{merge} values) than intensities measured by X-ray diffraction. The feasibility of obtaining phases by indirect methods was demonstrated by the structure determination of AQP0 and later on that of AQP4. Only electron diffraction patterns were collected, and the phases were obtained by molecular replacement using the structure of AQP1 as the search model (Gonen *et al.*, 2004b; Hiroaki *et al.*, 2006). The programs originally developed to process X-ray data could be used without modifications, but to find meaningful solutions, it proved crucial to first refine the unit cell size because the

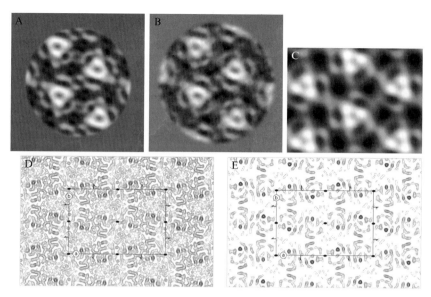

Figure 5.6 New developments in the processing of images of 2D crystals. (A–C) Three projection maps calculated from an image of a negatively stained, partially disordered 2D crystal of the ammonium transporter AmtB using three different algorithms implemented in the *2dx* software package. The final, twofold averaged projection maps were calculated using (A) maximum likelihood, (B) cross correlation alignment, and (C) standard crystallographic processing. The maximum likelihood approach opens a new avenue for processing images of partially disordered crystals, which are not optimally processed with the standard crystallographic image processing procedures. (D, E) $p22_12_1$-symmetrized projection maps of the L-arginine/agmatine antiporter AdiC calculated from images that were unbent using the standard crystallographic approach in *IPLT* (D) and a refined approach (E), which was fine-tuned for this particular project and yielded a more reliable projection map. Note that the noise level is significantly higher in the projection map obtained with the conventional approach (D) than in the projection map obtained with the refined approach (E). Panels (A)–(C) adapted from Zeng *et al.* (2007); panel (E) adapted from Casagrande *et al.* (2008).

magnification of an electron microscope cannot be calibrated very accurately. As the resolution of the AQP0 data increased from 3 to 1.9 Å, it also became important to use the appropriate scattering factors for refinement calculations (Gonen *et al.*, 2005). X-rays sample the electron density in the specimen, whereas electrons interact with the shielded Coulomb potential of the specimen. This difference is reflected in the atomic scattering factors. Calculations with the AQP0 data set showed that for refinement of the 3 Å structure, it did not make a significant difference whether scattering factors for X-rays or electrons were used. For the refinement of the 1.9 Å resolution structure, however, the best result was obtained when the calculations were performed with the scattering factors for electrons at the acceleration voltage used for data collection (300 kV).

3.5. Electron crystallography versus X-ray crystallography

Unlike X-rays, electrons carry a charge and thus interact differently with neutral and charged atoms in the specimen. The difference in the scattering of electrons by neutral and charged atoms is most pronounced in the low-resolution range. Theoretically, it should thus be possible to visualize the charge states of amino acid residues by calculating the difference between a map that contains data over the entire resolution range and a map that lacks the low-resolution data (< 5 Å). The situation is, however, complicated by the fact that solute effects also affect the low-resolution range. Nevertheless, studies on bR (Mitsuoka *et al.*, 1999a) as well as mathematical simulations (Hirai *et al.*, 2007) suggest that it may be possible to see the ionization state of amino acid residues in electron crystallographic density maps. Further calculations revealed that the resulting difference peaks strongly depend on the accuracy of the atomic coordinates. To identify charge states, it is likely necessary that the structure of the protein first be determined at a resolution of about 1.5 Å. While it is unlikely that this resolution can be achieved for bR, for which data is collected from fused purple membranes, *in vitro* assembled 2D crystals of AQPs may reach sufficient order to determine a density map at such a resolution. An interesting test specimen would be AQP0, not only because it has formed the best-ordered 2D crystals to date but also because its water conductance is pH-regulated. Visualization of the charge state of key residues in AQP0 may thus reveal the mechanism underlying its pH regulation.

For every AQP structure determined by electron crystallography, there is now also an X-ray structure available, which provides the opportunity to compare structures produced by the two techniques.

The electron crystallographic structure of AQP1 was the first structure of an AQP (Murata *et al.*, 2000), but the crystals were not of very high quality and data collection was very challenging. The density map thus only had an in-plane resolution of 3.8 Å and even less in *z*-direction. Another electron crystallographic structure of AQP1 was based on a density map with similar resolution (in-plane resolution of 3.7 Å) (Ren *et al.*, 2001). Molecular dynamics simulations proved to be an excellent tool to evaluate the quality of the two atomic models and to refine the structure of AQP1. To produce a more accurate structure, the X-ray structure of GlpF (Fu *et al.*, 2000) was used to create a homology model for AQP1, which was then refined against the 3.8 Å resolution electron crystallographic data, first by rigid body refinement and then by a molecular dynamics simulated annealing procedure that refined the torsion angles (de Groot *et al.*, 2001). When X-ray crystallography produced the structure of bovine AQP1 (Sui *et al.*, 2001), which has a sequence identity of over 90% with the human protein, it was possible to compare the X-ray structure with the refined and the two original EM structures (de Groot *et al.*, 2003). All the AQP1 structures showed a high degree of similarity, with root mean square deviations (RMSDs) between the backbone atoms of less than 2.5 Å. The two original EM-based models for human AQP1, however, also showed a number of significant differences to the X-ray structure of the bovine protein due to low resolution of the map and the resulting inaccuracies in model building (Fig. 5.7A). The refined EM structure, on the other hand, agreed very well with the X-ray structure, except for the loop regions, and also showed better agreement with the X-ray structure of AQP1 than with that of GlpF, indicating that refinement—if at all—only introduced a small bias toward the GlpF structure. These findings thus demonstrated the power of molecular dynamics calculations for the refinement of EM structures.

Comparison of the AQP0 structures determined by electron crystallography at 1.9 Å (Gonen et al., 2005) and by X-ray crystallography at 2.2 Å provided not only biological information on junction formation and channel gating but also showed differences between AQP0 in a lipidic environment and solubilized by a detergent micelle. With the exception of the conformation of the extracellular loops, the two structures are extremely similar (RMSD of 0.61 Å between the backbone atoms) (Fig. 5.7B) and could thus be compared on an atom-to-atom basis (Hite et al., 2008). Analysis of the B-factors showed that atoms in the electron crystallographic AQP0 structure that contacted lipid molecules had generally lower B-factors than the corresponding atoms in the X-ray structure that are

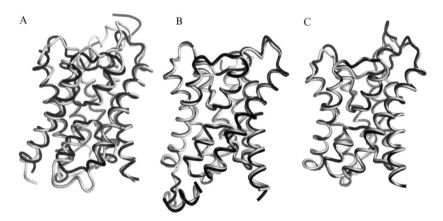

Figure 5.7 Comparison of atomic models of AQP1, AQP0, and AQP4 determined by electron and X-ray crystallography. The ribbon diagrams of the structures determined by electron crystallography are shown in white and those of the corresponding structures determined by X-ray crystallography are shown in black. (A) AQP1 (PDB entries 1FQY and 1J4N). (B) AQP0 (PDB entries 2B6O and 1YMG). (C) AQP4 (PDB entries 2ZZ9 and 3GD8). All ribbon diagrams only show the amino acid residues that are present in both the X-ray and the electron crystallographic models. Figures were created using OpenStructure (www.openstructure.org).

presumably interacting with detergent molecules. The same trend was found in other structures of membrane proteins that resolved associated lipid or detergent molecules (Hite *et al.*, 2008). This finding suggests that lipids have a stabilizing effect on protein residues they contact. While this effect may not be important for the very stable AQPs, it may make a difference for more labile membrane proteins such as G-protein coupled receptors (Rosenbaum *et al.*, 2009), for which lipids may help stabilize their structures.

Like with AQP0, the electron crystallographic and X-ray structures of AQP4 are very similar (RMSD of 0.61 Å between the backbone atoms) (Fig. 5.7C). Nevertheless, there is a remarkable difference, as the densities for the water molecules in the 2.8 Å resolution EM map appear much better defined than in the 1.8 Å resolution X-ray map (Fig. 5.8). Although this remains to be proven, the better definition of the waters in the EM map may be attributed to the dipole moments of the α-helices, especially those of the two short pore helices HB and HE, which may be influenced by the "nonglobular" environment of the lipid bilayer (Sengupta *et al.*, 2005). As dipole moments of α-helices may be stronger in lipid bilayers than in detergent micelles, they may more strongly affect the localization of water molecules in the AQP4 channel, providing a more accurate view of the physiological situation.

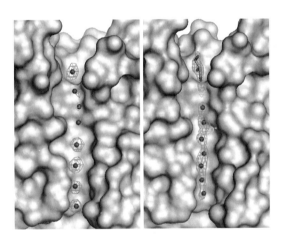

Figure 5.8 Comparison of the solute densities in the maps of AQP4 determined by electron and X-ray crystallography. The EM structure (PDB entry 2ZZ9) (A) and the X-ray structure (PDB entry 3GD8) (B) of AQP4 are represented as molecular surfaces. The water molecules in the channel are shown as red spheres. The densities for water and glycerol molecules are displayed as blue ($2F_o - F_c$) and green ($F_o - F_c$) wireframes. The water densities appear better defined in the EM map than in the X-ray map. (See Color Insert.)

ACKNOWLEDGMENTS

We thank Dr. Kazutoshi Tani for preparing Fig. 5.8. Electron crystallographic work on AQPs in the Walz laboratory is supported by NIH grants P01 GM062580 (to Stephen C. Harrison) and R01 EY015107 and R01 GM082927 (to T. W.). Electron crystallographic work on AQPs in the Fujiyoshi laboratory is supported by Grants-in-Aid for Specially Promoted Research, and the Japan New Energy and Industrial Technology Development Organization (NEDO). Electron crystallographic work on AQPs in the Engel laboratory was supported by the Swiss National Science Foundation (SNF Grant 3100A0-108299 to A. E.), the National Center of Competence in Research (NCCR) of Structural Biology, the European Union (EU projects LSHG-CT-2004-502828 and LSHG-CT-2005-018811), and the Maurice E. Müller Foundation of Switzerland. A. D. S. is supported by a fellowship from the Swiss National Science Foundation (PA00P3_126253). T. W. is an investigator in the Howard Hughes Medical Institute.

REFERENCES

Agre, P., Sasaki, S., and Chrispeels, M. J. (1993). Aquaporins: A family of water channel proteins. *Am. J. Physiol.* **265,** F461.

Amos, L. A., Henderson, R., and Unwin, P. N. (1982). Three-dimensional structure determination by electron microscopy of two-dimensional crystals. *Prog. Biophys. Mol. Biol.* **39,** 183–231.

Andrews, S., Reichow, S. L., and Gonen, T. (2008). Electron crystallography of aquaporins. *IUBMB Life* **60,** 430–436.

Bok, D., Dockstader, J., and Horwitz, J. (1982). Immunocytochemical localization of the lens main intrinsic polypeptide (MIP26) in communicating junctions. *J. Cell Biol.* **92,** 213–220.

Booy, F. P., and Pawley, J. B. (1993). Cryo-crinkling: What happens to carbon films on copper grids at low temperature. *Ultramicroscopy* **48,** 273–280.

Braun, T., Philippsen, A., Wirtz, S., Borgnia, M. J., Agre, P., Kühlbrandt, W., Engel, A., and Stahlberg, H. (2000). The 3.7 Å projection map of the glycerol facilitator GlpF: A variant of the aquaporin tetramer. *EMBO Rep.* **1,** 183–189.

Calamita, G., Bishai, W. R., Preston, G. M., Guggino, W. B., and Agre, P. (1995). Molecular cloning and characterization of AqpZ, a water channel from *Escherichia coli*. *J. Biol. Chem.* **270,** 29063–29066.

Carbrey, J. M., Gorelick-Feldman, D. A., Kozono, D., Praetorius, J., Nielsen, S., and Agre, P. (2003). Aquaglyceroporin AQP9: Solute permeation and metabolic control of expression in liver. *Proc. Natl. Acad. Sci. USA* **100,** 2945–2950.

Casagrande, F., Ratera, M., Schenk, A. D., Chami, M., Valencia, E., Lopez, J. M., Torrents, D., Engel, A., Palacin, M., and Fotiadis, D. (2008). Projection structure of a member of the amino acid/polyamine/organocation transporter superfamily. *J. Biol. Chem.* **283,** 33240–33248.

Casagrande, F., Harder, D., Schenk, A., Meury, M., Ucurum, Z., Engel, A., Weitz, D., Daniel, H., and Fotiadis, D. (2009). Projection structure of DtpD (YbgH), a prokaryotic member of the peptide transporter family. *J. Mol. Biol.* **394,** 708–717.

Cheng, Y., and Walz, T. (2009). The advent of near-atomic resolution in single-particle electron microscopy. *Annu. Rev. Biochem.* **78,** 723–742.

Cheng, A., van Hoek, A. N., Yeager, M., Verkman, A. S., and Mitra, A. K. (1997). Three-dimensional organization of a human water channel. *Nature* **387,** 627–630.

Crowther, R. A., Henderson, R., and Smith, J. M. (1996). MRC image processing programs. *J. Struct. Biol.* **116,** 9–16.

Daniels, M. J., Chrispeels, M. J., and Yeager, M. (1999). Projection structure of a plant vacuole membrane aquaporin by electron cryo-crystallography. *J. Mol. Biol.* **294,** 1337–1349.

de Groot, B. L., Engel, A., and Grubmüller, H. (2001). A refined structure of human aquaporin-1. *FEBS Lett.* **504,** 206–211.

de Groot, B. L., Engel, A., and Grubmüller, H. (2003). The structure of the aquaporin-1 water channel: A comparison between cryo-electron microscopy and X-ray crystallography. *J. Mol. Biol.* **325,** 485–493.

Deen, P. M., Verdijk, M. A., Knoers, N. V., Wieringa, B., Monnens, L. A., van Os, C. H., and van Oost, B. A. (1994). Requirement of human renal water channel aquaporin-2 for vasopressin-dependent concentration of urine. *Science* **264,** 92–95.

Downing, K. H. (1991). Spot-scan imaging in transmission electron microscopy. *Science* **251,** 53–59.

Dunia, I., Manenti, S., Rousselet, A., and Benedetti, E. L. (1987). Electron microscopic observations of reconstituted proteoliposomes with the purified major intrinsic membrane protein of eye lens fibers. *J. Cell Biol.* **105,** 1679–1689.

Elkjaer, M., Vajda, Z., Nejsum, L. N., Kwon, T., Jensen, U. B., Amiry-Moghaddam, M., Frøkiaer, J., and Nielsen, S. (2000). Immunolocalization of AQP9 in liver, epididymis, testis, spleen, and brain. *Biochem. Biophys. Res. Commun.* **276,** 1118–1128.

Engel, A., Fujiyoshi, Y., Gonen, T., and Walz, T. (2008). Junction-forming aquaporins. *Curr. Opin. Struct. Biol.* **18,** 229–235.

Fotiadis, D., Hasler, L., Müller, D. J., Stahlberg, H., Kistler, J., and Engel, A. (2000). Surface tongue-and-groove contours on lens MIP facilitate cell-to-cell adherence. *J. Mol. Biol.* **300,** 779–789.

Fraysse, L. C., Wells, B., McCann, M. C., and Kjellbom, P. (2005). Specific plasma membrane aquaporins of the PIP1 subfamily are expressed in sieve elements and guard cells. *Biol. Cell* **97,** 519–534.

Fu, D., Libson, A., Miercke, L. J., Weitzman, C., Nollert, P., Krucinski, J., and Stroud, R. M. (2000). Structure of a glycerol-conducting channel and the basis for its selectivity. *Science* **290,** 481–486.

Fujiyoshi, Y. (1989). High resolution cryo-electron microscopy for biological macromolecules. *J. Electron Microsc.* **38,** 97–101.

Fujiyoshi, Y. (1998). The structural study of membrane proteins by electron crystallography. *Adv. Biophys.* **35,** 25–80.

Fujiyoshi, Y., and Unwin, N. (2008). Electron crystallography of proteins in membranes. *Curr. Opin. Struct. Biol.* **18,** 587–592.

Furman, C. S., Gorelick-Feldman, D. A., Davidson, K. G., Yasumura, T., Neely, J. D., Agre, P., and Rash, J. E. (2003). Aquaporin-4 square array assembly: Opposing actions of M1 and M23 isoforms. *Proc. Natl. Acad. Sci. USA* **100,** 13609–13614.

Fushimi, K., Uchida, S., Hara, Y., Hirata, Y., Marumo, F., and Sasaki, S. (1993). Cloning and expression of apical membrane water channel of rat kidney collecting tubule. *Nature* **361,** 549–552.

Gipson, B., Zeng, X., Zhang, Z. Y., and Stahlberg, H. (2007a). 2dx—User-friendly image processing for 2D crystals. *J. Struct. Biol.* **157,** 64–72.

Gipson, B., Zeng, X., and Stahlberg, H. (2007b). 2dx_merge: Data management and merging for 2D crystal images. *J. Struct. Biol.* **160,** 375–384.

Glaeser, R. M. (1992). Specimen flatness of thin crystalline arrays: Influence of the substrate. *Ultramicroscopy* **46,** 33–43.

Gomes, D., Agasse, A., Thiébaud, P., Delrot, S., Gerós, H., and Chaumont, F. (2009). Aquaporins are multifunctional water and solute transporters highly divergent in living organisms. *Biochim. Biophys. Acta* **1788**, 1213–1228.

Gonen, T., and Walz, T. (2006). The structure of aquaporins. *Q. Rev. Biophys.* **39**, 361–396.

Gonen, T., Cheng, Y., Kistler, J., and Walz, T. (2004a). Aquaporin-0 membrane junctions form upon proteolytic cleavage. *J. Mol. Biol.* **342**, 1337–1345.

Gonen, T., Sliz, P., Kistler, J., Cheng, Y., and Walz, T. (2004b). Aquaporin-0 membrane junctions reveal the structure of a closed water pore. *Nature* **429**, 193–197.

Gonen, T., Cheng, Y., Sliz, P., Hiroaki, Y., Fujiyoshi, Y., Harrison, S. C., and Walz, T. (2005). Lipid-protein interactions in double-layered two-dimensional AQP0 crystals. *Nature* **438**, 633–638.

Gorin, M. B., Yancey, S. B., Cline, J., Revel, J. P., and Horwitz, J. (1984). The major intrinsic protein (MIP) of the bovine lens fiber membrane: Characterization and structure based on cDNA cloning. *Cell* **39**, 49–59.

Gyobu, N., Tani, K., Hiroaki, Y., Kamegawa, A., Mitsuoka, K., and Fujiyoshi, Y. (2004). Improved specimen preparation for cryo-electron microscopy using a symmetric carbon sandwich technique. *J. Struct. Biol.* **146**, 325–333.

Han, B. G., Wolf, S. G., Vonck, J., and Glaeser, R. M. (1994). Specimen flatness of glucose-embedded biological materials for electron crystallography is affected significantly by the choice of carbon evaporation stock. *Ultramicroscopy* **55**, 1–5.

Han, B. G., Guliaev, A. B., Walian, P. J., and Jap, B. K. (2006). Water transport in AQP0 aquaporin: Molecular dynamics studies. *J. Mol. Biol.* **360**, 285–296.

Harries, W. E., Akhavan, D., Miercke, L. J., Khademi, S., and Stroud, R. M. (2004). The channel architecture of aquaporin 0 at a 2.2-Å resolution. *Proc. Natl. Acad. Sci. USA* **101**, 14045–14050.

Hasler, L., Walz, T., Tittmann, P., Gross, H., Kistler, J., and Engel, A. (1998a). Purified lens major intrinsic protein (MIP) forms highly ordered tetragonal two-dimensional arrays by reconstitution. *J. Mol. Biol.* **279**, 855–864.

Hasler, L., Heymann, J. B., Engel, A., Kistler, J., and Walz, T. (1998b). 2D crystallization of membrane proteins: Rationales and examples. *J. Struct. Biol.* **121**, 162–171.

Hayward, S. B., and Glaeser, R. M. (1979). Radiation damage of purple membrane at low temperature. *Ultramicroscopy* **4**, 201–210.

Hayward, S. B., and Glaeser, R. M. (1980). High resolution cold stage for the JEOL 100B and 100C electron microscopes. *Ultramicroscopy* **5**, 3–8.

Henderson, R., and Unwin, P. N. (1975). Three-dimensional model of purple membrane obtained by electron microscopy. *Nature* **257**, 28–32.

Henderson, R., Baldwin, J. M., Ceska, T. A., Zemlin, F., Beckmann, E., and Downing, K. H. (1990). Model for the structure of bacteriorhodopsin based on high-resolution electron cryo-microscopy. *J. Mol. Biol.* **213**, 899–929.

Henderson, R., Raeburn, C., and Vigers, G. (1991). A side-entry cold holder for cryo-electron microscopy. *Ultramicroscopy* **35**, 45–53.

Hirai, T., Murata, K., Mitsuoka, K., Kimura, Y., and Fujiyoshi, Y. (1999). Trehalose embedding technique for high-resolution electron crystallography: Application to structural study on bacteriorhodopsin. *J. Electron Microsc.* **48**, 653–658.

Hirai, T., Mitsuoka, K., Kidera, A., and Fujiyoshi, Y. (2007). Simulation of charge effects on density maps obtained by high-resolution electron crystallography. *J. Electron Microsc.* **56**, 131–140.

Hiroaki, Y., Tani, K., Kamegawa, A., Gyobu, N., Nishikawa, K., Suzuki, H., Walz, T., Sasaki, S., Mitsuoka, K., Kimura, K., Mizoguchi, A., and Fujiyoshi, Y. (2006). Implications of the aquaporin-4 structure on array formation and cell adhesion. *J. Mol. Biol.* **355**, 628–639.

Hite, R. K., Raunser, S., and Walz, T. (2007). Revival of electron crystallography. *Curr. Opin. Struct. Biol.* **17**, 389–395.

Hite, R. K., Gonen, T., Harrison, S. C., and Walz, T. (2008). Interactions of lipids with aquaporin-0 and other membrane proteins. *Pflugers Arch.* **456**, 651–661.

Hite, R. K., Li, Z., and Walz, T. (2010a). Principles of membrane protein interactions with annular lipids deduced from aquaporin-0 2D crystals. *EMBO J.* **29**, 1652–1658.

Hite, R. K., Schenk, A. D., Li, Z., Cheng, Y., and Walz, T. (2010b). Collecting electron crystallographic data of two-dimensional protein crystals. *Methods Enzymol.* **483**.

Ho, J. D., Yeh, R., Sandstrom, A., Chorny, I., Harries, W. E., Robbins, R. A., Miercke, L. J., and Stroud, R. M. (2009). Crystal structure of human aquaporin 4 at 1.8 Å and its mechanism of conductance. *Proc. Natl. Acad. Sci. USA* **106**, 7437–7442.

Hub, J. S., Grubmüller, H., and de Groot, B. L. (2009). Dynamics and energetics of permeation through aquaporins. What do we learn from molecular dynamics simulations? *Handb. Exp. Pharmacol.* **190**, 57–76.

Iacovache, I., Biasini, M., Kowal, J., Kukulski, W., Chami, M., Van der Goot, F. G., Engel, A., and Rémigy, H. W. (2010). The 2DX robot: A membrane protein 2D crystallization Swiss army knife. *J. Struct. Biol.* **169**, 370–378.

Jap, B. K., and Li, H. (1995). Structure of the osmo-regulated H_2O-channel, AQP-CHIP, in projection at 3.5 Å resolution. *J. Mol. Biol.* **251**, 413–420.

Jap, B. K., Zulauf, M., Scheybani, T., Hefti, A., Baumeister, W., Aebi, U., and Engel, A. (1992). 2D crystallization: From art to science. *Ultramicroscopy* **46**, 45–84.

Jegerschöld, C., Pawelzik, S., Purhonen, P., Bhakat, P., Gheorghe, K. R., Gyobu, N., Mitsuoka, K., Morgenstern, R., Jakobsson, P.-J., and Hebert, H. (2008). Structural basis for induced formation of the inflammatory mediator prostaglandin E2. *Proc. Natl. Acad. Sci. USA* **105**, 11110–11115.

Jensen, M. Ø., Dror, R. O., Xu, H., Borhani, D. W., Arkin, I. T., Eastwood, M. P., and Shaw, D. E. (2008). Dynamic control of slow water transport by aquaporin 0: Implications for hydration and junction stability in the eye lens. *Proc. Natl. Acad. Sci. USA* **105**, 14430–14435.

Jiang, J., Daniels, B. V., and Fu, D. (2006). Crystal structure of AqpZ tetramer reveals two distinct Arg-189 conformations associated with water permeation through the narrowest constriction of the water-conducting channel. *J. Biol. Chem.* **281**, 454–460.

Jung, J. S., Bhat, R. V., Preston, G. M., Guggino, W. B., Baraban, J. M., and Agre, P. (1994). Molecular characterization of an aquaporin cDNA from brain: Candidate osmoreceptor and regulator of water balance. *Proc. Natl. Acad. Sci. USA* **91**, 13052–13056.

Khandelia, H., Jensen, M.Ø., and Mouritsen, O. G. (2009). To gate or not to gate: Using molecular dynamics simulations to morph gated plant aquaporins into constitutively open conformations. *J. Phys. Chem. B* **113**, 5239–5244.

Kimura, Y., Vassylyev, D. G., Miyazawa, A., Kidera, A., Matsushima, M., Mitsuoka, K., Murata, K., Hirai, T., and Fujiyoshi, Y. (1997). Surface of bacteriorhodopsin revealed by high-resolution electron crystallography. *Nature* **389**, 206–211.

Kühlbrandt, W., Wang, D. N., and Fujiyoshi, Y. (1994). Atomic model of plant light-harvesting complex by electron crystallography. *Nature* **367**, 614–621.

Kukulski, W., Schenk, A. D., Johanson, U., Braun, T., de Groot, B. L., Fotiadis, D., Kjellbom, P., and Engel, A. (2005). The 5 Å structure of heterologously expressed plant aquaporin SoPIP2;1. *J. Mol. Biol.* **350**, 611–616.

Landis, D. M., and Reese, T. S. (1981). Astrocyte membrane structure: Changes after circulatory arrest. *J. Cell Biol.* **88**, 660–663.

Maurel, C., Verdoucq, L., Luu, D. T., and Santoni, V. (2008). Plant aquaporins: Membrane channels with multiple integrated functions. *Annu. Rev. Plant Biol.* **59**, 595–624.

Mitra, A. K., Yeager, M., van Hoek, A. N., Wiener, M. C., and Verkman, A. S. (1994). Projection structure of the CHIP28 water channel in lipid bilayer membranes at 12-Å resolution. *Biochemistry* **33,** 12735–12740.

Mitra, A. K., van Hoek, A. N., Wiener, M. C., Verkman, A. S., and Yeager, M. (1995). The CHIP28 water channel visualized in ice by electron crystallography. *Nat. Struct. Biol.* **2,** 726–729.

Mitsuoka, K., Hirai, T., Murata, K., Miyazawa, A., Kidera, A., Kimura, Y., and Fujiyoshi, Y. (1999a). The structure of bacteriorhodopsin at 3.0 Å resolution based on electron crystallography: Implication of the charge distribution. *J. Mol. Biol.* **286,** 861–882.

Mitsuoka, K., Murata, K., Walz, T., Hirai, T., Agre, P., Heymann, J. B., Engel, A., and Fujiyoshi, Y. (1999b). The structure of aquaporin-1 at 4.5-Å resolution reveals short α-helices in the center of the monomer. *J. Struct. Biol.* **128,** 34–43.

Mulders, S. M., Preston, G. M., Deen, P. M., Guggino, W. B., van Os, C. H., and Agre, P. (1995). Water channel properties of major intrinsic protein of lens. *J. Biol. Chem.* **270,** 9010–9016.

Murata, K., Mitsuoka, K., Hirai, T., Walz, T., Agre, P., Heymann, J. B., Engel, A., and Fujiyoshi, Y. (2000). Structural determinants of water permeation through aquaporin-1. *Nature* **407,** 599–605.

Nedvetsky, P. I., Tamma, G., Beulshausen, S., Valenti, G., Rosenthal, W., and Klussmann, E. (2009). Regulation of aquaporin-2 trafficking. *Handb. Exp. Pharmacol.* **190,** 133–157.

Nielsen, S., Nagelhus, E. A., Amiry-Moghaddam, M., Bourque, C., Agre, P., and Ottersen, O. P. (1997). Specialized membrane domains for water transport in glial cells: High-resolution immunogold cytochemistry of aquaporin-4 in rat brain. *J. Neurosci.* **17,** 171–180.

Nogales, E., Wolf, S. G., and Downing, K. H. (1998). Structure of the αβ tubulin dimer by electron crystallography. *Nature* **391,** 199–203.

Nyblom, M., Frick, A., Wang, Y., Ekvall, M., Hallgren, K., Hedfalk, K., Neutze, R., Tajkhorshid, E., and Törnroth-Horsefield, S. (2009). Structural and functional analysis of SoPIP2;1 mutants adds insight into plant aquaporin gating. *J. Mol. Biol.* **387,** 653–668.

Oshima, A., Tani, K., Hiroaki, Y., Fujiyoshi, Y., and Sosinsky, G. E. (2007). Three-dimensional structure of a human connexin26 gap junction channel reveals a plug in the vestibule. *Proc. Natl. Acad. Sci. USA* **104,** 10034–10039.

Pao, G. M., Wu, L. F., Johnson, K. D., Höfte, H., Chrispeels, M. J., Sweet, G., Sandal, N. N., and Saier, M. H. J. (1991). Evolution of the MIP family of integral membrane transport proteins. *Mol. Microbiol.* **5,** 33–37.

Philippsen, A., Schenk, A. D., Stahlberg, H., and Engel, A. (2003). Iplt—Image processing library and toolkit for the electron microscopy community. *J. Struct. Biol.* **144,** 4–12.

Philippsen, A., Schenk, A. D., Signorell, G. A., Mariani, V., Berneche, S., and Engel, A. (2007). Collaborative EM image processing with the IPLT image processing library and toolbox. *J. Struct. Biol.* **157,** 28–37.

Preston, G. M., and Agre, P. (1991). Isolation of the cDNA for erythrocyte integral membrane protein of 28 kilodaltons: Member of an ancient channel family. *Proc. Natl. Acad. Sci. USA* **88,** 11110–11114.

Preston, G. M., Carroll, T. P., Guggino, W. B., and Agre, P. (1992). Appearance of water channels in *Xenopus* oocytes expressing red cell CHIP28 protein. *Science* **256,** 385–387.

Rash, J. E., Yasumura, T., Hudson, C. S., Agre, P., and Nielsen, S. (1998). Direct immunogold labeling of aquaporin-4 in square arrays of astrocyte and ependymocyte plasma membranes in rat brain and spinal cord. *Proc. Natl. Acad. Sci. USA* **95,** 11981–11986.

Raunser, S., and Walz, T. (2009). Electron crystallography as a technique to study the structure of membrane proteins in a lipidic environment. *Annu. Rev. Biophys.* **38,** 89–105.

Rémigy, H. W., Caujolle-Bert, D., Suda, K., Schenk, A., Chami, M., and Engel, A. (2003). Membrane protein reconstitution and crystallization by controlled dilution. *FEBS Lett.* **555,** 160–169.

Ren, G., Reddy, V. S., Cheng, A., Melnyk, P., and Mitra, A. K. (2001). Visualization of a water-selective pore by electron crystallography in vitreous ice. *Proc. Natl. Acad. Sci. USA* **98,** 1398–1403.

Rigaud, J. L., Mosser, G., Lacapere, J. J., Olofsson, A., Levy, D., and Ranck, J. L. (1997). Bio-Beads: An efficient strategy for two-dimensional crystallization of membrane proteins. *J. Struct. Biol.* **118,** 226–235.

Ringler, P., Borgnia, M. J., Stahlberg, H., Maloney, P. C., Agre, P., and Engel, A. (1999). Structure of the water channel AqpZ from *Escherichia coli* revealed by electron crystallography. *J. Mol. Biol.* **291,** 1181–1190.

Rosenbaum, D. M., Rasmussen, S. G., and Kobilka, B. K. (2009). The structure and function of G-protein-coupled receptors. *Nature* **459,** 356–363.

Sanner, M. F., Olson, A. J., and Spehner, J. C. (1996). Reduced surface: An efficient way to compute molecular surfaces. *Biopolymers* **38,** 305–320.

Savage, D. F., Egea, P. F., Robles-Colmenares, Y., O'Connell, J. D. R., and Stroud, R. M. (2003). Architecture and selectivity in aquaporins: 2.5 Å X-ray structure of aquaporin Z. *PLoS Biol.* **1,** E72.

Schenk, A. D., Werten, P. J., Scheuring, S., de Groot, B. L., Müller, S. A., Stahlberg, H., Philippsen, A., and Engel, A. (2005). The 4.5 Å structure of human AQP2. *J. Mol. Biol.* **350,** 278–289.

Sengupta, D., Behera, R. N., Smith, J. C., and Ullmann, G. M. (2005). The α helix dipole: Screened out? *Structure* **13,** 849–855.

Signorell, G. A., Kaufmann, T. C., Kukulski, W., Engel, A., and Rémigy, H.-W. (2007). Controlled 2D crystallization of membrane proteins using methyl-β-cyclodextrin. *J. Struct. Biol.* **157,** 321–328.

Silberstein, C., Bouley, R., Huang, Y., Fang, P., Pastor-Soler, N., Brown, D., and van Hoek, A. N. (2004). Membrane organization and function of M1 and M23 isoforms of aquaporin-4 in epithelial cells. *Am. J. Physiol. Renal. Physiol.* **287,** F501–F511.

Stahlberg, H., Braun, T., de Groot, B., Philippsen, A., Borgnia, M. J., Agre, P., Kühlbrandt, W., and Engel, A. (2000). The 6.9-Å structure of GlpF: A basis for homology modeling of the glycerol channel from *Escherichia coli*. *J. Struct. Biol.* **132,** 133–141.

Sui, H., Han, B. G., Lee, J. K., Walian, P., and Jap, B. K. (2001). Structural basis of water-specific transport through the AQP1 water channel. *Nature* **414,** 872–878.

Suzuki, H., Nishikawa, K., Hiroaki, Y., and Fujiyoshi, Y. (2008). Formation of aquaporin-4 arrays is inhibited by palmitoylation of N-terminal cysteine residues. *Biochim. Biophys. Acta* **1778,** 1181–1189.

Tani, K., Mitsuma, T., Hiroaki, Y., Kamegawa, A., Nishikawa, K., Tanimura, Y., and Fujiyoshi, Y. (2009). Mechanism of aquaporin-4's fast and highly selective water conduction and proton exclusion. *J. Mol. Biol.* **389,** 694–706.

Törnroth-Horsefield, S., Wang, Y., Hedfalk, K., Johanson, U., Karlsson, M., Tajkhorshid, E., Neutze, R., and Kjellbom, P. (2006). Structural mechanism of plant aquaporin gating. *Nature* **439,** 688–694.

Tsukaguchi, H., Shayakul, C., Berger, U. V., Mackenzie, B., Devidas, S., Guggino, W. B., van Hoek, A. N., and Hediger, M. A. (1998). Molecular characterization of a broad selectivity neutral solute channel. *J. Biol. Chem.* **273,** 24737–24743.

Unger, V. M., Kumar, N. M., Gilula, N. B., and Yeager, M. (1999). Three-dimensional structure of a recombinant gap junction membrane channel. *Science* **283**, 1176–1180.

Unwin, P. N., and Henderson, R. (1975). Molecular structure determination by electron microscopy of unstained crystalline specimen. *J. Mol. Biol.* **94**, 425–440.

Verbavatz, J. M., Ma, T., Gobin, R., and Verkman, A. S. (1997). Absence of orthogonal arrays in kidney, brain and muscle from transgenic knockout mice lacking water channel aquaporin-4. *J. Cell Sci.* **110**, 2855–2860.

Viadiu, H., Gonen, T., and Walz, T. (2007). Projection map of aquaporin-9 at 7 Å resolution. *J. Mol. Biol.* **367**, 80–88.

Vonck, J. (2000). Parameters affecting specimen flatness of two-dimensional crystals for electron crystallography. *Ultramicroscopy* **85**, 123–129.

Walz, T., Smith, B. L., Zeidel, M. L., Engel, A., and Agre, P. (1994a). Biologically active two-dimensional crystals of aquaporin CHIP. *J. Biol. Chem.* **269**, 1583–1586.

Walz, T., Smith, B. L., Agre, P., and Engel, A. (1994b). The three-dimensional structure of human erythrocyte aquaporin CHIP. *EMBO J.* **13**, 2985–2993.

Walz, T., Typke, D., Smith, B. L., Agre, P., and Engel, A. (1995). Projection map of aquaporin-1 determined by electron crystallography. *Nat. Struct. Biol.* **2**, 730–732.

Walz, T., Hirai, T., Murata, K., Heymann, J. B., Mitsuoka, K., Fujiyoshi, Y., Smith, B. L., Agre, P., and Engel, A. (1997). The three-dimensional structure of aquaporin-1. *Nature* **387**, 624–627.

Walz, T., Fujiyoshi, Y., and Engel, A. (2009). The AQP structure and functional implications. *Handb. Exp. Pharmacol.* **190**, 31–56.

Wang, D. N., and Kühlbrandt, W. (1991). High-resolution electron crystallography of light-harvesting chlorophyll a/b-protein complex in three different media. *J. Mol. Biol.* **217**, 691–699.

Williams, R. C., and Glaeser, R. M. (1972). Ultrathin carbon support films for electron microscopy. *Science* **175**, 1000–1001.

Zeidel, M. L., Ambudkar, S. V., Smith, B. L., and Agre, P. (1992). Reconstitution of functional water channels in liposomes containing purified red cell CHIP28 protein. *Biochemistry* **31**, 7436–7440.

Zeng, X., Stahlberg, H., and Grigorieff, N. (2007). A maximum likelihood approach to two-dimensional crystals. *J. Struct. Biol.* **160**, 362–374.

Zhao, G., Johnson, M. C., Schnell, J. R., Kanaoka, Y., Haase, W., Irikura, D., Lam, B. K., and Schmidt-Krey, I. (2010). Two-dimensional crystallization conditions of human leukotriene C(4) synthase requiring adjustment of a particularly large combination of specific parameters. *J. Struct. Biol.* **169**, 450–454.

Zhou, Z. H. (2008). Towards atomic resolution structural determination by single-particle cryo-electron microscopy. *Curr. Opin. Struct. Biol.* **18**, 218–228.

CHAPTER SIX

Cryoelectron Microscopy Applications in the Study of Tubulin Structure, Microtubule Architecture, Dynamics and Assemblies, and Interaction of Microtubules with Motors

Kenneth H. Downing* *and* Eva Nogales*,†,‡

Contents

Abstract

Cryo-EM is ideally suited for the study of cytoskeleton polymers and their interaction with cellular partners. Our understanding of microtubule (MT) structure and interactions has benefited tremendously from the application of different EM techniques, from the use of electron crystallography to determine the first high-resolution structure of tubulin, to electron tomographic reconstructions of unique MT-based organelles; from molecular details governing the regulated interaction of MTs with kinesin motors, to an atomic description of

* Life Sciences Division, Donner Laboratory, Lawrence Berkeley National Laboratory, Berkeley, California, USA
† Life Sciences Division, Lawrence Berkeley National Laboratory, Berkeley, California, USA
‡ Department of Molecular and Cell Biology, Howard Hughes Medical Institute, UC Berkeley, Berkeley, California, USA

Methods in Enzymology, Volume 483
ISSN 0076-6879, DOI: 10.1016/S0076-6879(10)83006-X

how antimitotic agents bind to tubulin and affect MT stability. In this chapter, we review these structural findings with an emphasis on how cryo-EM enables our studies.

1. Introduction: The Role of Electron Microscopy in Tubulin Studies

Microtubules (MTs) are hollow, fibrous polymers present as part of the cytoskeleton in all eukaryotic cells. They play many essential roles, including intracellular transport, chromosome segregation, and cell movement. To carry out these functions, MTs should be very stable or highly dynamic, growing, and shrinking to allow the ends to explore the cell volume. In these activities, they interact with a large number of other proteins that affect their stability, act as motors using MTs as a track to move cargo, or make use of MT's dynamic character to localize or carry out work.

Electron microscopy (EM) has been intimately involved with many aspects of the study of MTs and their principal component protein, tubulin, and a great deal of what we know about structural aspects of MTs has come from EM. It is interesting to review how developments in EM technology have been frequently followed by new information about MTs. Soon after the development of ultramicrotomes for cutting resin sections suitable for EM, MTs were discovered by examination of sections of plant tissue (Ledbetter and Porter, 1963). The overall architecture containing 13 protofilaments of tubulin monomers was soon revealed (Ledbetter and Porter, 1964). Shortly after the development of techniques for deriving 3D models from EM images of helical structures, the arrangement of the subunits was resolved in negative stain (Amos and Klug, 1974). Soon after, it was discovered that 2D crystals could be produced by polymerizing tubulin in the presence of zinc ions (Gaskin and Kress, 1977; Larsson *et al.*, 1976), and the nascent technique of electron crystallography was soon applied to obtain a slightly higher resolution map of the structure (Baker and Amos, 1978; Crepeau *et al.*, 1978). Further work on these crystals was limited by their small size and high disorder. As single-particle methods began to develop, methods were proposed for combining crystallographic and single-particle approaches to improve the data extracted from the tubulin crystals (Crepeau and Fram, 1981). This in turn led to the development of techniques for "lattice unbending" that proved to be key to extending EM resolution to the atomic level with protein crystals (Henderson *et al.*, 1986). At about this time, the technology of cryo-EM was being developed (Adrian *et al.*, 1984). The combination of these two advances led to the first protein structures solved by electron crystallography, including bacteriorhodopsin (bR; Henderson *et al.*, 1990) and the plant light harvesting complex (LHC; Kühlbrandt *et al.*, 1994). These successes inspired us to begin work on the tubulin crystals, starting with the demonstration that they could be prepared to produce electron diffraction to a resolution sufficient to build an atomic model (Downing and Jontes, 1992). At about the same time, cryo-EM was being used to study the structure and assembly/disassembly processes of MTs (Mandelkow *et al.*, 1991). The crystallographic approach eventually led to the full atomic structure of tubulin (Löwe *et al.*, 2001; Nogales *et al.*, 1998), and the characterization of the interactions of tubulin with various anticancer ligands (Nettles *et al.*, 2004; Snyder *et al.*, 2001). More recently, new approaches in cryo-EM have been developed to visualize MTs and other tubulin assemblies at improved resolution, and interacting with a variety of cellular partners, most remarkably kinesin (Sindelar and Downing, 2010). Electron tomography has been used to study larger and much more complex tubulin-based complexes (Bui *et al.*, 2009; Nicastro *et al.*, 2006; Sui and Downing, 2006). In this chapter, we focus on the various applications of cryo-EM in the crystallographic determination of the tubulin structure, the characterization of assembly and disassembly intermediates in MT dynamics, and on the interaction of MTs with motor proteins.

2. TUBULIN STUDIES BY ELECTRON CRYSTALLOGRAPHY

Our efforts to study tubulin by electron crystallography began with work to improve the crystals themselves and to prepare them for high-resolution study. It had long been known that under many conditions tubulin loses its ability to polymerize with a half life of just hours, so the early attempts to grow crystals used incubation times on the order of 20 min (Baker and Amos, 1978; Larsson *et al.*, 1976). The crystals were a few tenths of a micron in size, and apparently sufficiently disordered to limit the resolution even in negatively stained preparations. We eventually found that in the presence of salt and protease inhibitors and at slightly subphysiological temperatures incubation for up to 24 h was possible, yielding crystals over a micron in size (Nogales *et al.*, 1995). Since the time of the earlier experiments, Taxol had been discovered as the first of a series of small ligands that stabilize MTs. We found that Taxol also stabilizes the crystals, and it was routinely added after crystal formation to avoid any problems of crystals depolymerizing, for example, from decrease of the temperature during sample preparation. The first attempts to embed these crystals in glucose, simply following the approach that had been successful with bR, gave somewhat promising results (Downing and Jontes, 1992). Inspired by the use of tannin embedment with catalase and LHC, we obtained even better diffraction patterns extending to 3.5 Å with tannin. We also applied the "back injection" technique that had been worked out with bR and then applied to LHC, modified for a combination of tannin and glucose. A small square of carbon film was floated from mica onto a 1% tannin solution and picked up on a bare molybdenum grid. The crystal suspension was injected into the lens of liquid adhering to the grid. An equal volume of 1% glucose was then added and the solution thoroughly but gently mixed. The grid was then simply blotted, briefly air dried, and frozen in liquid nitrogen. The high sugar concentration avoids crystallization of the water on freezing, and samples prepared this way have shown diffraction spots up to 2.5 Å. Figure 6.1 shows an example of a diffraction pattern from a tubulin crystal embedded in the tannin–glucose mixture. While this may not be considered to be a cryo-EM sample in the most common sense, all of our work has been done with the specimen at low temperature using all of the protocols that are generally applied with frozen-hydrated specimens.

As has been common with all the other samples studied by electron crystallography, the tubulin crystals have been plagued by issues of specimen flatness (see Chapter 11 in MIE volume 481). This problem is manifested with tilted crystals in the rapid loss of diffraction spot sharpness and intensity with increasing distance from the tilt axis (Glaeser *et al.*, 1991).

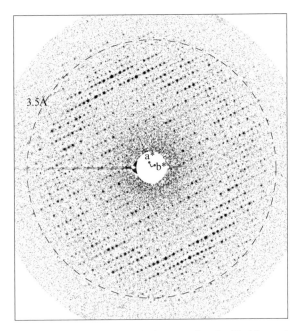

Figure 6.1   Diffraction pattern from a tubulin crystal embedded in a mixture of tannin and glucose. Diffraction spots are frequently seen well beyond 3.5 Å.

The interpretation of this observation is that the specimen is not perfectly flat, and variations in tilt angle cause sampling of different parts of the reciprocal lattice rods in different areas of the crystal. Depending on the length scale of the variation in tilt, the diffraction spots may blur increasingly with increasing distance from the tilt axis or may simply disappear. The use of molybdenum grids rather than copper reduces the effect somewhat because molybdenum has a lower thermal expansion coefficient, closer to that of carbon, leading to less wrinkling of the carbon when the sample is frozen. While it is still not possible to generalize the complete solution to the problem, use of a particular type of carbon rod, evaporation conditions, age of the carbon film, and other factors seem to play critical roles in obtaining good diffraction. These factors may vary significantly from one type of specimen to another. We have found that, as with bR, it is important to use carbon films that are on the order of 1–2 weeks old for best results with tubulin. The "double carbon film" specimen preparation technique (Gyobu et al., 2004), in which a second carbon film is placed on the droplet adhering to the grid before blotting and freezing, appears to provide a very substantial improvement in crystal flatness. For the tubulin crystals, the success rate dramatically increased by limiting the amount of blotting, leaving a meniscus of tannin–glucose on the grid with only small areas in each square of the right thickness, which eliminated the deleterious effect of the liquid removal and added mechanical strength to the grid.

One can estimate the number of diffraction patterns or images that are required to fill reciprocal space with sufficient sampling to reach a target resolution of d as $N = \pi D/d$, where D is the protein thickness. For tubulin at a resolution of 3.5 Å, this number would be about 50. Factors such as the nonuniformity of angular sampling, limited range of tilt angles, and especially the noise level influence the number actually needed. The initial data set for diffraction amplitudes used about 100 diffraction patterns (Nogales et al., 1998), and this was subsequently expanded to 200 with most of the new patterns collected at higher tilts (Löwe et al., 2001). About 130 images were used to obtain structure factor phases. Data collection and processing generally followed the protocols of the MRC image processing system, which for the images included several cycles of lattice unbending (Crowther et al., 1996).

One enhancement for image recording developed during the course of this work was the implementation of a procedure for dynamic focus correction with spot-scan imaging. All of the images were collected using a spot-scan protocol, in which the beam is focused to a diameter of 300–1000 Å and stepped over the image in a 2D raster rather than illuminating the entire area at once (Downing, 1991). This approach had been shown to give greatly improved data quality, presumably by reducing beam-induced specimen movement during imaging. With tilted specimens, the beam was scanned in the direction of the tilt axis and the focus was adjusted by an amount between scan lines to keep the entire image at the same focus (Downing, 1992). This greatly reduced the effort involved in image processing by removing the focus ramp that normally occurs across the image of a tilted sample, in addition to eliminating any loss of quality that might occur in areas of high defocus.

The initial 3D density map was calculated at a nominal resolution of 3.7 Å, and while the quality was at least as good as that of a typical X-ray map at the same resolution, fitting the polypeptide into the map was a challenge. Nonetheless, a full atomic model was built which left very little ambiguity and revealed essentially all the details of the long-sought tubulin structure (Nogales *et al.*, 1998). With additional diffraction data and application of structure refinement procedures from X-ray crystallography, the quality of the model improved along with extension of the resolution to 3.5 Å (Löwe *et al.*, 2001). Figure 6.2 is a section of the density map used to derive the final structure, showing the high quality of the data and strong constraints that it provided for structure building.

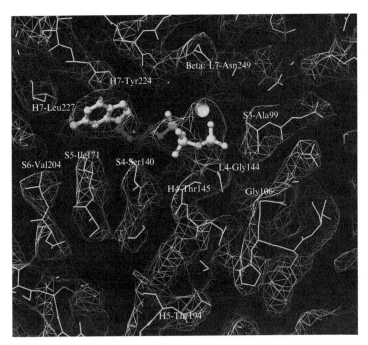

Figure 6.2 Section of the refined electron crystallographic density map used to derive the final structure of the tubulin dimer. This section includes the GTP bound to alpha-tubulin, which is shown as a ball-and-stick model, along with a sphere representing the associated Mg ion. The atomic model of the protein is shown as a wire structure, with several of the residues in the neighborhood of the nucleotide marked.

Figure 6.3 is a ribbon diagram of the tubulin dimer structure. The alpha- and beta-tubulin monomers share a high degree of structure similarity, with a core of beta sheet surrounded by alpha helices. The N-terminal part of the chain forms a Rossmann fold, having alternating strand and helix segments quite similar in topology to the fold found in proteins such as GAPDH. Each of the loops connecting a beta strand to the following helix has some contact with the nucleotide, a significant departure from many other GTP- and ATP-binding proteins, which coordinate most of the nucleotide via the so-called "P-loop." A second domain is formed by a series of beta strands and helices, with the "core helix" providing a strong interface between the domains. A third domain is identified as containing a pair of helices running along the surface of the monomer on the side opposite from the core helix. The essential C-terminal extension of tubulin, containing the last 12–15 residues in alpha and beta-tubulin was unstructured.

It is important to realize that the zinc-induced sheets we refer to here as "crystals" are in fact a polymer form of tubulin, involving large intradimer

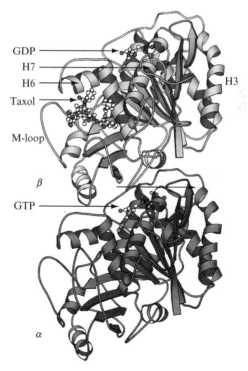

Figure 6.3 Ribbon diagram of the tubulin dimer structure. This view shows the surface facing the lumen of the microtubule, with the plus end toward the top. Several of the secondary structure features and cofactors are marked.

and interdimer contacts along protofilaments that are conserved with the MT (see later). Just like MTs, these polymers can go through cycles of assembly and disassembly with temperature, they depend on GTP, and they are stabilized by Taxol. Thus the tubulin electron crystallographic structure goes beyond defining the fold of the tubulin molecule and carries functional information with direct relevance to our understanding of MT assembly.

Important for understanding the properties of nucleotide-binding, hydrolysis, and exchange, which are central to the process of MT dynamic instability, the structure shows that the nucleotide sits at the surfaces of the monomers that are buried in intermonomer contacts along protofilaments. The GTP bound to alpha-tubulin within the alpha–beta dimer is firmly trapped at the intradimer interface. While the GDP bound to beta-tubulin appears just as trapped within the protofilament, its position at the inter-dimer interface renders it exchangeable at the surface of a free dimer. This arrangement thus explains the differences in exchangeability of the nucleo-tide in the tubulin monomers and the change in exchangeability upon polymerization: the nucleotide in beta is exchangeable only in the dimer state but becomes buried and thus nonexchangeable, like in alpha, within the polymer. The position of the exchangeable nucleotide and the structure of the interdimer interface readily tells us about how hydrolysis of the nucleotide in beta is stimulated by residues from alpha concomitant with polymerization.

The biggest difference between alpha and beta-tubulins is found in the loop between helices 9 and 10, which is 10 residues shorter in beta-tubulin. Most interestingly, filling the space left by this deletion is the binding site of the broadly used cancer therapeutic Taxol. This location of Taxol suggested ways in which it might stabilize MTs, either by favoring contacts between protofilaments or by inhibiting relative movement of the nucleotide-binding and intermediate domains (Amos and Löwe, 1999; Nogales *et al.*, 1998). Smaller structural differences between the GDP-bound beta-tubulin and the GTP-bound alpha-tubulin are located in the T5 loop (involved in ribose binding), its neighboring helix H6, helix H10, and the M and T7 loops. These are all essential regions in longitudinal (T5, H6–H7 loop, T7) and lateral (M-loop, H10).

3. MICROTUBULE STRUCTURE

From the birth of cryo-EM, application of this technique to the study of MTs has been highly revealing about the structure and behavior of these polymers. As early as 1985, the MT lattice was resolved by cryo-EM at a resolution at least as good as the earlier work with negative stain, allowing a 3D reconstruction that showed the arrangement of monomers (Mandelkow and Mandelkow, 1985). In the course of this work, MTs sometimes disassembled due to cooling of the sample prior to freezing. Depolymerization occurred with peeling of protofilaments away from the MT into spirals and rings with a narrow distribution of diameters. Such rings had been seen earlier in stain and characterized by X-ray scattering, and the cryo-EM work now suggested that they represented the natural, curved form of protofilaments containing GDP. This in turn led to models for storage of the energy of hydrolysis as strain within the MT, where the protofilaments are constrained by lattice contacts to be straight (Caplow et al., 1994). A series of papers used the improved preservation of the cylindrical shape of frozen-hydrated MTs to characterize variations in protofilament numbers and the associated surface lattice (Chrétien and Wade, 1991; Song and Mandlekow, 1995; Wade et al., 1990). Furthermore, helical reconstruction techniques were being applied to the study of naked and kinesin-bound MTs, with resolutions approaching 20 Å. By the time the tubulin crystal structure became available, reconstructions were thus available to dock the crystal structure into the MT density.

The protofilaments in the zinc-induced crystals are antiparallel, while those in MTs are parallel. Thus, while the crystal structure provided a clear view of the interactions within dimers and between dimers along the protofilaments, the actual interaction between protofilaments, as well as the orientation of the dimer in the MT, was still unknown. The asymmetry of the tubulin dimer was sufficient to allow docking into an MT map having a resolution around 20 Å with enough precision to clearly define the general orientation and regions of contacts between protofilaments. The C-terminal domain faces the outside of the MT, defining a crest for motor binding, while the Taxol binding site faces the lumen of the MT, near the site of lateral contact between protofilaments (Nogales et al., 1999). The loop connecting H7 to S9 along the side of the monomers, named the M-loop, plays a major role in lateral contacts between protofilaments via its interaction with the loop between H1 and S2 in the N-terminal domain of the adjacent subunit.

In an effort to improve understanding of the MT structure, we developed a protocol for MT image processing based more on a combination of single-particle and crystallographic methods than on helical reconstruction. The MT images were boxed into small regions, typically about 10 dimers in length, and each subsection was aligned to a reference. We focused our analysis on 13-protofilament MTs, which are not strictly helical because of a discontinuity or "seam" in the surface lattice where alpha and beta monomers meet, but are amenable to averaging along the axis. With a target resolution in the 5–10 Å range the differences between alpha and beta could be ignored, giving a structure which is an average of the two monomers. This approach gave better tolerance to the distortions that are inevitable in a structure like that of the MT and allowed extension of the resolution to about 8 Å, sufficient to visualize most of the secondary structure (Li *et al.*, 2002). This improved resolution led to a far greater precision in docking the crystal structure into the density map and showed more clearly the interaction of the M-loop with loops from the adjacent monomers. This work has recently been extended to the derivation of maps with comparable resolution of MTs with 11, 12, 13, 14, 15, and 16 protofilaments (Sui and Downing, 2010). These new results confirm that the variation in protofilament number is accommodated by changes in the loop regions involved in interprotofilament contacts. Figure 6.4 shows a comparison of the six 3D density maps from this work.

While our approach for processing MT images and the inclusion of more data than in previous work certainly contributed much to the resolution extension, it may also be that the nature of the specimen preparations provided a substantial benefit as well. All of the bare MT images, as well as those of kinesin-decorated MTs described later, have been recorded at 400 kV. While this decreases the contrast below what is more commonly obtained at 120 or 200 kV, it allows working with thicker ice, which provides a more rigid structure and may produce images with higher inherent resolution. On the other hand, there are now well over a dozen reconstructions of MT complexes that all reach the 8–9 Å resolution range but not beyond, suggesting that the inherent flexibility of MTs may currently limit resolution. Conformational heterogeneity could easily arise from distortions that produce an oval cross-section. If these distortions are in fact the source of the resolution limit, it should not be too difficult to use classification techniques to identify them and computationally achieve significantly higher resolution.

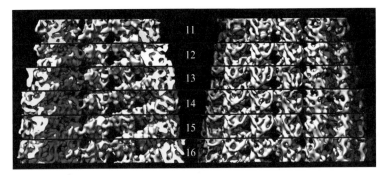

Figure 6.4 Three-dimensional reconstruction of microtubule structures. This figure shows the inner (left) and outer (right) surfaces of six 3D density maps obtained from microtubules with 11, 12, 13, 14, 15, and 16 protofilaments described in Sui and Downing (2010).

4. STRUCTURE OF MICROTUBULE ASSEMBLY/ DISASSEMBLY INTERMEDIATES

A very interesting facet of MT dynamics is that the nucleotide-regulated tubulin assembly and disassembly processes occur via structural intermediates, rather than individual tubulin subunits directly adding to or leaving the MT lattice (Figs. 6.5A and 6.6A). These structural intermediates have been directly observed at the ends of MTs using cryo-EM. Mandelkow and colleagues first characterized the peeling of depolymerizing MTs via "ram's-horn" intermediates (Mandelkow *et al.*, 1991). These structures are outwardly curved protofilaments that break down into ring-like structures closely resembling

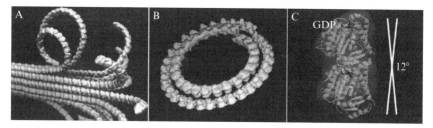

Figure 6.5 Structure of microtubule disassembly intermediates and their relationship to tubulin conformation and nucleotide state. (A) Artistic rendition of the microtubule depolymerization process via peeling of protofilaments concomitant with the relaxation of GDP–tubulin into its low energy state. (B) Cryo-EM reconstruction of GDP–tubulin assembled into double-layered tubes closely corresponding to GDP–tubulin rings and a surrogate for depolymerizing protofilaments. (C) Schematic of the kinked dimer conformation observed in the cryo-EM reconstruction of GDP–tubulin.

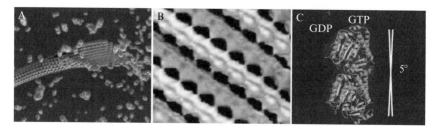

Figure 6.6 Structure of microtubule assembly intermediates and their relationship to tubulin conformation and nucleotide state. (A) Artistic rendition of the microtubule polymerization process via open sheets that later close into a tube. (B) Cryo-EM reconstruction of GMPCPP-tubulin assembled at low temperatures as a surrogate for the assembly sheets. All the protofilaments run in the same direction, but every other one is slightly rotated around its axis. (C) Schematic of the partially straightened dimer conformation observed in the cryo-EM reconstruction of GMPCPP-tubulin.

the double-rings formed *de novo* from GDP–tubulin (Frigon and Timasheff, 1975). On the other hand, Chrétien *et al.* (1995) described the presence of open sheet structures with slight outward curvature at the end of rapidly growing MTs. How do these distinct oligomeric states of tubulin relate to the structure and interactions of tubulin subunits?

The curved protofilaments at the ends of shortening MTs constitute a structural intermediate in the disassembly process where GDP–tubulin is in its relaxed state, clearly distinct from its constrained state in the body of the MT wall. We used the fact that high concentrations of divalent cations stabilize tubulin ring assembly, likely by a charge shielding mechanism that may be shared with MAPs (microtubule-associated proteins), to form tubular assemblies of GDP–tubulin in which ring closure does not occur and the protofilaments form a tight, double-layer helix (Fig. 6.5B). This helix, which can form a tube containing tens of turns, appears to recapitulate the shape of the horn-like protofilament structures at depolymerizing MT ends, and its intrinsic order facilitated its structural characterization by cryo-EM. The double-layer nature of these tubes results in systematic overlap of the Bessel terms from the inner and outer layers on all the layer lines, making impossible to obtain the orientation relationship between different images of the tubes directly using traditional helical reconstruction methods (see Chapter 5 in MIE volume 482). To overcome this issue and still take advantage of the helical symmetry of the tubes, we generated a modified Fourier space-based, helical reconstruction algorithm to deal with Bessel order overlap and the need of multiple projection views of the structure (Wang and Nogales, 2005a). This method relies on an iterative approach to generate accurate values for the relative orientations between different projection images. The overall procedure is as follows: first, we calculate initial reconstructions for the inner-layer and outer-layer helices using an initial input for the relative orientations of all the projection images (we demonstrated that these could be arbitrary values). We then use those initial 3D reconstructions as references, calculate their sum with different orientation combinations (for the inner and outer layers), and cross-correlate them with the raw projection images to find a new set of orientations for each of the projection images. At this point, we apply the new set of orientations to produce an improved 3D reconstruction of the inner and outer helices. The cycle of cross-correlation search and reconstruction is repeated until the result converges, that is, the orientations do not change and the resolution of the reconstruction does not improve.

We used the above methodology to produce independent 3D reconstructions of the inner and outer layers of the tube, using data up to 10 Å resolution (Wang and Nogales, 2005b). Our structures showed, for the first time, distinctive intra- and interdimer interactions and thus a distinction between the GTP and GDP interfaces (Fig. 6.5B). While both interfaces are kinked, the bending angles are clearly different, and one interface is dramatically more flexible than the other (Wang and Nogales, 2005b). The cryo-EM, in agreement with X-ray studies of disassembled tubulin bound to the RB3 stathmin fragment (Ravelli *et al.*, 2004), showed that, irrespective of the presence or absence of a depolymerizer, the bending of the intra- and interdimer interfaces in GDP–tubulin (Fig. 6.5C) is incompatible with the formation of the lateral contacts which are present in MTs.

A two-step MT assembly process, involving a structural intermediate, was directly suggested by the cryo-EM studies of Chrétien and colleagues about a decade ago. They showed that under conditions of fast tubulin assembly, growth occurs via open sheets at the ends of MTs that later close into a cylinder (Chrétien *et al.*, 1995). A model of this process is shown in Fig. 6.6A. As a stable, surrogate assembly of those sheets we studied a tubulin polymer that forms in the presence of the nonhydrolysable GTP analogue GMPCPP when MT assembly is inhibited by low temperatures. These structures showed protofilaments to be slightly and smoothly curved, with small indistinguishable intra- and interdimer kinks between tubulin monomers (Wang and Nogales, 2005b; Fig. 6.6C). Most importantly, the structure showed the presence of alternating lateral contacts between protofilaments, which otherwise preserved the precise stagger between protofilaments seen in the MT (Fig. 6.6B). This means that the structures would be able to convert into MTs without the need of longitudinal sliding between protofilaments, but simply by a rotation of the protofilaments. This type of arrangement, involving alternative lateral contacts without longitudinal displacements between protofilaments, is fully compatible with the extended sheets observed by Chretien and colleagues at the growing end of MTs, and their direct conversion into MTs by closure.

Altogether, the studies above support the separation of the process of straightening from the curved, depolymerized state to the straight protofilaments in MTs into two stages: one, nucleotide-dependent, that allows for lateral association into a curved sheet, and a later one that occurs upon MT closure.

5. MECHANISM OF KINESIN MOVEMENT ALONG MICROTUBULE

During the time that the crystal structure was being determined, substantial progress was being made in the study of motor-decorated MTs in order to understand the interaction of these essential motor proteins with MTs. A series of papers resolved the basic pattern of monomer and dimer binding of conventional kinesins and the minus end-directed ncd, most using helical reconstruction and leading to a resolution around 20 Å

(Arnal *et al.*, 1996; Hirose *et al.*, 1995; Hoenger *et al.*, 1995; Kikkawa *et al.*, 1995). Conformational changes in the motors were identified as a consequence of altering the nucleotide state (Rice *et al.*, 1999). To some extent, these changes could be correlated with structural changes in the motors seen by X-ray crystallography. However, there was no absolute correlation of the X-ray structure conformations with the nucleotide, so it remained unclear how ATP binding and hydrolysis were associated with MT binding.

Following our determination of the 8-Å MT structure, we applied the same basic methods to the study of kinesin-decorated MTs. One enhancement of the approach was required by the fact that we could not average over the seam, so identification of the seam location in the 13-protofilament MTs became necessary. Again, a reference-based approach allowed identification of the proper orientation. Plotting the correlation between each segment of the MT image and a series of projections calculated for a full rotation of the reference model produces 13 peaks as the tubulin protofilaments align. The height of the peaks increases with increasing alignment of the kinesins, so that the proper orientation is identified from the highest peak. The series of correlation peaks provides a particularly robust determination of the angle along with a measure of confidence in its determination. Using this approach we reached the 8–9 Å resolution range, allowing precise docking of a kinesin crystal structure into the map and identification of structural rearrangements that occur upon binding of the motor to the MT (Sindelar and Downing, 2007). Subsequent cryo-EM work with kinesin in a series of nucleotide states that reflect the full hydrolysis cycle has led to a fairly complete understanding of the conformational changes that drive motor stepping (Sindelar and Downing, 2010). Figure 6.7 shows density maps for the nucleotide-free and ATP states with the fitted atomic model, along with a cartoon representation of the conformational changes associated with nucleotide-binding and the power stroke.

One essential feature revealed by this work is that binding to the MT causes the stabilization of an extension of the switch-II helix of kinesin. This extension is absent in almost all of the dozens of X-ray crystal structures of kinesins in the relevant nucleotide states. While the conformational changes that had been seen in the previous cryo-EM studies could be correlated with different conformations seen in the X-ray structures, the lack of correlation between the conformation and the nucleotide state in these structures made it difficult to obtain much functional insight. However, from the cryo-EM work the paradigm became clear that formation of the helix extension stabilized components of the switch-I and -II regions which complete the binding pocket for the nucleotide. Binding of ATP then triggers tilting of the core of the motor with respect to the MT-attached domain, which in turn exposes the neck-linker binding region on the opposite side of the molecule. Docking of the neck-linker to this region then pulls the lagging motor head forward in a mechanical step along the MT.

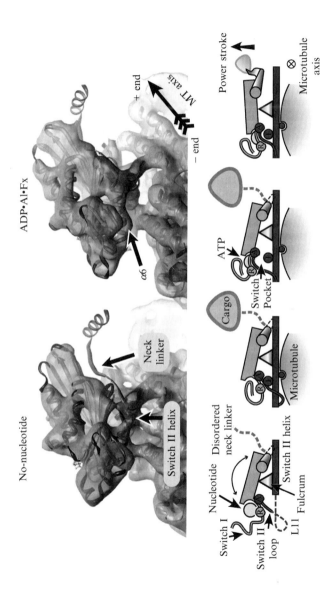

Figure 6.7 Kinesin–microtubule interaction. *Top:* Density maps of the kinesin–microtubule complex with no nucleotide (left); with ADP and aluminum fluoride as an analogue of the ATP-bound state (right). Atomic models have been docked into the density. The most notable difference between these states is a tilt of the core of the motor with respect to the microtubule. *Bottom:* Cartoon representation of the conformational changes associated with nucleotide-binding and the power stroke. In unbound kinesin, the core is free to tilt back and forth independent of nucleotide. Once bound to a microtubule, the switch-II helix forms a base for residues that either prop the core to the right, with no nucleotide, or pull it to the left with ATP. Reprinted from Sindelar and Downing (2010).

An interesting finding of the higher resolution cryo-EM work is that the switch-I regions of the kinesin X-ray structures were not compatible with the experimental density of the cryo-EM maps, because the structure in this region changes upon MT binding and formation of the switch-II extension. However, the corresponding region from the actin-associated motor myosin in the ATP-bound state fits very well, as might have been predicted based on the evolutionary relation between the two motors.

6. DRUG BINDING STUDIED BY DIFFRACTION AND MODELING

The tubulin–Taxol complex of the zinc-induced crystals served as an excellent starting point for understanding the interaction and mechanism of a number of potential anticancer agents and other ligands that affect MTs' roles in the cell cycle. The Taxol conformation itself and its binding mode have been the subject of intense interest as a number of groups have worked toward development of novel derivatives with improved properties. The conformation derived from electron crystallography differs substantially from conformations resolved by NMR measurements in either polar or nonpolar solvents. Because of the limited resolution of our data, there were also limits on the confidence in details of the conformation. It was thus of interest to synthesize a Taxol derivative that was locked in the conformation of the crystals to see if it would in fact bind and to identify differences in binding constants. A compound with this conformation did in fact yield higher activity than the native Taxol molecule, lending support to the crystallographic result (Ganesh *et al.*, 2007).

Diffraction methods have been well worked out in X-ray crystallography for identifying small differences in ligands or structure between related protein structures. Briefly, one can use differences in diffraction amplitudes to identify differences in two closely related structures. A change in structure will, of course, change both amplitudes and phases of structure factors, but for sufficiently small changes the differences in either one can be ignored. The method has been widely used in X-ray diffraction for studying changes in ligand binding or structure. Some success has also been obtained in electron crystallography, for example, in characterizing changes in bR during its photocycle, where tilts in alpha helices could be visualized at moderate resolution (Subramaniam *et al.*, 1993; Vonck, 1996). In the case of MT-stabilizing ligands bound to the tubulin sheets, we found that difference maps based on data sets of about 200 electron diffraction patterns could indeed confirm that several of the compounds bind in the same location. However, the data quality, including resolution, completeness, and signal-to-noise ratio, were not sufficient to unambiguously determine the orientation and conformation of the highly flexible compounds.

Extending the diffraction approach by incorporation of computational modeling led to a plausible model for the complex of epothilone-A with tubulin (Nettles *et al.*, 2004). An extensive library of candidate conformations of the epothilone molecule was derived, starting with a set of conformations that had been determined by crystallography and NMR. Each of these conformations was computationally docked onto tubulin in an initial energy-minimized orientation, and the complex was then refined against the electron diffraction data using the crystallographic programs refmac and cns. Only a very small subset of the trial complexes could be successfully refined, and these converged to one structure that was then further refined. The resultant structure is consistent with a wealth of structure–activity data that has been collected from variants of the epothilone motif (Nettles and Downing, 2009). On the other hand, it differs from a number of previous proposals derived by different computational modeling methods as well as subsequent models derived by NMR and SAR-based modeling (Reese *et al.*, 2007). Increasing the quality of the diffraction data to more strongly constrain the modeling should help to resolve whatever ambiguities remain.

7. TOMOGRAPHY FOR LARGER STRUCTURES

As with the other aspects of cryo-EM, the development of cryoelectron tomography has paralleled advances in our studies of tubulin and the complexes it forms. Tomography gives access to some larger MT-based superstructures, and one of the more intensively studied of these is the ubiquitous axoneme found in eukaryotic flagella and cilia. It had long been recognized that this structure is formed by nine MT doublets—themselves complexes of one complete and one incomplete MT—along with (usually) two singlet MTs and a host of other proteins. It has been estimated that axonemes are made of over 600 proteins. By far the best characterized of these are the dyneins, motor proteins providing the forces that give the axoneme its characteristic undulating movement seen, for example, in the tails of swimming sperm and the flagella of *Chlamydomonas*. Tomography of relatively intact axonemes has been particularly productive in identifying the locations and interactions of several isoforms of dyneins that bridge adjacent MT doublets and cause them to move with respect to each other, as well as some of the closely associated subunits (Bui *et al.*, 2009; Heuser *et al.*, 2009). Because the axoneme is such a large structure, on the order of 2500 Å in diameter, the resolution in such tomographic studies is somewhat limited. The signal-to-noise ratio, and thus the resolution, can be substantially enhanced by volume averaging of segments along the axoneme.

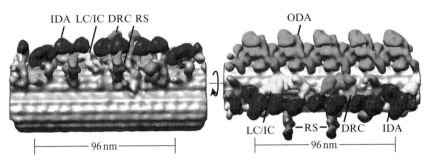

Figure 6.8   Tomography of *Chlamydomonas* axoneme. A section of a microtubule doublet obtained by averaging densities around and along axonemes is shown, encompassing the basic 96 nm repeat of dyneins and associated proteins. Views are as seen from the center of the axoneme (left) and tangential to the doublets (right). Densities attached to the microtubule are mainly components of the inner dynein arms, aside from the radial spoke components seen pointing down in the view at right. ODA, outer dynein arms (light blue in color figure); IDA, (red) inner dynein arms; DRC (green), dynein regulatory complex; LC (yellow), light chain of dynein 7; RS (blue), radial spokes. This view is similar to that in Bui *et al.* (2009) but represents the majority of dynein arrangements, not the uncommon one between doublets 1 and 9. Figure kindly provided by Drs K. H. Bui and T. Ishikawa. (See Color Insert.)

Figure 6.8 shows the results of averaging subvolumes from tomograms of *Chlamydomonas* axonemes, revealing the locations and interactions of dyneins and other identifiable components (Bui *et al.*, 2009).

Studies of smaller components, notably the isolated doublets, have the potential to reach higher resolution. Again, the use of averaging is essential to boost the SNR to take advantage of the resolution inherent in the tomogram. In studies of sea urchin sperm doublets, the resolution was sufficient to construct a model of the tubulin arrangement within the doublet and then to use this model to identify features of the 3D density map corresponding to some of the nontubulin components (Sui and Downing, 2006). Figure 6.9 shows a section of the density map derived from tomography of isolated doublets, along with the pseudoatomic model of the tubulin component and nontubulin components identified by difference mapping. As in much of the other work currently underway using EM to study tubulin-based structures, the focus is now on the mechanisms and functional aspects of the interactions of tubulin with the vast array of its binding partners.

Electron tomography also gives access to studies of macromolecules in their native environment of the cell. This often involves specimen areas that are much too thick for EM even with intermediate voltage microscopes. Thus sectioning of vitreous frozen materials is required. While this has proven to be a difficult technology to master, recent developments have led to some notable achievements (see Chapter 8). MTs are among the most easily visualized substructures in typical eukaryotic cells. One of the surprises in several studies that focused on MTs in such sections was the

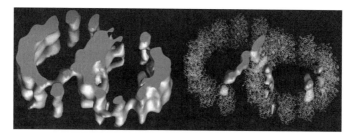

Figure 6.9 Tomographic reconstruction of sea urchin microtubule doublet. *Left*: Isosurface view of a section of the doublet reconstruction determined in Sui and Downing (2006). *Right*: Pseudoatomic model of the tubulin component is shown as a wire structure, and the component of the difference map showing nontubulin components is shown as a solid surface.

appearance of material within the lumen of the MT (Cyrklaff *et al.*, 2007). In stained resin sections, densely staining material is often seen inside MTs, but it was never entirely clear whether or not this was a staining artifact. Observation of material in cryosections removed this doubt, since these specimens should be free of such artifacts. The appearance of material with periodic spacings within the MTs has been particularly intriguing, although there is as yet no clear evidence of what such material might be.

 ## 8. Conclusion

Over the last 35 years, EM and particularly cryo-EM have proven crucial for our structural and functional knowledge of tubulin, its assembly into MTs and higher order structures, as well as for our understanding of how cellular factors interact with MTs and how anticancer agents bind to tubulin and affect its assembly properties. As in the study of other polymers of biological relevance, EM has offered the unique possibility of studying the large, functionally active assemblies that constitute the biologically relevant unit. With structural information spanning from the high-resolution electron crystallographic studies of tubulin to the tomographic analysis of the cellular MT cytoskeleton, there is no question that EM has provided, and will continue to provide, vivid, direct information on the complex and essential MT system and the molecular mechanisms that underlie its cellular functions.

ACKNOWLEDGMENTS

This work was supported by NIH grant GM51487 and by the U.S. Department of Energy under Contract No. DE–AC02–05CH11231. E. N. is a Howard Hughes Medical Institute Investigator.

REFERENCES

Adrian, M., Dubochet, J., Lepault, J., and McDowall, A. W. (1984). Cryo-electron micros-copy of viruses. *Nature* **308,** 32–36.

Amos, L. A., and Klug, A. (1974). Arrangement of subunits in flagellar microtubules. *J. Cell Sci.* **14,** 523–549.

Amos, L., and Löwe, J. (1999). How Taxol stabilizes microtubule structure. *Chem. Biol.* **6,** R65–R69.

Arnal, I., Metoz, F., DeBonis, S., and Wade, R. H. (1996). Three-dimensional structure of functional motor proteins on microtubules. *Curr. Biol.* **6,** 1265–1270.

Baker, T. S., and Amos, L. A. (1978). Structure of the tubulin dimer in zinc-induced sheets. *J. Mol. Biol.* **123,** 89–106.

Bui, K. H., Sakakibara, H., Movassagh, T., Oiwa, K., and Ishikawa, T. (2009). Asymmetry of inner dynein arms and inter-doublet links in *Chlamydomonas flagella. J. Cell Biol.* **186,** 437–446.

Caplow, M., Ruhlen, R. L., and Shanks, J. (1994). The free energy of hydrolysis of a microtubule-bound nucleotide triphosphate is near zero: All of the free energy for hydrolysis is stored in the microtubule lattice. *J. Cell Biol.* **127,** 779–788.

Chrétien, D., and Wade, R. (1991). New data on the microtubule surface lattice. *Biol. Cell* **71,** 161–174.

Chrétien, D., Fuller, S. D., and Karsenti, E. (1995). Structure of growing microtubule ends: Two-dimensional sheets close into tubes at variable rates. *J. Cell Biol.* **129,** 1311–1328.

Crepeau, R. H., and Fram, E. K. (1981). Reconstruction of imperfectly ordered zinc-induced sheets using cross-correlation and real space averaging. *Ultramicroscopy* **6,** 7–18.

Crepeau, R. H., McEwen, B., and Edelstein, S. J. (1978). Differences in alpha and beta polypeptide chains of tubulin resolved by electron microscopy with image reconstruc-tion. *Proc. Natl. Acad. Sci. USA* **75,** 5006–5010.

Crowther, R. A., Henderson, R., and Smith, J. M. (1996). MRC image processing programs. *J. Struct. Biol.* **116,** 9–16.

Cyrklaff, M., *et al.* (2007). Cryoelectron tomography reveals periodic material at the inner side of subpellicular microtubules in apicomplexan parasites. *J. Exp. Med.* **204,** 1281–1287.

Downing, K. H. (1991). Spot-scan imaging in transmission electron microscopy. *Science* **251,** 53–59.

Downing, K. H. (1992). Automatic focus correction for spot-scan imaging of tilted speci-mens. *Ultramicroscopy* **46,** 199–206.

Downing, K. H., and Jontes, J. (1992). Projection map of tubulin in zinc-induced sheets at 4 Å resolution. *J. Struct. Biol.* **109,** 152–159.

Frigon, R. P., and Timasheff, S. N. (1975). Magnesium-induced self-association of calf brain tubulin. I. Stoichiometry. *Biochemistry* **14,** 4559–4566.

Ganesh, T., *et al.* (2007). Evaluation of the tubulin-bound paclitaxel conformation: Synthe-sis, biology, and SAR studies of C-4 to C-3′ bridged paclitaxel analogues. *J. Med. Chem.* **50,** 713–725.

Gaskin, F., and Kress, Y. (1977). Zinc ion-induced assembly of tubulin. *J. Biol. Chem.* **252,** 6918–6924.

Glaeser, R. M., Zilker, A., Radermacher, M., Gaub, H. E., Hartmann, T., and Baumeister, W. (1991). Interfacial energies and surface-tension forces involved in the preparation of thin, flat crystals of biological macromolecules for high-resolution electron microscopy. *J. Microsc.* **161,** 21–45.

Gyobu, N., Tani, K., Hiroaki, Y., Kamegawa, A., Mitsuoka, K., and Fujiyoshi, Y. (2004). Improved specimen preparation for cryo-electron microscopy using a symmetric carbon sandwich technique. *J. Struct. Biol.* **146,** 325–333.

Henderson, R., Baldwin, J. M., Downing, K. H., Lepault, J., and Zemlin, F. (1986). Structure of purple membrane from *Halobacterium halobium*: Recording, measurement and evaluation of electron micrographs at 3.5 Å resolution. *Ultramicroscopy* **19**, 147–178.

Henderson, R., Baldwin, J. M., Ceska, T. A., Zemlin, F., Beckman, E., and Downing, K. H. (1990). Model for the structure of bacteriorhodopsin based on high-resolution electron cryo-microscopy. *J. Mol. Biol.* **213**, 899–929.

Heuser, T., Raytchev, M., Krell, J., Porter, M. E., and Nicastro, D. (2009). The dynein regulatory complex is the nexin link and a major regulatory node in cilia and flagella. *J. Cell Biol.* **187**, 921–933.

Hirose, K., Lockhart, A., Cross, R. A., and Amos, L. A. (1995). Nucleotide-dependent angular change in kinesin motor domain bound to tubulin. *Nature* **376**, 277–279.

Hoenger, A., Sablin, E. P., Vale, R. D., Fletterick, R. J., and Milligan, R. A. (1995). Three-dimensional structure of a tubulin–motor-protein complex. *Nature* **376**, 271–274.

Kikkawa, M., Ishikawa, T., Wakabayashi, T., and Hirokawa, N. (1995). Three-dimensional structure of the kinesin head–microtubule complex. *Nature* **376**, 274–277.

Kühlbrandt, W., Wang, D. N., and Fujiyoshi, Y. (1994). Atomic model of plant light-harvesting complex by electron crystallography. *Nature* **367**, 614–621.

Larsson, H., Wallin, M., and Edstrom, A. (1976). Induction of a sheet polymer of tubulin by Zn^{2+}. *Exp. Cell Res.* **100**, 104–110.

Ledbetter, M. C., and Porter, K. R. (1963). A "microtubule" in plant fine structure. *J. Cell Biol.* **19**, 239–250.

Ledbetter, M. C., and Porter, K. R. (1964). Morphology of microtubules of plant cell. *Science* **144**, 872–874.

Li, H., DeRosier, D., Nicholson, W., Nogales, E., and Downing, K. (2002). Microtubule structure at 8 Å resolution. *Structure* **10**, 1317–1328.

Löwe, J., Li, H., Downing, K. H., and Nogales, E. (2001). Refined structure of alpha beta-tubulin at 3.5 Å resolution. *J. Mol. Biol.* **313**, 1045–1057.

Mandelkow, E.-M., and Mandelkow, E. (1985). Unstained microtubules studied by cryo-electron microscopy. Substructure, supertwist and diassembly. *J. Mol. Biol.* **181**, 123–135.

Mandelkow, E.-M., Mandelkow, E., and Milligan, R. A. (1991). Microtubule dynamics and microtubule caps: A time-resolved cryo-electron microscopy study. *J. Cell Biol.* **114**, 977–991.

Nettles, J. H., and Downing, K. H. (2009). The tubulin binding mode of microtubule stabilizing agents studied by electron crystallography. *In* "Topics in Current Chemistry: Microtubule Stabilizing and Destabilizing Agents: Synthetic, Structural and Mechanistic Insights," (T. Carlomagno, ed.).Springer, Heidelberg.

Nettles, J. H., Li, H. L., Cornett, B., Krahn, J. M., Snyder, J. P., and Downing, K. H. (2004). The binding mode of epothilone A on α,β-tubulin by electron crystallography. *Science* **205**, 866–869.

Nicastro, D., Schwartz, C., Pierson, J., Gaudette, R., Porter, M. E., and McIntosh, J. R. (2006). The molecular architecture of axonemes revealed by cryoelectron tomography. *Science* **313**, 944–948.

Nogales, E., Wolf, S. G., Zhang, S. X., and Downing, K. H. (1995). Preservation of 2-D crystals of tubulin for electron crystallography. *J. Struct. Biol.* **115**, 199–208.

Nogales, E., Wolf, S. G., and Downing, K. H. (1998). Structure of the $\alpha\beta$ tubulin dimer by electron crystallography. *Nature* **391**, 199–203.

Nogales, E., Whittaker, M., Milligan, R. A., and Downing, K. H. (1999). High resolution model of the microtubule. *Cell* **96**, 79–88.

Ravelli, R. B. G., Gigant, B., Curmi, P. A., Jourdain, I., Lachkar, S., Sobel, A., and Knossow, M. (2004). Insight into tubulin regulation from a complex with colchicine and a stathmin-like domain. *Nature* **428**, 198–202.

Reese, M., Sanchez-Pedregal, V. M., Kubicek, K., Meiler, J., Blommers, M. J., Griesinger, C., and Carlomagno, T. (2007). Structural basis of the activity of the microtubule-stabilizing agent epothilone A studied by NMR spectroscopy in solution. *Angew. Chem. Int. Ed. Engl.* **46,** 1864–1868.

Rice, S., *et al.* (1999). A structural change in the kinesin motor protein that drives motility. *Nature* **402,** 778–784.

Sindelar, C. V., and Downing, K. H. (2007). The beginning of kinesin's force-generating cycle visualized at 9-Å resolution. *J. Cell Biol.* **177,** 377–385.

Sindelar, C. V., and Downing, K. H. (2010). An atomic-level mechanism for activation of the kinesin molecular motors. *Proc. Natl. Acad. Sci. USA* **107,** 4111–4116.

Snyder, J. P., Nettles, J. H., Cornett, B., Downing, K. H., and Nogales, E. (2001). The binding conformation of Taxol in beta-tubulin: A model based on electron crystallographic density. *Proc. Natl. Acad. Sci. USA* **98,** 5312–5316.

Song, Y. H., and Mandlekow, E. (1995). The anatomy of flagellar microtubules: Polarity, seams, junctions, and lattice. *J. Cell Biol.* **128,** 81–94.

Subramaniam, S., Gerstein, M., Oesterhelt, D., and Henderson, R. (1993). Electron diffraction analysis of structural changes in the photocycle of bacteriorhodopsin. *EMBO J.* **12,** 1–8.

Sui, H., and Downing, K. H. (2006). Molecular architecture of axonemal microtubule doublets revealed by cryo-electron tomography. *Nature* **442,** 475–478.

Sui, H., and Downing, K. H. (2010). Structural Basis of Inter-Protofilament Interaction and Lateral Deformation of Microtubules. *Structure* **18,** 1022–1031.

Vonck, J. (1996). A three-dimensional difference map of the N intermediate in the bacteriorhodopsin photocycle: Part of the F helix twists in the M to N transition. *Biochemistry* **35,** 5870–5878.

Wade, R. H., Chrétien, D., and Job, D. (1990). Characterization of microtubule protofilament numbers. How does the surface lattice accomodate? *J. Mol. Biol.* **212,** 775–786.

Wang, H. W., and Nogales, E. (2005a). An iterative Fourier–Bessel algorithm for reconstruction of helical structures with severe Bessel overlap. *J. Struct. Biol.* **149,** 65–78.

Wang, H. W., and Nogales, E. (2005b). Nucleotide-dependent bending flexibility of tubulin regulates microtubule assembly. *Nature* **435,** 911–915.

CHAPTER SEVEN

HELICAL CRYSTALLIZATION OF TWO EXAMPLE MEMBRANE PROTEINS: MSBA AND THE CA^{2+}-ATPASE

John Paul Glaves,*,[†] Lauren Fisher,[‡] Andrew Ward,[‡] and Howard S. Young*,[†]

Contents

Abstract

Helical crystallization is a powerful tool for the moderate resolution structure determination of integral membrane proteins, where the insight gained often includes domain architecture and the disposition of α-helical segments. A necessary first step toward helical crystallization involves membrane protein reconstitution, which itself is a powerful technique for structure–function studies of integral membrane proteins. The correct insertion of a detergent-solubilized, purified membrane protein into lipid vesicles (proteoliposomes) can facilitate the functional characterization of the protein in a well-defined, chemically pure environment without interference from other membrane-associated components. In addition, the lipid-to-protein ratio can be controlled during reconstitution to

* Department of Biochemistry, School of Molecular and Systems Medicine, University of Alberta, Edmonton, Alberta, Canada
† National Institute for Nanotechnology, University of Alberta, Edmonton, Alberta, Canada
‡ The Scripps Research Institute, La Jolla, California, USA

Methods in Enzymology, Volume 483
ISSN 0076-6879, DOI: 10.1016/S0076-6879(10)83007-1

generate a high concentration of a particular membrane protein in the proteoliposomes, which are then suitable for both functional assays and crystallization trials.
Traditional approaches to two-dimensional crystallization for electron microscopy
rely on dialysis methods for the simultaneous reconstitution and crystallization of
a membrane protein [Kühlbrandt, W. (1992). Two-dimensional crystallization of
membrane proteins. *Q. Rev. Biophys.* **25**, 1–49.], yet some systems allow these
two steps to be experimentally separated and independently considered. Some
examples of integral membrane proteins that have been reconstituted and crystallized in a helical lattice include cytochrome bc1 complex from bovine heart [Akiba,
T., *et al.* (1996). Three-dimensional structure of bovine cytochrome bc_1 complex by
electron cryomicroscopy and helical image reconstruction. *Nat. Struct. Biol.* **3**,
553–561.], *Escherichia coli* melibiose permease [Rigaud, J. L., *et al.* (1997).
Bio-beads: An efficient strategy for two-dimensional crystallization of membrane
proteins. *J. Struct. Biol.* **118**, 226–235.], a bacterial ATP-binding cassette transporter MsbA [Ward, A., *et al.* (2009). Nucleotide dependent packing differences in
helical crystals of the ABC transporter MsbA. *J. Struct. Biol.* **165**, 169–175.], and
the sarcoplasmic reticulum Ca^{2+}-ATPase [Young, H. S., *et al.* (1997). How to make
tubular crystals by reconstitution of detergent-solubilized Ca^{2+}-ATPase. *Biophys.
J.* **72**, 2545–2558.]. The reconstitution and helical crystallization of MsbA and
Ca^{2+}-ATPase will be the focus of this chapter.

1. HELICAL CRYSTALLIZATION OF THE BACTERIAL INTEGRAL MEMBRANE PROTEIN, MsbA

MsbA is an ATP-binding cassette (ABC) transporter that couples ATP
hydrolysis to export lipid A and various substrates across the inner membrane of Gram-negative bacteria. MsbA is expressed as a 64.5-kDa half-
transporter and forms a symmetric homodimer in its functional state. Each
dimer has two soluble ATP-binding domains and two transmembrane
domains (TMDs) with a total of 12 membrane-spanning helices. MsbA is
a structural and functional homologue of the human multidrug resistant
transporter, P-glycoprotein (Pgp), which is widely studied for its ability to
pump chemotherapeutics and a variety of drugs out of cells. Several conformational states of MsbA have now been studied by X-ray crystallography
(Ward *et al.*, 2007). However, these structures represent proteins bound to a
detergent micelle, not its native environment embedded within a lipid
bilayer. Structural information about membrane embedded MsbA was
achieved by reconstituting purified protein into a lipid bilayer, which
resulted in the formation of helical crystals. Electron cryomicroscopy and
helical image analysis of these crystals has yielded a variety of moderate
resolution maps (Ward *et al.*, 2009).

1.1. Methods — MsbA

MsbA from *Vibrio cholerae* (VC-MsbA) was expressed and purified according to Ward *et al.* (2007). Purified MsbA was stored in 20 mM Tris–HCl (pH8.0), 20 mM NaCl, and 0.04–0.1% Cymal-7 for reconstitution experiments. Additional detergents, such as *n*-dodecyl β-D-maltoside (βDDM) and *n*-undecyl β-D-maltoside (UDM), were used for purification with the same methods and evaluated during crystallization. Purified MsbA was reconstituted into proteoliposomes by mixing it with a lipid suspension followed by detergent removal. Synthetic lipids with a fatty acid chain length of 16 or 18 carbon atoms were selected to model conditions commonly found in biological membranes. 1,2-dioleoyl-*sn*-glycero-3-phospho-L-serine (18:1 DOPS), 1,2-dioleoyl-*sn*-glycero-3-phosphoethanolamine (18:1 DOPE), 1,2-dioleoyl-*sn*-glycero-3-phosphocholine (18:1 DOPC), 1-palmitoyl-2-oleoyl-*sn*-glycero-3-phosphoethanolamine (16:0–18:1 POPE), and 1-palmitoyl-2-oleoyl-*sn*-glycero-3-phosphocholine (16:0–18:1 POPC) were purchased from Avanti Polar Lipids, Inc. (Alabaster, AL), solubilized in chloroform, dried down under nitrogen gas, and stored at -20 °C. On the day of experiment, crystallization buffer was added to the dried lipids and sonicated for approximately 20 min to completely resuspend the lipids in aqueous solution. Once resuspended, the lipid was combined with the purified protein at an appropriate lipid-to-protein ratio (w/w). Various protein concentrations ranging from 0.2 to 1.0 mg/ml and lipid-to-protein ratios ranging from 1:6 to 4:1 were tested. Later trials tested combinations of the 18:1 lipids with 1,2-diheptanoyl-*sn*-glycero-3-phosphocholine (7:0 DHPC) and 1,2-dimyristoyl-*sn*-glycero-3-phosphocholine (14:0 DMPC), which have shorter fatty acid chains.

Typically, the next step requires the removal of detergent; however, this depends on the choice of detergent and the dilution factor in adding the membrane protein (MsbA) to the lipid suspension. Each detergent has a characteristic critical micellar concentration (CMC), which is the concentration above which micelles will form. The CMC is approximately equal to the number of detergent monomers in the solution (Kühlbrandt, 1992). In order to remove the protein from the detergent micelle so it can incorporate into the lipid bilayer, the detergent concentration in solution must be brought below the CMC. There are a variety of detergent removal methods, which include dilution, adsorption, and dialysis. Adsorption via BioBeads (Bio-Rad Laboratories, Inc., Hercules, CA) was the most utilized method for MsbA trials because of the speed of detergent removal and the ease of use. Finally, these suspensions were incubated at a constant temperature to induce crystal formation, and temperatures above the phase transition temperature of the respective lipid were selected. Various temperatures ranging from 4 to 37 °C were evaluated. The best experimental temperature was found to be room temperature (22–25 °C).

A wide variety of conditions were tested during the crystallization trials of VC–MsbA. Variables such as final protein concentration, lipid-to-protein ratio, lipid type, and acyl-chain length, pH, detergent removal method, and temperature had to be adjusted in order to establish conditions that promoted protein reconstitution (i.e., insertion into a membrane bilayer) as opposed to aggregation. Initial experiments evaluated only a few variables, contained minimal components, and explored broad concentration ranges. Subsequent trials were based on the results from the previous experiment and incorporated more parameters and narrower ranges. We arrived at the following conditions for the optimal formation of helical crystals, which were suitable for electron cryomicroscopy. Chloroform solubilized DOPS was dried down under nitrogen gas and resuspended in crystallization buffer (20 mM citric acid buffer, pH 5.5, and 50 mM NaCl). Cymal-7 solubilized VC–MsbA was combined with the DOPS suspension in crystallization buffer at a final concentration of 0.6 mg/ml and a 1:1 lipid-to-protein ratio. Approximately 5–10% isopropyl alcohol was then added to the mixture. The optimal conditions for growing helical crystals of VC–MsbA required the addition of a nucleotide transition-state mimetic that trapped VC–MsbA in a single conformation. Nucleotide was added to the solution by combining 5 mM MgCl$_2$ and 5 mM ATP. In order to trap MsbA in the enzyme–substrate complex, 2 mM AlCl$_3$ and 2 mM NaF were also added. The mixture was incubated at room temperature (25 °C) and the protein formed well-ordered helical arrays as it reconstituted into lipid bilayers (Fig. 7.1). Notably, detergent removal was not ultimately required because the addition of buffer was sufficient to dilute the solution below the CMC of Cymal-7 (0.19 mM).

After 5 days incubation, the MsbA helical crystals were harvested. An aliquot of the crystallization solution (4 μl) was placed on a glow-discharged Quantifoil holey carbon EM grid (SPI Supplies, West Chester, PA), blotted with filter paper, and flash-frozen in liquid ethane (Fig. 7.2). The grids were

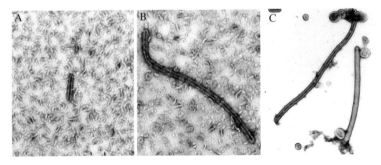

Figure 7.1   Images of negatively stained specimens showing VC–MsbA crystal growth overtime. (A) Nucleation of a helical crystal at 2 h, (B) extension of a helical crystal at 6 h, and (C) the final preparation of helical crystals with clean background at 24 h.

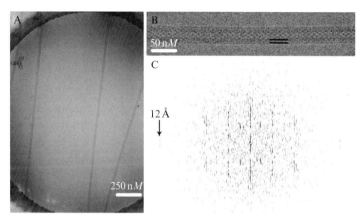

Figure 7.2 Electron cryomicroscopy of VC–MsbA helical crystals. (A) Electron micrograph of three vitrified helical VC–MsbA crystals suspended over a hole on a Quantifoil holey carbon EM grid. (B) High magnification image of a VC–MsbA crystal. The parallel black lines denote the approximate location of the membrane bilayer. (C) Computed Fourier transform of the helical crystal shown in (B) with layer lines visible to 12 Å resolution.

subsequently stored in liquid nitrogen. The vitrified helical crystals were imaged at liquid nitrogen temperature on a Philips CM120 microscope at an accelerating voltage of 120 kV and 34,480× magnification. Images were recorded on Kodak SO163 film under low dose conditions with defocus values ranging from 0.7 to 1.5 μm. The best crystals were selected by optical diffraction, and the micrographs were scanned using a Nikon Super Coolscan 9000 ED at 4000 dpi. The scanning interval and step size were calculated to be 6.35 μm and 1.8 Å per pixel, respectively. MRC/PHOELIX-based helical programs and custom scripts were used for helical reconstruction (Fig. 7.3).

2. HELICAL CRYSTALLIZATION OF THE SARCOPLASMIC RETICULUM CA^{2+}-ATPASE

The sarcoplasmic reticulum (SR) Ca^{2+}-ATPase is a P-type ion pump that couples ATP hydrolysis to the transport of cytosolic calcium into the lumen of the SR, in exchange for luminal protons. Ca^{2+}-ATPase is a 110-kDa protein with three cytoplasmic domains that interact with bound nucleotide and 10 transmembrane helices that form the ion transport sites. Ca^{2+}-ATPase provides an excellent model system because of the long history of reconstitution and two-dimensional crystallization of this protein (e.g., see Dux and Martonosi, 1983; Dux et al., 1985, 1987; Lacapere et al., 1998; Levy et al., 1990b; Martonosi et al., 1987; Misra et al., 1991). Indeed, the knowledge

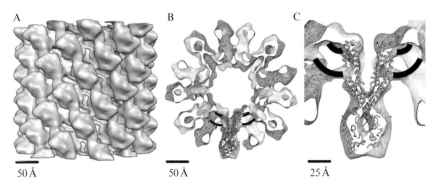

Figure 7.3   The density maps of VC-MsbA generated by helical reconstruction to 15 Å resolution. (A) Longitudinal view of the density map (gray surface), which is equivalent to the direction of view of the helical crystal shown in Fig. 7.2B. Each subunit in the density map corresponds to a dimer of MsbA. (B) Cross-sectional view of the helical reconstruction (gray mesh). The X-ray structure of the MsbA dimer (yellow cartoon) from *Salmonella typhimurium* (PDB 3B60) has been docked into the EM density map. (C) Close-up of the fitted MsbA X-ray structure and corresponding EM density map. The putative location of the lipid bilayer is delineated by the black lines. (See Color Insert.)

generated from early two-dimensional crystallization studies of the SR Ca^{2+}-ATPase (Misra *et al.*, 1991) ultimately contributed to the first X-ray crystallographic structure of this P-type ion pump (Toyoshima *et al.*, 2000). The last decade has seen a revolution in our understanding of P-type ion pumps—there are now structures for every intermediate in the Ca^{2+}-ATPase transport cycle (Olesen *et al.*, 2004, 2007; Sorensen *et al.*, 2004; Toyoshima and Mizutani, 2004; Toyoshima and Nomura, 2002; Toyoshima *et al.*, 2000, 2004, 2007), as well as initial structures for two other P-type ion pumps (Morth *et al.*, 2007; Pedersen *et al.*, 2007). Despite this wealth of structural information, the Ca^{2+}-ATPase remains a valuable experimental system for understanding the physical and chemical principles that link functional reconstitution with two- and three-dimensional crystallization. While reconstitution and crystallization remain empirical processes, there is often observed a correlation between the functional preservation of a particular membrane protein and the crystallization potential under particular experimental conditions. For Ca^{2+}-ATPase, a functional readout allowed us to prioritize the screening of conditions for reconstitution and crystallization trials, particularly in regard to the choice of detergents and lipids (Young *et al.*, 1997).

2.1. Methods—Ca^{2+}-ATPase

Skeletal muscle SR membranes were prepared from rabbit hind leg muscle (Eletr and Inesi, 1972), and detergent-solubilized Ca^{2+}-ATPase was purified by Reactive Red affinity chromatography (Stokes and Green, 1990).

Elution from the affinity column took advantage of functional properties of Ca^{2+}-ATPase, and required 4 mM ADP in the presence of calcium [20 mM MOPS, pH 7.0, 1 mM $MgCl_2$, 1 mM $CaCl_2$, 20% glycerol, and 0.1% detergent ($C_{12}E_8$; n-dodecyl-octaethylene-glycol monoether)]. The peak fractions were highly enriched for Ca^{2+}-ATPase ($>98\%$ pure by SDS-PAGE) and contained a minimum of 3–4 mg/ml protein. The specific ATPase activity of this Ca^{2+}-ATPase preparation was typically 10–12 μmol/mg/min as measured at 25 °C in the detergent-solubilized state using a coupled enzyme assay (Warren et al., 1974). The protein was stored at -80 °C, though the retention of activity after subsequent thawing and refreezing steps required the addition of 0.5–1 mg lipid (typically, egg yolk phosphatidylcholine; EYPC) per mg of protein.

The reconstitution of Ca^{2+}-ATPase has been described in detail (Levy et al., 1990a,c; Young et al., 1997). The appropriate amounts of lipids in organic solvent (Avanti Polar Lipids, Inc.) were dried to a thin film under nitrogen gas and lyophilized. For maximal Ca^{2+}-ATPase activity and helical crystallization, EYPC was combined with egg yolk phosphatidylethanolamine (EYPE) and egg yolk phosphatidic acid (EYPA) in an 8:1:1 weight ratio. However, for crystallization screening, a variety of lipid types and mixtures were substituted in this step. Buffer (20 mM imidazole, pH 7.0, 100 mM KCl) and detergent ($C_{12}E_8$) were added to the thin lipid film, followed by vigorous vortexing to achieve complete solubilization. Detergent-solubilized, affinity-purified Ca^{2+}-ATPase (~ 1 mg/ml) was then combined with this detergent–lipid suspension to achieve a weight ratio of 1 protein to 1 lipid to 2 detergents (a molar ratio of 1 Ca^{2+}-ATPase to ~ 150 lipids). The reconstitution of Ca^{2+}-ATPase into proteoliposomes was accomplished by the slow removal of detergent over a 4-h time course at room temperature, via the incremental addition of SM2-BioBeads. Finally, the reconstituted proteoliposomes containing Ca^{2+}-ATPase were removed from BioBeads and purified on a sucrose step-gradient. The density gradient centrifugation separated the two populations of reconstituted vesicles that result from slow detergent removal—a population of lipid vesicles and a population of proteoliposomes that are densely packed with Ca^{2+}-ATPase.

The proteoliposomes were used for crystallization screening directly after the sucrose step-gradient purification. The step-gradient removed excess lipid and any aggregated protein that did not reconstitute into the proteoliposomes, thereby providing a well-defined system for crystallization. After step-gradient purification, the reconstituted proteoliposomes typically had a molar ratio of one Ca^{2+}-ATPase to 120 lipids, and all Ca^{2+}-ATPase molecules were oriented with their cytoplasmic domains on the outside of the proteoliposomes (Young et al., 1997). However, the proteoliposomes can also be used without sucrose step-gradient purification, since there was no detectable protein aggregation during Ca^{2+}-ATPase reconstitution. In this case, the excess lipid may ultimately be useful for crystallization.

For crystallization, an aliquot of proteoliposomes (~ 50 μg Ca^{2+}-ATPase) was diluted into crystallization buffer (500 μl of 20 mM imidazole, pH 7.4, 100 mM KCl, 5 mM MgCl$_2$, 0.5 mM EGTA, and 0.5 mM Na$_3$VO$_4$ at 4 °C), and the proteoliposomes were then collected by centrifugation at $\sim 40,000 \times g$ for 30 min (Allegra 64R, Beckman Coulter). The resultant supernatant was gently aspirated away and 25 μl of fresh crystallization buffer was layered on top of the pellet. The point of this step was twofold—the sample was transferred into buffer conditions suitable for crystal formation and sucrose was removed so that proteoliposome fusion could occur for the growth of two-dimensional crystals. Once in crystallization buffer, the pellet was subjected to two freeze–thaw cycles in liquid N$_2$, and the pellet was gently resuspended with a micropipette followed by two additional cycles of freeze–thaw. Although crystallization within proteoliposomes usually occurred after 1 day, incubation at 4 °C for 4–7 days was optimal for a high frequency of well-ordered crystals. The key elements of this crystallization protocol included (i) freeze–thaw cycles to promote fusion of the proteolipo-somes, (ii) the removal of calcium by EGTA to promote a calcium-free state of Ca^{2+}-ATPase, and (iii) a decameric species of vanadate, which promotes the formation of antiparallel dimer ribbons of Ca^{2+}-ATPase. In addition, crystal formation was enhanced by a specific inhibitor or an endogenous regulator of Ca^{2+}-ATPase, thapsigargin (Sagara and Inesi, 1991) or phospho-lamban (Kirchberger *et al.*, 1975), respectively. However, neither thapsigar-gin nor phospholamban were absolutely required for crystal formation (Young *et al.*, 2001).

For screening crystallization conditions, samples (3–5 μl) were pipeted onto glow-discharged, carbon-coated grids and negatively stained by rinsing with two to three drops of cold uranyl acetate (2%). Once suitable crystals were identified, frozen-hydrated samples were prepared on holey carbon support films (300 mesh grids) that had been glow-discharged in air. An aliquot of the crystal suspension (typically 5 μl) was placed on a grid, blotted to a thin layer of residual solution, and then plunged into liquid ethane (-180 °C). Crystals were imaged in a Tecnai F20 electron microscope (FEI Company, Eindhoven, Netherlands) in the Microscopy and Imaging Facility (University of Calgary) or a JEOL 2000FS electron microscope (JEOL Ltd., Tokyo, Japan) in the Electron Microscopy Facility (National Institute for Nanotechnology, Uni-versity of Alberta and National Research Council of Canada). A standard room-temperature holder was used for negatively stained samples and a Gatan 626 cryoholder (Gatan Inc., Pleasanton, CA) was used for frozen-hydrated samples. Low-dose images were recorded either on film at 50,000× magnification (Tecnai F20) or on image plates at 35,800× magnifi-cation (JEOL 2200FS). Films were digitized at a scanning resolution of 6.35 μm with a Nikon Super Coolscan 9000 (the final pixel size was 2.54 Å after a two-by-two compression). The image plates were digitized at a scan-ning resolution of 15 μm (the final pixel size was 4.44 Å). All data were recorded with defocus levels between 0.5 and 2 μm, with an emphasis on low defocus images (0.5 and 1 μm) for frozen-hydrated samples.

3. Discussion

3.1. Ca^{2+}-ATPase: A case study

The Ca^{2+}-ATPase is an excellent model system for understanding helical crystallization, because the parameters that govern reconstitution and crystallization have been experimentally separated and independently considered in detail. Importantly, there is generally observed a positive correlation between the measured activity of Ca^{2+}-ATPase postreconstitution and the crystallization potential of the reconstituted proteoliposomes (see Table 1 in Young *et al.*, 1997). It had long been known that vanadate induces two-dimensional arrays in standard preparations of native SR membranes (Taylor *et al.*, 1984), with small crystalline vesicles ($\sim$0.2 μm diameter) and the occasional short, helical crystal being observed. The same was true for initial preparations of reconstituted proteoliposomes containing purified Ca^{2+}-ATPase. The question arose, what parameters control the preferential formation of helical crystals? Since the small SR vesicles or reconstituted proteoliposomes ($\sim$0.2 μm diameter) must somehow become larger helical crystals (typically 5–10 μm in length), we hypothesized that membrane fusion is a primary determinant for the formation of a high frequency of well-ordered helical crystals. We decided to attempt to influence membrane fusion of the reconstituted proteoliposomes using treatments that mildly destabilize the lipid bilayer—the addition of small amounts of detergent to the proteoliposomes or the addition of nonbilayer forming lipids in the reconstitution step.

In the first approach, the reconstituted Ca^{2+}-ATPase proteoliposomes were transferred to crystallization buffer and subjected to freeze–thaw cycles as described above. Immediately following the freeze–thaw cycles, various detergents including C$_{12}$E$_8$, *n*-octyl-β-D-glucoside (OG) and dihexanoyl-phosphatidylcholine (DHPC) were added to the samples. In the example shown in Fig. 7.4, the reconstituted proteoliposomes contained only Ca^{2+}-ATPase and the lipid EYPC. These proteoliposomes were resistant to forming helical crystals, yet the addition of 0.04 mg/ml C$_{12}$E$_8$ significantly facilitated crystal formation. This concentration is approximately equal to the CMC of this detergent, and higher concentrations begin to solubilize the proteoliposomes giving rise to crystalline membrane fragments (Young *et al.*, 1997). DHPC was also effective at producing helical crystals, whereas the addition of OG was far less effective (for all detergents, we tested concentrations ranging from half to twice the detergent CMC; data not shown). Therefore, we had identified preliminary reconstitution and crystallization conditions that give rise to rare or poorly ordered crystals,

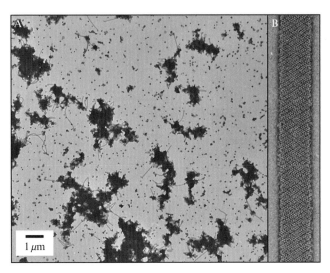

Figure 7.4   The effect of detergent on the helical crystallization of reconstituted Ca^{2+}-ATPase proteoliposomes. Small amounts of $C_{12}E_8$ (0.04–0.16 mg/ml) were added following reconstitution and freeze–thaw cycles in crystallization buffer. In this method, the formation of helical crystals does not depend on the lipid composition, since EYPC alone is sufficient. The detergent produces a higher background composed of small membrane fragments, which is more prevalent at higher detergent concentrations. (A) A low magnification image of a negatively stained specimen is shown in order to convey the overall appearance of the sample. There are large aggregates that result from the freeze–thaw process, and most of the helical crystals (~ 70 nm diameter) extend from these aggregates. Small proteoliposomes are also visible (~ 0.2 μm diameter). (B) A higher magnification image is shown for a single negatively stained helical crystal (~ 70 nm diameter) in order to convey the overall appearance of the helical lattice.

and the use of detergent additives enhanced crystal formation. However, the helical crystals of Ca^{2+}-ATPase produced by this method were difficult to preserve in the frozen-hydrated state. The blotting and vitrification process caused disordering of the helical crystals, indicating that the detergent must be removed prior to specimen preparation for electron cryomicroscopy.

The second approach of modifying the lipid composition during the Ca^{2+}-ATPase reconstitution step was much more successful at producing well-ordered helical crystals for electron cryomicroscopy. This concept is based on the molecular shape of lipids, which contribute to the overall physical state of the membrane (Cullis and de Kruijff, 1979). In cases where the lipid head group and acyl-chains have similar cross-sectional areas, the molecule has a cylindrical shape (phosphatidylcholine and phosphatidylserine), whereas lipids with a small head group have a cone shape (phosphatidylethanolamine and phosphatidic acid). Such lipid polymorphism plays a clear role in membrane fusion (Zimmerberg and Chernomordik, 1999) because the lipid bilayer formed by cylindrical shaped lipids is destabilized by cone-shaped lipids. We found that the ATPase activity of our reconstituted proteoliposomes was maximal when EYPC was supplemented with 10% EYPE and 10% EYPA, and this also gave rise to the best helical crystallization conditions (Fig. 7.5). This effect on Ca^{2+}-ATPase activity and crystal formation occurred over a relatively narrow range of lipid concentrations (5–15% each of EYPE and EYPA). While the negatively charged head groups of EYPS and EYPA could substitute for one another, EYPA was far more effective. In contrast, either cholesterol or a more unsaturated distribution of acyl-chains (soy phosphatidylcholine) had a negative impact on Ca^{2+}-ATPase activity and helical crystal formation. Therefore, a critical factor in crystal formation was the specific mixture of lipids added during Ca^{2+}-ATPase reconstitution (Young et al., 1997). Although crystalline patches could be observed in proteoliposomes composed of most lipids explored, small amounts of EYPE and EYPA (or EYPS) were required for extensive formation of helical crystals.

3.2. A lipid dependent switch between helical and two-dimensional crystals

The helical crystals formed from reconstituted proteoliposomes containing mixtures of EYPC, EYPE, and EYPA were suitable for high-resolution electron cryomicroscopy, and they have been used for structure determination (Young and Stokes, 2004; Young et al., 2001). In fact, the principles described above have also been used in the structure determination of the Na^+/K^+-ATPase from duck salt gland (Rice et al., 2001) and the Kdp-ATPase from E. coli (Hu et al., 2008). For the Ca^{2+}-ATPase helical reconstruction, the underlying crystal symmetry and molecular packing were similar to those found for native SR preparations. The helical crystals formed from the reconstituted proteoliposomes have a variable diameter that is slightly larger than those formed from native SR membranes, and this reflects significant variability in the helical symmetry (Toyoshima and Unwin, 1990; Toyoshima et al., 1993). To limit the choices in helical symmetry, the narrowest tubes are chosen for imaging and image processing, and these narrow helical crystals are often better preserved in frozen-hydrated specimens. As observed for native SR membranes (Zhang et al., 1998), the helical crystals are composed of antiparallel dimer ribbons of Ca^{2+}-ATPase molecules arranged with p2 symmetry (lattice parameters of $a \cong 62$ Å, $b \cong 115$ Å, and $\gamma \cong 73°$ in the middle of the membrane). The helical crystals generated from this method are of sufficient quality that individual images often possess visible layer line data to 10 Å resolution (Fig. 7.5), and higher resolution is readily attainable with data averaging (Xu et al., 2002).

While helical crystals are an ideal specimen type for electron cryomicroscopy, large planar two-dimensional crystals are the standard for high resolution

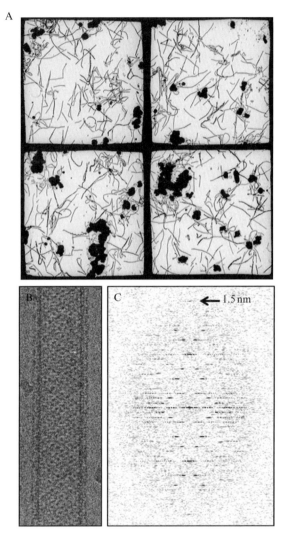

Figure 7.5 The effect of a lipid mixture on the helical crystallization of reconstituted Ca^{2+}-ATPase proteoliposomes. EYPC, EYPE, and EYPA were used in the reconstitution step, followed by freeze–thaw cycles in crystallization buffer. In this method, the formation of helical crystals is dramatically impacted by the lipid composition. (A) A very low magnification image of a negatively stained specimen is shown in order to convey the overall appearance of the sample. There are large aggregates that result from the freeze–thaw process, but most of the sample is in the form of helical crystals ($\sim$70 nm diameter). The entire 400 mesh grid had this same appearance. (B) A higher magnification image is shown for a single frozen-hydrated helical crystal and (C) the computed Fourier transform is shown in order to convey the overall appearance and quality of the helical lattice. The helical crystal is $\sim$70 nm in diameter, and the computed diffraction pattern shows layer line data to at least 15 Å resolution (individual crystals often have layer line data up to 10 Å resolution).

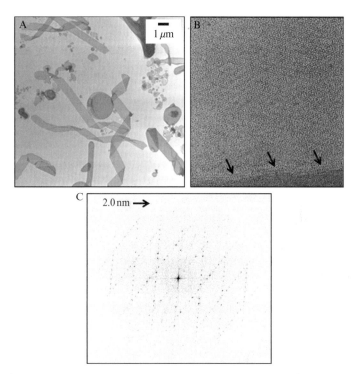

Figure 7.6   The effect of magnesium on the crystallization behavior of reconstituted Ca^{2+}-ATPase proteoliposomes. EYPC, EYPE, and EYPA were used in the reconstitution step, followed by freeze–thaw cycles in crystallization buffer. In this method, the magnesium concentration was increased from 5 to 30 mM, which altered the morphology from p2 helical crystals to $p22_12_1$ two-dimensional crystals. (A) A low magnification image of a negatively stained specimen is shown in order to convey the overall appearance of the sample. There are often large aggregates and vesicles that result from the freeze–thaw process, but most of the sample is in the form of two-dimensional crystals (0.5–3.0 μm wide). (B) A higher magnification image is shown for a region of a single frozen-hydrated crystal and (C) its computed diffraction pattern is shown in order to convey the overall appearance and quality of the crystal lattice. The crystal edge is indicated by arrows in the lower part of panel (B), and the computed diffraction pattern shows amplitude data to at least 15 Å resolution (a diffraction peak at 20 Å resolution is indicated).

structure determination. With this in mind, we serendipitously discovered conditions that generate large two-dimensional crystals from reconstituted Ca^{2+}-ATPase proteoliposomes (Fig. 7.6; Stokes *et al.*, 2006). The helical and two-dimensional crystals were grown using the same conditions described above, but with one small change—the magnesium concentration determines the crystal type (5 mM magnesium gives rise to helical crystals, and 30 mM magnesium gives rise to large two-dimensional crystals). The two-dimensional crystals are comprised of antiparallel dimer ribbons of Ca^{2+}-ATPase molecules as seen in the helical crystals, but packed into a lattice with $p22_12_1$ symmetry (lattice parameters of $a \cong 350$ Å, $b \cong 71$ Å, and $\gamma = 90°$). There are several interesting points to consider here. First, the two-dimensional crystals are typically 0.5–3 μm wide and up to $\sim$50 μm in length, which are much larger than either the starting proteoliposomes or the helical crystals. This suggests that membrane fusion is enhanced under conditions that permit two-dimensional crystallization. Second, the Ca^{2+}-ATPase molecules in our reconstituted proteoliposomes are all oriented with their cytoplasmic domains on the exterior of the vesicles (Young *et al.*, 1997). This facilitates the p2 symmetry seen in the helical crystals, yet it should be an impediment to the $p22_12_1$ symmetry seen in the two-dimensional crystals (Ca^{2+}-ATPase molecules protrude from both sides of the membrane). Nonetheless, the Ca^{2+}-ATPase molecules are able to reorient with respect to the membrane bilayer in the formation of this plane group. These effects are absolutely dependent on the presence of magnesium and EYPA, suggesting an electrostatic interaction between the divalent cation and the negatively charged lipid head group. While the underlying mechanism remains unknown, we have preliminary evidence that magnesium causes a lateral reorganization of EYPA in the reconstituted proteoliposomes (H. S. Young, unpublished observations). Perhaps lipid aggregation in the plane of the bilayer creates defects that ultimately promote fusion and allow Ca^{2+}-ATPase molecules to reorient relative to the membrane.

Irrespective of the mechanism, a particular lipid mixture was shown to be beneficial for the ATPase activity, as well as the helical and two-dimensional crystallization of reconstituted Ca^{2+}-ATPase proteoliposomes. This same lipid mixture (EYPC, EYPE, and EYPA) even proved useful in enhancing the quality of crystals for X-ray crystallographic studies (Moncoq *et al.*, 2007). While this lipid mixture is specific to Ca^{2+}-ATPase, the underlying principles may help to prioritize the screening of conditions for structure–function studies, which are often empirically based. The time frame for functional measurements (hours) is often more rapid than crystallization trials (days, weeks, or months), which makes a functional assay an effective means of prioritizing "crystallization space." Otherwise, the vast number of detergents and lipids that are commercially available make it impossible to conduct factorial screens, where individual components of crystallization conditions are varied in a systematic way. Instead, a functional readout allows the development of sparse matrices that retain conditions satisfying one criterion for the system under study.

4. Conclusions

MsbA and Ca^{2+}-ATPase exemplify the detailed insight that results from combining two-dimensional crystallization in a membrane environment with X-ray crystallographic studies in detergents. As a result, we gain a better understanding of the parameters that control crystal formation, as well as the conformational progression through the catalytic cycles of these transport proteins. While X-ray crystallography provides high resolution information in a detergent environment, helical crystallization can provide moderate resolution information in a native membrane. In both cases, crystallization is dependent on the conformation of the transporter, yet helical crystallization is influenced by additional parameters such as the choice of lipids and the reconstitution procedure. Following reconstitution, MsbA formed helical arrays under conditions that stabilized either an enzyme–substrate complex (trapped by ADP aluminum fluoride or AMPPNP) or an enzyme–product complex (trapped by ADP-vanadate). Reconstitution into a defined lipid mixture (50% DMPC and 50% DOPS) was optimal for helical crystallization. This is not that dissimilar from the approach for Ca^{2+}-ATPase. The presence of EGTA, deca-, and *ortho*-vanadate promoted a calcium-free transition-state conformation that is necessary for the formation of antiparallel dimer arrays of Ca^{2+}-ATPase molecules. The specific lipid mixture (80% EYPC, 10% EYPE, and 10% EYPA) and reconstitution procedure facilitated the arrangement of these dimer arrays into a helical lattice. Moreover, these findings are applicable for functional studies (Trieber *et al.*, 2005, 2009), electron microscopy of helical crystals (Young and Stokes, 2004; Young *et al.*, 1997, 2001), electron crystallography of two-dimensional crystals (Stokes *et al.*, 2006), and X-ray crystallography (Moncoq *et al.*, 2007).

ACKNOWLEDGMENTS

This work was supported by grants from the Canadian Institutes of Health Research to H.S.Y. (MOP53306) and the National Institutes of Health to Ron Milligan (GM75820; Ward *et al.*, 2009). J. P. G. is the recipient of a Canada Graduate Scholarship Doctoral Award from the Canadian Institutes of Health Research and the Alberta Ingenuity Fund. H. S. Y. is a Senior Scholar of the Alberta Heritage Foundation for Medical Research.

REFERENCES

Akiba, T., *et al.* (1996). Three-dimensional structure of bovine cytochrome bc₁ complex by electron cryomicroscopy and helical image reconstruction. *Nat. Struct. Biol.* **3,** 553–561.

Cullis, P. R., and de Kruijff, B. (1979). Lipid polymorphism and the functional roles of lipids in biological membranes. *Biochim. Biophys. Acta* **559,** 399–420.

Dux, L., and Martonosi, A. (1983). Two-dimensional arrays of proteins in sarcoplasmic reticulum and purified Ca^{2+}-ATPase vesicles treated with vanadate. *J. Biol. Chem.* **258,** 2599–2603.

Dux, L., *et al.* (1985). Crystallization of the Ca^{2+}-ATPase of sarcoplasmic reticulum by calcium and lanthanide ions. *J. Biol. Chem.* **260,** 11730–11743.

Dux, L., *et al.* (1987). Crystallization of Ca²⁺-ATPase in detergent-solubilized sarcoplasmic reticulum. *J. Biol. Chem.* **262**, 6439–6442.

Eletr, S., and Inesi, G. (1972). Phospholipid orientation in sarcoplasmic reticulum membranes: Spin-label ESR and proton NMR studies. *Biochim. Biophys. Acta* **282**, 174–179.

Hu, G. B., *et al.* (2008). Three-dimensional structure of the KdpFABC complex of *Escherichia coli* by electron tomography of two-dimensional crystals. *J. Struct. Biol.* **161**, 411–418.

Kirchberger, M., *et al.* (1975). Phospholamban: A regulatory protein of the cardiac sarcoplasmic reticulum. *Recent Adv. Stud. Cardiac Struct. Metab.* **5**, 103–115.

Kühlbrandt, W. (1992). Two-dimensional crystallization of membrane proteins. *Q. Rev. Biophys.* **25**, 1–49.

Lacapere, J., *et al.* (1998). Two-dimensional crystallization of Ca-ATPase by detergent removal. *Biophys. J.* **75**, 1319–1329.

Levy, D., *et al.* (1990a). A systematic study of liposome and proteoliposome reconstitution involving Bio-Bead-mediated Triton X-100 removal. *Biochim. Biophys. Acta* **1025**, 179–190.

Levy, D., *et al.* (1990b). Phospholipid vesicle solubilization and reconstitution by detergents. Symmetrical analysis of the two processes using octaethylene glycol mono-n-dodecyl ether. *Biochemistry* **29**, 9480–9488.

Levy, D., *et al.* (1990c). Evidence for proton countertransport by the sarcoplasmic reticulum Ca²⁺-ATPase during calcium transport in reconstituted proteoliposomes with low ionic permeability. *J. Biol. Chem.* **265**, 19524–19534.

Martonosi, A., *et al.* (1987). Structure of Ca²⁺-ATPase in sarcoplasmic reticulum. *Soc. Gen. Physiol. Ser.* **41**, 257–286.

Misra, M., *et al.* (1991). Effect of organic anions on the crystallization of the Ca{+2+}-ATPase or muscle sarcoplasmic reticulum. *Biochim. Biophys. Acta* **1077**, 107–118.

Moncoq, K., *et al.* (2007). The molecular basis for cyclopiazonic acid inhibition of the sarcoplasmic reticulum calcium pump. *J. Biol. Chem.* **282**, 9748–9757.

Morth, J. P., *et al.* (2007). Crystal structure of the sodium–potassium pump. *Nature (London)* **450**, 1043–1049.

Olesen, C., *et al.* (2004). Dephosphorylation of the calcium pump coupled to counterion occlusion. *Science* **306**, 2251–2255.

Olesen, C., *et al.* (2007). The structural basis of calcium transport by the calcium pump. *Nature (London)* **450**, 1036–1042.

Pedersen, B. P., *et al.* (2007). Crystal structure of the plasma membrane proton pump. *Nature (London)* **450**, 1111–1114.

Rice, W. J., *et al.* (2001). Structure of Na+, K+-ATPase at 11 Angstroms resolution: Comparison with Ca2+-ATPase in E1 and E2 states. *Biophys. J.* **80**, 2187–2197.

Rigaud, J. L., *et al.* (1997). Bio-beads: An efficient strategy for two-dimensional crystallization of membrane proteins. *J. Struct. Biol.* **118**, 226–235.

Sagara, Y., and Inesi, G. (1991). Inhibition of the sarcoplasmic reticulum Ca{+2+} transport ATPase by thapsigargin at subnanomolar concentrations. *J. Biol. Chem.* **266**, 13503–13506.

Sorensen, T., *et al.* (2004). Phosphoryl transfer and calcium ion occlusion in the calcium pump. *Science* **304**, 1672–1675.

Stokes, D. L., and Green, N. M. (1990). Three-dimensional crystals of Ca-ATPase from sarcoplasmic reticulum: Symmetry and molecular packing. *Biophys. J.* **57**, 1–14.

Stokes, D., *et al.* (2006). Interactions between Ca²⁺-ATPase and the pentameric form of phospholamban in two-dimensional co-crystals. *Biophys. J.* **90**, 4213–4223.

Taylor, K., *et al.* (1984). Structure of the vanadate-induced crystals of sarcoplasmic reticulum Ca²⁺-ATPase. *J. Mol. Biol.* **174**, 193–204.

Toyoshima, C., and Mizutani, T. (2004). Crystal structure of the calcium pump with a bound ATP analogue. *Nature* **430**, 529–535.

Toyoshima, C., and Nomura, H. (2002). Structural changes in the calcium pump accompanying the dissociation of calcium. *Nature* **418**, 605–611.

Toyoshima, C., and Unwin, N. (1990). Three-dimensional structure of the acetylcholine receptor by cryoelectron microscopy and helical image reconstruction. *J. Cell Biol.* **111**, 2623–2635.

Toyoshima, C., *et al.* (1993). Three-dimensional cryo-electron microscopy of the calcium ion pump in the sarcoplasmic reticulum membrane. *Nature* **362**, 469–471.

Toyoshima, C., *et al.* (2000). Crystal structure of the calcium pump of sarcoplasmic reticulum at 2.6 A resolution. *Nature* **405**, 647–655.

Toyoshima, C., *et al.* (2004). Lumenal gating mechanism revealed in calcium pump crystal structures with phosphate analogues. *Nature* **432**, 361–368.

Toyoshima, C., *et al.* (2007). How processing of aspartylphosphate is coupled to lumenal gating of the ion pathway in the calcium pump. *Proc. Natl. Acad. Sci. USA* **104**, 19831–19836.

Trieber, C., *et al.* (2005). The effects of mutation on the regulatory properties of phospholamban in co-reconstituted membranes. *Biochemistry* **44**, 3289–3297.

Trieber, C., *et al.* (2009). The effects of phospholamban transmembrane mutants on the calcium affinity, maximal activity and cooperativity of the sarcoplasmic reticulum calcium pump. *Biochemistry* **48**, 9287–9296.

Ward, A., *et al.* (2007). Flexibility in the ABC transporter MsbA: Alternating access with a twist. *Proc. Natl. Acad. Sci. USA* **104**, 19005–19010.

Ward, A., *et al.* (2009). Nucleotide dependent packing differences in helical crystals of the ABC transporter MsbA. *J. Struct. Biol.* **165**, 169–175.

Warren, G. B., *et al.* (1974). Reconstitution of a calcium pump using defined membrane components. *Proc. Natl. Acad. Sci. USA* **71**, 622–626.

Xu, C., *et al.* (2002). A structural model for the catalytic cycle of Ca(2+)-ATPase. *J. Mol. Biol.* **316**, 201–211.

Young, H., and Stokes, D. (2004). The mechanics of calcium transport. *J. Membr. Biol.* **198**, 55–63.

Young, H. S., *et al.* (1997). How to make tubular crystals by reconstitution of detergent-solubilized Ca^{2+}-ATPase. *Biophys. J.* **72**, 2545–2558.

Young, H. S., *et al.* (2001). Locating phospholamban in co-crystals with Ca^{2+}-ATPase by cryoelectron microscopy. *Biophys. J.* **81**, 884–894.

Zhang, P., *et al.* (1998). Structure of the calcium pump from sarcoplasmic reticulum at 8 Angstroms resolution. *Nature* **392**, 835–839.

Zimmerberg, J., and Chernomordik, L. V. (1999). Membrane fusion. *Adv. Drug Deliv. Rev.* **38**, 197–205.

CHAPTER EIGHT

MULTIPARTICLE CRYO-EM OF RIBOSOMES

Justus Loerke, Jan Giesebrecht, *and* Christian M. T. Spahn

Contents

Abstract

As the resolution of cryo-EM reconstructions has improved to the subnanometer range, conformational and compositional heterogeneity have become increasing problems in cryo-EM, limiting the resolution of reconstructions. Since further purification is not feasible, the presence of several conformational states of ribosomal complexes in thermodynamic equilibrium requires methods for separating these states *in silico*.

We describe a procedure for generating subnanometer resolution cryo-EM structures from large sets of projection images of ribosomal complexes. The incremental K-means-like method of unsupervised 3D sorting discussed here allows separation of classes in the dataset by exploiting intrinsic divisions in the data. The classification procedure is described in detail and its effectiveness is illustrated using current examples from our work. Through a good separation of conformational modes, higher resolution reconstructions can be calculated. This increases information gained from single states, while exploiting the coexistence of multiple states to gather comprehensive mechanistic insight into biological processes like ribosomal translocation.

Institut für medizinische Physik und Biophysik, Charité, Universitätsmedizin Berlin, Berlin, Germany

Methods in Enzymology, Volume 483
ISSN 0076-6879, DOI: 10.1016/S0076-6879(10)83008-3

1. INTRODUCTION

Electron microscopy has experienced a prolific development in recent years through a combination of data collection from vitrified samples (see chapter 3 in MIE volume 481) and advanced digital image processing (see chapter 8 in MIE volume 482). Many large macromolecular complexes essential to every living cell are either too large or too heterogeneous to be analyzed by X-ray crystallography or nuclear magnetic resonance (NMR). Cryoelectron microscopy (cryo-EM) is now a rapidly emerging biophysical method for visualizing these interesting cellular components. Ribosomes have been ideal samples to develop single-particle-EM methodology due to their size, their dense RNA core, and their spherical shape (Frank, 2009). In the past 20 years, the structures of ribosomal complexes advanced considerably in resolution, from low-resolution 45 Å reconstructions based on only several hundred images (Frank *et al.*, 1991) to subnanometer resolution cryo-EM maps from hundreds of thousands of particle projections (Connell *et al.*, 2007; Halic *et al.*, 2004; Schuette *et al.*, 2009; Seidelt *et al.*, 2009; Schüler *et al.*, 2006; Villa *et al.*, 2009). This progress was the result of a constant improvement of microscopes, image processing software plus major growth in computational power and biochemical preparation techniques. A new generation of electron microscopes with advanced cryostage technology and better vacuum systems have been coupled with field emission electron guns (FEG) using acceleration voltages of 200–300 kV. These modern microscopes such as the *Tecnai Polara* can deliver resolution on the length scale of the carbon–carbon covalent bond. However, during imaging in the electron microscope, high spatial frequency information is attenuated more severely than low-frequency information (Saad *et al.*, 2001; Wade, 1992; Wade and Frank, 1977). A careful compensation for this effect by high-frequency enhancement is important in order to improve the resolution of our reconstructions into the subnanometer range.

During our recent structural work on the translation apparatus in bacterial, fungal, and mammalian systems, it became clear that all preparations of ribosomal complexes are mixed with a fraction of vacant or degraded particles. These projections of "bad" particles have to be removed from the data or otherwise they are averaged together with projections of the target complex, resulting in an obscured reconstruction. In addition, even relatively stable macromolecular machines such as ribosomes undergo large conformational changes during their functional cycle and therefore can occur in a mixture of conformational states (Spahn and Penczek, 2009). The heterogeneity of states appears to be intrinsic to ribosomal complexes (Munro *et al.*, 2009), so a further separation on the biochemical level is not feasible. Thus, in order to address these issues, improved image processing methods for separating ribosomal functional modes need to be developed.

2. DEALING WITH HETEROGENEITY: 3D SORTING

Heterogeneity in projection images of complexes results from (i) compositional heterogeneity, which may be caused by a mixture of aggregated, malformed, damaged, or dissociated complexes, etc. in addition to the functional complexes, (ii) intrinsic flexibility of the complexes under investigation, and (iii) conformational heterogeneity, caused by the existence of different states of the complex in a thermodynamic equilibrium before the freezing process. Heterogeneity may manifest itself in the form of low densities when visualizing a reconstructed volume. While the problems with composition of the sample can sometimes be solved during biochemical sample preparation or the particle selection step, flexibility, and conformational heterogeneity are not always readily apparent in the images and cannot be avoided (see chapter 12 in MIE volume 482 and chapter 9 in this volume). The problem is particularly challenging for ribosomal complexes because the ribosome is a highly dynamic machine and oscillates between many functional states during protein synthesis, for example, the pretranslocational (PRE) and the posttranslocational (POST) state during the elongation phase. The transition between the POST to the PRE state is catalyzed by the elongation factor EF-TU/eEF1A and the transition from the PRE back to the POST state called translocation is facilitated by the elongation factor EF-G/eEF2. Thus, already four major intermediate states of the elongation cycle are known, which can exist in various substates defined, for example, by the positions of the bound tRNA molecules or by the conformation of the bound elongation factor before and after GTP hydrolysis (Frank and Spahn, 2006; Munro et al., 2009). The structures of these intermediates are essential for a comprehensive understanding of the complex chain of events of the elongation cycle. Though the options for a further biochemical separation of ribosomal subcomplexes at reasonable cost seem to be exhausted, an in silico purification approach via 3D sorting can be an effective alternative.

Three-dimensional sorting introduces multiple references into the projection matching procedure (Penczek et al., 1994). As recently described in a review by Spahn and Penczek (2009), there are several methods available. The procedures most commonly used for ribosomes are (i) supervised classification (Allen et al., 2005; Gao et al., 2004; Valle et al., 2001), where template structures are selected manually and reflect a priori knowledge or assumptions of the user; and (ii) unsupervised classification, or clustering, which relies on the competition between templates for particles to drive selection. Unsupervised classification methods try to find clusters or separations inherent to the data itself; the name refers to the fact that an algorithm will try to determine divisions within the data without a priori knowledge or user intervention. The unsupervised method described below is based on an incremental K-means-like algorithm (Likas *et al.*, 2003); also, unsupervised procedures based on maximum-likelihood methods have been suggested (Scheres *et al.*, 2007; see chapter 11 in MIE volume 482).

In the first step of the sorting procedure, templates compete for particles (K-means belongs to the class of "competitive learning" algorithms); particles are assigned to the cluster closest to them, where, for comparison of particle images and projections of reference structures, the distance is usually measured by correlation between the two. In the second step, the templates are updated by reconstructing a volume from all the particles assigned to it in the first step. This simple two-step procedure is repeated in an iterative manner either for a given number of iterations or until convergence, that is stability of spatial orientation and class assignments of particles, is detected.

This three-dimensional sorting procedure is run in an incremental scheme (see Fig. 8.1). In the initial phase of alignment, the whole dataset will have to be aligned against one common reference. This choice of only one initial reference is a deviation of the incremental K-means in contrast to a classical K-means procedure, which starts with at least two clusters. Since the multireference alignment scheme is very sensitive toward off-center particles, the initial alignment step is crucial. Over a number of iterations, orientation parameters of particles are refined using projection matching with one reference only. This refinement is run with decreasing step size and search range for both angular and translation parameter search, and the reference is updated to reflect the new particle orientations. Once particles are aligned and orientation parameters are stable, a second neutral reference will be added, starting the first round of multireference sorting. If this new seeding structure is not too dissimilar, it will start attracting particles that do not fit well into the first, stable volume. These particles will be reassigned to the new reference and will be used for updating the new reference. Again, this two-reference alignment will be run with decreasing step size and search range to refine the alignment of all particles with respect to the two references. After a number of iterations, the orientations and the assignment of particles to the references will stabilize. At this point, a third reference may be added, starting the next round of sorting, and the procedure will repeat. Consecutive addition of new references will lead to the formation of new clusters in the dataset; if the number of references in the refinement exceeds the number of intrinsic divisions within the dataset, the addition of a new reference will usually lead to the formation of a very small subpopulation of particles that result in a poor reconstruction. At this point, the populations may be assessed and subpopulations may be split from the data and refined separately to improve the separation of the individual populations. Likewise, particle images that are assigned to corrupt reconstructions and can be assumed to contain degraded particles may be removed from the main dataset.

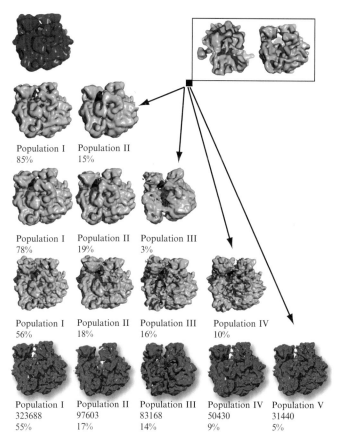

Figure 8.1 Multiparticle refinement strategy. In order to account for conformational heterogeneity of the *T. thermophilus* 70S·EF-Tu·Phe-tRNA·GDP·kirromycin complex, the multiparticle refinement was used to sort the dataset into up to five subpopulations (Penczek *et al.*, 2006; Connell *et al.*, 2008). During the refinement, the number of reference structures was successively increased by adding the structure of a vacant *E. coli* 70S ribosome as a new reference. The respective cryo-EM maps of the ribosomal complexes are depicted in the side view. Numbers and percentages indicate total particle projections and fractions of the whole dataset, respectively. Population I embodies the majority of particle images and represents the cryo-EM map analyzed in this study.

3. SUBNANOMETER MULTIPARTICLE CRYO-EM OF RIBOSOMES IN PRACTICE

3.1. Sample preparation, microscopy, and initial image processing

Samples of ribosomal complexes are flash-frozen in liquid ethane on carbon coated *Quantifoil* grids using an *FEI Vitrobot*. Typically, our data is collected on an *FEI Tecnai G2 Polara* microscope operating at 300 kV, at a nominal magnification of $39,000\times$ and under low-dose conditions (~ 20 e$^-$/Å^2) on film (*Kodak*). Micrographs are digitized on a *Heidelberger* drum-scanner with a step size of 4.76 μm corresponding to a pixel size of 1.26 Å on the object scale. Visual inspection of the micrographs and their power spectra is followed by template-based screening for particles using the software *Signature* (Chen and Grigorieff, 2007) and visual or automatic preselection of particles. The full resolution particles are boxed out and decimated for initial image processing. To determine the defocus values for the micrographs, the package *CTFfind* is employed (Mindell and Grigorieff, 2003). Particles are sorted into groups by defocus values and CTF correction is later applied separately to the intermediate volumes for each of these defocus groups (Frank *et al.*, 2000).

3.2. Multireference refinement

The typical size of our datasets is currently from several hundred thousand to over a million particle images from over a thousand micrographs. Boxing and all subsequent image processing steps are done using *SPIDER* (Frank *et al.*, 1996; see chapter 15 in MIE volume 482). The refinement is run in several phases, with decreasing pixel size, from 5.04 Å in the initial alignment phase to 3.78 and 2.52 Å for sorting and 1.26 Å for high-detail refinement and reconstruction.

The general procedure for multireference refinement incorporates a regular angular refinement (Gabashvili *et al.*, 2000; Penczek *et al.*, 1994) within an incremental K-means-like loop and is shown schematically in Fig. 8.2. The procedure runs as follows:

1. Start with a single neutral reference ($N = 1$) to align all particles.
2. Reference volumes are multiplied by CTF for each defocus group and K reference projections (depending on the angular step size) are generated for each of the N reference volumes, leading to a stack of $K \cdot N$ reference projections.
3. Particles are aligned per defocus group to the best matching reference projection.
4. Particles from each defocus group are divided into N classes according to the reference volume the best matching reference projection was derived from. N reconstructions corresponding to the N classes are computed from the aligned particles of each defocus group.
5. Reconstructions from individual defocus groups are CTF-corrected by Wiener filtration and merged to yield N intermediate volumes.
6. Intermediate volumes from step 5 are low-pass filtered according to the current resolution estimate. Optionally, intermediate volumes are thresholded, high-frequency enhanced, and masked before low-pass filtration. The resulting N volumes are used as references for the next iteration.

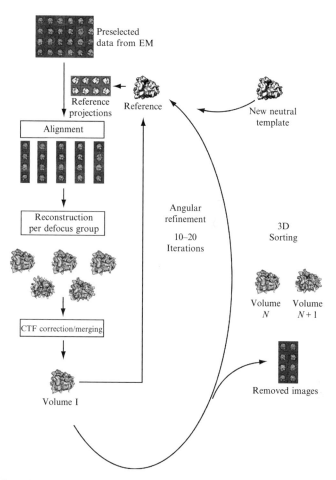

Figure 8.2   Schematic overview of the 3D sorting procedure. The 3D sorting procedure consists of two nested loops. The outer loop consists of the classification procedure. Starting with one reference, all particles are aligned to the reference in an iterative angular refinement procedure, comprising the inner loop. Particles are aligned to the reference volume sorted into defocus groups. Individual reconstructions are computed for each defocus group, merged and filtered to yield a new reference volume. Search parameters may be reduced to increase the precision of the search and the alignment is restarted with the updated reference volume and search parameters. When orientation parameters in the refinement have stabilized, a neutral template is added as an additional new reference. Search parameters are reset and the angular refinement is restarted with two reference volumes. The number of reference volumes is increased until there is no further separation. To increase the quality of the sorting, bad reference volumes, and the corresponding subpopulations of particle images may have to be removed from the dataset.

7. Search parameters can be decreased and the angular refinement continues by jumping back to step 2, until convergence is reached.

8. At the end of the refinement cycle (depending on convergence or a fixed schedule), add a new reference ($N \rightarrow N + 1$). Reset search parameters to the initial values and jump to step 2. Alternatively, classes from bad particles can be eliminated by removing the particle images from the dataset and the corresponding volumes from the set of reference volumes. Particles belonging to a specific subclass can also be isolated and again subjected to a separate multiparticle refinement.

9. If N exceeds a predetermined threshold or assessment of the computed volumes shows no further separation, the procedure is stopped.

In the initial alignment and classification phase, particle images used are fourfold decimated, corresponding to a pixel size of 5.04 Å. Since the center coordinates generated by particle identification methods are usually not precise enough, particles first have to be centered using projection matching with a fairly large translational search range. If particles are not centered and the translational search range is kept too low, multireference alignment may return misaligned particles, and the classification will generate classes with different center positions instead of structural changes.

In addition to centering, projection matching is also used to align all particles to the reference. The initial reference for our refinements is usually the low-pass filtered volume of an empty 70S ribosome from *Escherichia coli* for data from *Thermus thermophilus*, for instance (Ratje *et al.*, 2010), or an empty 80S ribosome from *Saccharomyces cerevisiae* (Schüler *et al.*, 2006). The reference may also be used for internal monitoring of reference bias. The bias monitoring was used by Connell *et al.* (2008) and has become standard practice for our refinements (Schuette *et al.*, 2009). For projects using *T. thermophilus*, for instance, the development of features specific for *T. thermophilus* from an *E. coli* template during refinement (Ratje *et al.*, 2010), is an important indication of data consistency. A lack of change during refinement may indicate reference bias.

The initial alignment refines parameters with increasing precision over a number of iterations. Orientation search starts with a large step size and search range, for both translational and angular search, which are slowly decreased over a number of refinement iterations. Typically, the angular refinement step will take 10–20 iterations. Once particle assignment and orientations have stabilized, the next reference can be added (following the philosophy of the sorting method described in the previous section, see Fig. 8.1) and refinement will continue with the next alignment round. During this alignment, the relative changes of orientation parameters between iterations are monitored as an indicator for convergence. After a new reference has been added, reassignment and reorientation of particles should become stable within a few iterations, and the relative changes of individual particles should drop rapidly as orientations converge. Significant changes of the orientation parameters over many iterations of alignment are usually evidence for fundamental problems, such as incorrect filter parameters leading to overfitting, or misalignments of references and mask with respect to each other.

We noticed during our refinements that misalignment of particles may be caused by overfitting of particles and noise to the negative densities present in the intermediate volumes used as references. Moreover, this overfitting may also lead to an overestimation of the resolution. The removal of the negative densities reduces these effects, so we apply thresholding of intermediate volumes in step 6 to enforce positivity from the very beginning.

The strong low-pass filtering introduced by the decimation step emphasizes only very general particle features. At a pixel size of 5.04 Å, only general structural changes such as ratcheting or ligand binding will be detectable, so that they will dominate the initial classification. In the initial phase of a refinement, a new dataset will usually contain a subset of particles that poorly agree with the dominant reference(s). Provided that the new template and the dominant existing references are not too similar, poorly matching particles will be attracted by the new reference. In the first rounds of sorting the new classes formed will contain mainly bad particles (e.g., dissociated ribosomal subunits, aggregates, or other artifacts), which are not relevant for further refinement. Class size will usually be fairly low and volumes reconstructed from these classes will show strong artifacts and warped structures. Particles from these artifactual classes are then removed from the dataset before the refinement continues.

To reduce overfitting and to reduce the computation required, data should initially be refined at high pixel sizes. There are two indicators that the classification at this decimation factor has been exhausted and that the pixel size will have to be decreased for further progress: (i) resolution exceeds a bound of a normalized spatial frequency of 0.25–0.3, where a frequency of 0.5 corresponds to the Nyquist frequency; (ii) addition of new references leads to low-occupancy classes or to the formation of identical classes. The number of particles that need to be removed from the data before 3D sorting gives good results may be large. For a recent complex of 70S ribosomes with tRNAs and EF-G bound and stalled with fusidic acid (Ratje et al., 2010), roughly 283,000 particles, including vacant ribosomes without ligands, had to be removed from a total of 587,000 particles. Only 304,000 particles showing strong, albeit fragmented, density for EF-G were processed further, eventually splitting up into two subpopulations with 113,000 and 156,000 particles, with resolutions of 7.8 and 7.6 Å, respectively. Similar numbers appear in the recent reconstruction of a 70S·tRNA·EF-Tu·GDP·kyrromycin complex (Schuette et al., 2009), where the 6.4 Å resolution reconstruction of the complex of interest comprises roughly 163,000 particles out of a dataset with a total of 586,000 particles. It is noteworthy that the complex has been affinity purified and that essentially all ribosomes contained the ternary complex (tRNA·EF-Tu·GDP) as a ligand.

When the pixel size is decreased, all remaining particles will have to be centered again, which is necessary due to the possible errors of orientation parameters introduced during the pixel size change. Likewise, the relative alignment of reference volumes with respect to each other may have to be checked as well. If the dataset still contains degraded particle images at this point, it may be necessary to remove the classes forming artifactual reconstructions.

Interesting subpopulations of particles may be either separated from the dataset, refined and classified separately, or classes deemed to be irrelevant or warped may be removed and the remaining dataset is refined further. To keep the amount of computation time wasted on irrelevant particles low, it is usually the interesting populations that are separated from the rest and refined separately. This should be handled carefully, though, since separation of classes may remove good data prematurely and bias subsequent refinement steps.

Eventually, the attenuation of high-frequency information during imaging in the microscope and subsequent image processing steps will make subtle distinctions between particles difficult. So high-frequency enhancement is used, either in the form of B-factor, matched filters (Schüler *et al.*, 2006) or power spectrum adjustment (Ratje *et al.*, 2010) with the help of high-resolution X-ray data. This amplification increases the resolution again, allowing separation of small differences that are otherwise unnoticed. The effects of high-frequency enhancement can be illustrated using the data from our current 70S·EF-G·FA project (Ratje *et al.*, 2010). At a pixel size of 2.52 Å, separation of the classes was not progressing any further than at 3.78 Å, showing a dominant population with EF-G, but with a disordered head domain on the 30S subunit indicating further heterogeneity. Using high-frequency enhancement, this population split into two further subpopulations. Different amounts of intersubunit rotation, differing by about 3°, and head swiveling, differing about 3°, and subtle differences in tRNA binding between the two states could finally be detected.

Regrettably, high-frequency noise is also amplified by high-frequency enhancement. To reduce the effects of noise on the structure, real-space soft masking is employed. It can already be employed at higher pixel sizes, but becomes very important in conjunction with high-frequency enhancement, to reduce errors from amplified noise. Soft masks are created from low-pass filtered reference ribosome structures by first creating a binary mask from the structure (Frank, 2006). This binary mask is dilated a few pixels and multiplied along the new-grown edges with a cosine function, creating a smooth drop-off of the resulting mask.

Once relevant classes have stabilized, they will be separated from the remaining dataset and the orientation parameters are determined separately in an angular refinement, before reconstructions are calculated. To reduce the effect of interpolation errors on the final structure, reconstructions should be done either at the lowest pixel size of 1.26 Å or with a sophisticated algorithm like gridding (see chapter 1 in MIE volume 482).

4. INTERPRETATION

At barely subnanometer resolution (9–5 Å), it is in general not possible to model unknown structures *ab initio* in terms of identifying positions of single atoms. However, if sequences and secondary structure predictions are available tracing the backbone of nucleic acids comes into reach, because secondary structure elements of RNA/DNA (RNA double helices) are considerably larger than α-helices or β-sheets of proteins and usually the major and minor grooves are very well defined (see chapter 1 in this volume). As proof of concept *de novo* modeling of the nearly 200 nucleotide long CrPV IRES RNA (Hellen and Sarnow, 2001) has been done based on a 7.3-Å resolution cryo-EM map of the CrPV IRES-80S complex from *S. cerevisiae* (Schüler *et al.*, 2006). The map showed detailed density for the CrPV IRES in the intersubunit space of the ribosome (see Fig. 8.3), allowing *de novo* modeling of the complete sugar–phosphate backbone of the RNA. An alignment of this model with the X-ray structure of the ribosome binding domain of unbound dicistroviridae IRES (Pfingsten *et al.*, 2006) illustrates the general agreement of the two models derived from two different biophysical techniques (Fig. 8.3D).

Furthermore, if atomic models based on crystallographic data are available (Schmeing and Ramakrishnan, 2009), they can be used to substantially enhance the information deduced from lower resolution cryo-EM reconstructions. One recent example from our group is the structure of the decoding complex from *T. thermophilus*. The sorted major conformation containing 323,668 projections was refined at full pixel size to a final resolution of 6.4 Å by the conservative 0.5 cutoff of the FSC criterion (Schuette *et al.*, 2009). The map shows a substantial improvement in details compared to previous maps of the ribosomal decoding complex (Stark *et al.*, 1997; Valle *et al.*, 2003; see Fig. 8.4A and B). RNA helices and α-helical secondary structure elements of proteins are clearly resolved and distinct density is present even for low-molecular weight ligands like the antibiotic kirromycin (Fig. 8.4C). The reconstruction was analyzed further, beginning with density segmentation for the ribosomal subunits and bound ligands based on a clustering procedure using *SPIDER*. Then models were docked into the resulting density sections. If possible, we prefer to fit monolithic blocks of models as rigid bodies to our reconstructions, as in our work on the bacterial decoding complex. We fit 70S models derived from a 2.8-Å resolution X-ray crystallographic structure (Selmer *et al.*, 2006) as three rigid bodies, corresponding to the main body of the 50S subunit, and the

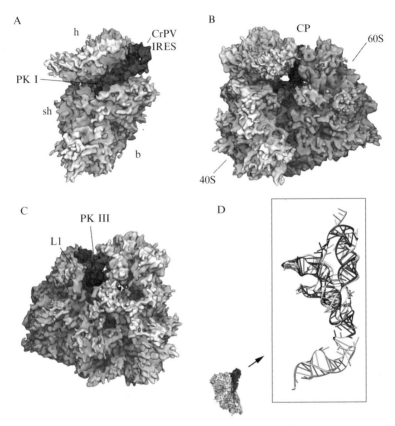

Figure 8.3 80S ribosome with bound CrPV IRES. 7.3 Å cryo-EM map of the CrPV·IRES·RNA in complex with the yeast ribosome. The map is shown (A) from the back if the 60S subunit with the 60S subunit removed, (B) from the A-site, and (C) from the L1 protuberance (40S subunit, yellow; 60S subunit, blue; CrPV·IRES, magenta). Landmarks for the 40S subunit: b, body; h, head; sh, shoulder. Landmarks for the 60S subunit: CP, central protuberance; L1, L1 protuberance; PKI, pseudoknot I; PKIII, pseudoknot III. In (D) the structure of the CrPV·IRES·RNA molecular model (gray and magenta ribbons) from the cryo-EM map is shown. Aligned to this structure is the model of the PSIV IRES based on a 3.1-Å X-ray structure (Pfingsten *et al.*, 2006; orange ribbons). (See Color Insert.)

head and the body/platform domains of the 30S subunit. For rigid body docking, we employ the *SITUS* software package (Wriggers *et al.*, 1999) or *UCSF Chimera* (Pettersen *et al.*, 2004). For more sophisticated modeling, molecular dynamics flexible fitting methods (Orzechowski and Tama, 2008; Trabuco *et al.*, 2008; Whitford *et al.*, 2009) can be employed for further insights.

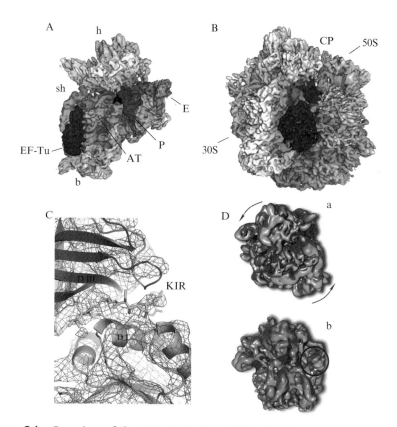

Figure 8.4 Overview of the 70S·EF-Tu·Phe-tRNA·GDP·kirromycin complex. A surface representation of the cryo-EM map is shown (A) from the 50S side, with 50S removed and (B) from the A-site (30S subunit, yellow; 50S subunit, blue; EF-Tu, red; A/T-tRNA, light magenta; P-tRNA, green; E-tRNA, orange; mRNA, dark blue). Landmarks for the 40S subunit: b, body; h, head; sh, shoulder. Landmarks for the 60S subunit: CP, central protuberance, L1, L1 protuberance. (C) Density for the low-molecular weight kirromycin seen between the domains of EF-Tu. (D) (a) Superposition of Population I (gold) compared to Population II (pink, transparent). The reconstructions are shown from the top with the 50S subunit above the 30S subunit. An arrow indicates an intersubunit rotation of the 30S subunit with respect to the 50S subunit. (b) Superposition of Population I (gold) compared to Population III (blue, transparent). The reconstructions are shown from the side. The stalk base of Population III is abbreviated (red circle). (See Color Insert.)

5. CONCLUSIONS

What started several years ago as a naïve attempt to remove projections of empty ribosomes or ribosomal complexes with a substochiometric occupancy of ligands from our data to avoid an averaging of ill-matching particle images, evolved over time to a complex hierarchy of refinement cycles. In combination with methods for exploiting as much of the high-frequency information available from the new generation of FEG driven cryomicroscopes as possible, we are now able to obtain routinely reconstructions of functional ribosomal complexes, for instance, intermediates of the elongation cycle, from bacteria, yeast, and mammalian systems with subnanometer resolution.

Today, we are convinced that ribosomes in living cells exist in a broad spectrum of functional states that can mostly be recovered only as mixtures (Alberts, 1998). This situation seemed to make the classical single-particle approach only applicable for rare occasions where a homogeneous complex could be biochemically isolated. Reconstructions with near atomic resolution for most of the intermediates of the elongation cycle appear to be still challenging for cryo-EM techniques. On the other hand, a broad variety of vital conformations are trapped during the vitrification process offering the opportunity to study not only one but many "snapshots" of an active complex. With 3D sorting strategies, it is possible to separate images of degraded complexes from those of intact ribosomes. It is also possible to isolate and reconstruct projections derived from functional subconformations of ribosomes and reconstruct them to subnanometer resolution. This allows structural access to each step in translation, such as ligand binding events, peptide bond formation, or GTP hydrolysis on an elongation factor from comparable samples to study the whole chain of actions. The ribosome can spontaneously adopt distinct native state conformations that are in equilibrium at room temperature. Accordingly, a metastable energy landscape view of the function of ribosomes is emerging similar to models that have been developed for folding and dynamics of smaller proteins (Munro *et al.*, 2009). The energy landscape view might be valid also for other large macromolecular complexes and could therefore constitute a basic principle. Cryo-EM, in its ability to capture projections of molecules in their close-to-native states in combination with 3D sorting procedures might become a key technology to study the dynamics of molecular machines at work.

ACKNOWLEDGMENTS

We thank J.-C. Schuette for his assistance with illustrations and Grant Jensen and Pawel A. Penczek for critical reading of the manuscript.

REFERENCES

Alberts, B. (1998). The cell as a collection of protein machines: Preparing the next generation of molecular biologists. *Cell* **92,** 291–294.
Allen, G. S., Zavialov, A., Gursky, R., Ehrenberg, M., and Frank, J. (2005). The cryo-EM structure of a translation initiation complex from *Escherichia coli. Cell* **121,** 703–712.

Chen, J. Z., and Grigorieff, N. (2007). SIGNATURE: A single-particle selection system for molecular electron microscopy. *J. Struct. Biol.* **157,** 168–173.

Connell, S. R., Takemoto, C., Wilson, D. N., Wang, H., Murayama, K., Terada, T., Shirouzu, M., Rost, M., Schüler, M., Giesebrecht, J., Dabrowski, M., Mielke, T., *et al.* (2007). Structure basis for interaction of the ribosome with the switch regions of GTP-bound elongation factors. *Mol. Cell* **25,** 751–764.

Connell, S. R., Topf, M., Qin, Y., Wilson, D. N., Mielke, T., Fucini, P., Nierhaus, K. N., and Spahn, C. M. T. (2008). A new tRNA intermediate revealed on the ribosome during EF4-mediated back-translocation. *Nat. Struct. Mol. Biol.* **15,** 910–915.

Frank, J. (2006). Three-Dimensional Electron Microscopy of Macromolecular Assemblies. Oxford University Press, New York.

Frank, J. (2009). Single-particle reconstruction of biological macromolecules in electron microscopy—30 years. *Q. Rev. Biophys.* **42,** 139–158.

Frank, J., and Spahn, C. M. T. (2006). The ribosome and the mechanism of protein synthesis. *Rep. Prog. Phys.* **69,** 1383–1417.

Frank, J., Penczek, P. A., Grassucci, R. A., and Srivastava, S. (1991). Three-dimensional reconstruction of the 70S *Escherichia coli* ribosome in ice: The distribution of ribosomal RNA. *J. Cell Biol.* **115,** 597–605.

Frank, J., Radermacher, M., Penczek, P. A., Zhu, J., Li, Y., Ladjadj, M., and Leith, A. (1996). SPIDER and WEB: Processing and visualization of images in 3D electron microscopy and related fields. *J. Struct. Biol.* **116,** 190–199.

Frank, J., Penczek, P. A., Agrawal, R. K., Grassucci, R. A., and Heagle, A. B. (2000). Three-dimensional cryoelectron microscopy of ribosomes. *Methods Enzymol.* **317,** 276–291.

Gabashvili, I. S., Agrawal, R. K., Spahn, C. M. T., Grassucci, R. A., Svergun, D. I., Frank, J., and Penczek, P. A. (2000). Solution structure of the *E. coli* 70S ribosome at 11.5 Å resolution. *Cell* **100,** 537–549.

Gao, H., Valle, M., Ehrenberg, M., and Frank, J. (2004). Dynamics of EF-G interaction with the ribosome explored by classification of a heterogeneous cryo-EM dataset. *J. Struct. Biol.* **147,** 283–290.

Halic, M., Becker, T., Pool, M. R., Spahn, C. M. T., Grassucci, R. A., Frank, J., and Beckmann, R. (2004). Structure of the signal recognition particle interacting with the elongation-arrested ribosome. *Nature* **427,** 806–814.

Hellen, C. U. T., and Sarnow, P. (2001). Internal ribosome entry sites in eukaryotic mRNA molecules. *Genes Dev.* **15,** 1593–1612.

Likas, A., Vlassis, N., and Verbeek, J. J. (2003). The global K-means clustering algorithm. *Pattern Recognit.* **36,** 451–461.

Mindell, J. A., and Grigorieff, N. (2003). Accurate determination of local defocus and specimen tilt in electron microscopy. *J. Struct. Biol.* **142,** 334–347.

Munro, J. B., Sanbonmatsu, K. Y., Spahn, C. M. T., and Blanchard, S. C. (2009). Navigating the ribosome's metastable energy landscape. *Trends Biochem. Sci.* **34,** 390–400.

Orzechowski, M., and Tama, F. (2008). Flexible fitting of high-resolution X-ray structures into cryoelectron microscopy maps using biased molecular dynamics simulations. *Biophys. J.* **95,** 5692–5705.

Penczek, P. A., Grassucci, R. A., and Frank, J. (1994). The ribosome at improved resolution: New techniques for merging and orientation refinement in 3D cryo-electron microscopy of biological particles. *Ultramicroscopy* **53,** 251–270.

Penczek, P. A., Frank, J. and Spahn, C. M. T. (2006). A method of focused classification, based on the bootstrap 3D variance analysis, and its application to EF-G-dependent translocation. *J. Struct. Biol.* **154,** 184–194.

Pettersen, E. F., Goddard, T. D., Huang, C. C., Couch, G. S., Greenblatt, D. M., Meng, E. C., and Ferrin, T. E. (2004). UCSF Chimera—A visualization system for exploratory research and analysis. *J. Comput. Chem.* **25**, 1605–1612.

Pfingsten, J. S., Costantino, D., and Kieft, J. S. (2006). Structural basis for ribosome recruitment and manipulation by a viral IRES RNA. *Science* **314**, 1450–1454.

Ratje, A. H., Loerke, J., Mikolajka, A., Brünner, M., Hildebrand, P. W., Starosta, A., Doenhoefer, A., Connell, S. R., Fucini, P., Mielke, T., Whitford, P. C., Onuchic, J. N., *et al.* (2010). Submitted.

Saad, A., Ludtke, S. J., Jakana, J., Rixon, F. J., Tsuruta, H., and Chiu, W. (2001). Fourier amplitude decay of electron cryomicroscopic images of single particles and effects on structure determination. *J. Struct. Biol.* **133**, 32–42.

Scheres, S. H. W., Gao, H., Valle, M., Herman, G. T., Eggermont, P. P. B., Frank, J. and Carazo, J.-M. (2007). Disentangling conformational states of macromolecules in 3D-EM through likelihood optimization. *Nat. Methods* **4**, 27–29.

Schmeing, T. M., and Ramakrishnan, V. (2009). What recent ribosome structures have revealed about the mechanism of translation. *Nature* **461**, 1234–1242.

Schuette, J.-C., Murphy, F. V., Kelley, A. C., Weir, J. R., Giesebrecht, J., Connell, S. R., Loerke, J., Mielke, T., Zhang, W., Penczek, P. A., Ramakrishnan, V., and Spahn, C. M. T. (2009). GTPase activation of elongation factor EF-Tu by the ribosome during decoding. *EMBO J.* **28**, 755–765.

Schüler, M., Connell, S. R., Lescoute, A., Giesebrecht, J., Dabrowski, M., Schroeer, B., Mielke, T., Penczek, P. A., Westhof, E., and Spahn, C. M. T. (2006). Structure of the ribosome-bound cricket paralysis virus IRES RNA. *Nat. Struct. Mol. Biol.* **13**, 1092–1096.

Seidelt, B., Innis, C. A., Wilson, D. N., Gartmann, M., Armache, J.-P., Villa, E., Trabuco, L. G., Becker, T., Mielke, T., Schulten, K., Steitz, T. A., and Beckmann, R. (2009). Structural insight into nascent polypeptide chain-mediated translational stalling. *Science* **326**, 1412–1415.

Selmer, M., Dunham, C., Murphy, F. V., Weixlbaumer, A., Petry, S., Kelley, A. C., Weir, J. R., and Ramakrishnan, V. (2006). Structure of the 70S ribosome complexed with mRNA and tRNA. *Science* **313**, 1935–1942.

Spahn, C. M. T., and Penczek, P. A. (2009). Exploring conformational modes of macromolecular assemblies by multiparticle cryo-EM. *Curr. Opin. Struct. Biol.* **19**, 623–631.

Stark, H., Rodnina, M. V., Rinke-Appel, J., Brimacombe, R., Wintermeyer, W., and van Heel, M. (1997). Visualization of elongation factor Tu on the *Escherichia coli* ribosome. *Nature* **389**, 403–406.

Trabuco, L. G., Villa, E., Mitra, K., Frank, J., and Schulten, K. (2008). Flexible fitting of atomic structures into electron microscopy maps using molecular dynamics. *Structure* **16**, 673–683.

Valle, M., Sengupta, J., Swami, N. K., Grassucci, R. A., Burkhardt, N., Nierhaus, K. N., Agrawal, R., and Frank, J. (2001). Cryo-EM reveals an active role for aminoacyl-tRNA in the accommodation process. *EMBO J.* **21**, 3557–3567.

Valle, M., Zavialov, A., Snegupta, J., Ehrenberg, M., and Frank, J. (2003). Locking and unlocking of ribosomal motions. *Cell* **114**, 123–134.

Villa, E., Sengupta, J., Trabuco, L. G., LeBarron, J., Baxter, W. T., Shaikh, T. R., Grassucci, R. A., Nissen, P., Ehrenberg, M., Schulten, K., and Frank, J. (2009). Ribosome-induced changes in elongation factor Tu conformation control GTP hydrolysis. *Proc. Natl. Acad. Sci. USA* **106**, 1063–1068.

Wade, R. H. (1992). A brief look at imaging and contrast transfer. *Ultramicroscopy* **46**, 145–156.

Wade, R. H., and Frank, J. (1977). Electron microscopic transfer functions for partially coherent axial illumination and chromatic defocus spread. *Optik* **49,** 81–92.

Whitford, P. C., Noel, J. K., Gosavi, S., Schug, A., Sanbonmatsu, K. Y., and Onuchic, J. N. (2009). An all-atom structure-based potential for proteins: Bridging minimal models with all-atom empirical forcefields. *Proteins* **75,** 430–441.

Wriggers, W., Milligan, R. A., and McCammon, J. A. (1999). Situs: A package for docking crystal structures into low-resolution maps from electron microscopy. *J. Struct. Biol.* **125,** 185–195.

CHAPTER NINE

SINGLE-PARTICLE ELECTRON MICROSCOPY OF ANIMAL FATTY ACID SYNTHASE: DESCRIBING MACROMOLECULAR REARRANGEMENTS THAT ENABLE CATALYSIS

Edward J. Brignole *and* Francisco Asturias

Contents

Department of Cell Biology, The Scripps Research Institute, La Jolla, California, USA

Methods in Enzymology, Volume 483
ISSN 0076-6879, DOI: 10.1016/S0076-6879(10)83009-5

Abstract

We have used macromolecular electron microscopy (EM) to characterize the conformational flexibility of the animal fatty acid synthase (FAS). Here we describe in detail methods employed for image collection and analysis. We also provide an account of how EM results were interpreted by considering a high-resolution static FAS X-ray structure and functional data to arrive at a molecular understanding of the way in which conformational pliability enables fatty acid synthesis.

1. ELECTRON MICROSCOPY AND THE NEXT FRONTIER IN STRUCTURAL BIOLOGY

Electron microscopy (EM) enables the visualization of biological structures ranging in size from individual macromolecules to whole cells. A variety of sample preparation, data acquisition, and image analysis methods make possible the use of EM projection images to reconstruct the 3D structures of biological entities spanning a 10^6 molecular weight range. For homogeneous, rigid, biological macromolecules, EM can be used to generate atomic resolution 3D structures (e.g., Chen *et al.*, 2009; Cong *et al.*, 2010; Henderson *et al.*, 1990; Kuhlbrandt *et al.*, 1994; Miyazawa *et al.*, 2003; Yonekura *et al.*, 2003; Zhang *et al.*, 2010). One of the most exciting and innovative uses of EM is the structural characterization of dynamic, asymmetric macromolecular complexes responsible for a variety of critical cellular processes. By virtue of its ability to visualize individual macromolecules, and through increasingly sophisticated image collection and analysis techniques, EM is emerging as the method of choice for structural analysis of macromolecules that display a heterogeneous mixture of conformational states and subunit compositions. Correlation of structural information from EM to functional data is proving essential for unraveling the mechanism of action of biological macromolecular machines.

2. THE CATALYTIC CYCLE OF FAS REQUIRES CONFORMATIONAL CHANGES

Cells synthesize fatty acids to build membrane phospholipids (Swinnen *et al.*, 2003; Vance and Vance, 2004), as posttranslational modifications that target proteins to membranes (Resh, 2006), and for energy storage as triglycerides (Coleman and Lee, 2004). Because of its central role in metabolic homeostasis, FAS has been targeted for treatment of infection (Zhang *et al.*, 2006), cancer (Kuhajda *et al.*, 1994; Menendez and Lupu, 2007), and obesity (Loftus *et al.*, 2000).

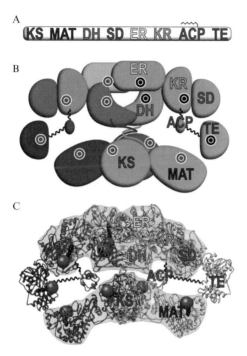

Figure 9.1 (A) The animal FAS polypeptide contains six catalytic domains: ketoacyl synthase (KS), malonyl/acetyl transferase (MAT), dehydrase (DH), enoyl reductase (ER), ketoacyl reductase (KR), and thioesterase (TE), as well as a nonenzymatic structural domain (SD) and an acyl-carrier domain (ACP). (B) Domain organization of the coiled FAS homodimer (one monomer in color, the other in gray) highlighting the active sites that can be visited by the red carrier domain (red targets) and those that the red carrier domain cannot access (black targets). Flexible linkage between KR-ACP and ACP-TE is indicated as wavy lines. (C) The N-terminal KS domain through the KR domain were observed in the FAS crystal structure (gray outline, PDB: 2vz9). Structures of the carrier domain (red, PDB: 2png) and TE (maroon, PDB: 1xkt) are positioned for illustrative purposes connected by flexible interdomain linkages indicated by wavy lines as in (B). Red balls indicate the docking site on each catalytic domain that can be accessed by the red carrier domain. The black structure extending from each red ball represents the substrate cargo delivered by the carrier domain into the active site. Two catalytic sites that can be contacted by the red ACP are located in the opposite reaction chamber, apparently beyond reach of the red carrier domain due to its short 10-residue tether to the KR domain. Using EM, we uncovered dramatic structural flexibility of FAS that permits contacts between carrier and catalytic domains. (See Color Insert.)

In animals, fatty acids are synthesized *de novo* in the cytosol by a 550-kDa FAS homodimer in which each single polypeptide monomer contributes all six enzymatic active sites required for production of fatty acids plus an acyl-carrier domain (Fig. 9.1A). In the active form of FAS, the monomers coil to form two separate reaction chambers, each possessing a full complement of catalytic domains (Asturias *et al.*, 2005; Maier *et al.*, 2006; Fig. 9.1B). A remarkable feature of FAS is that a growing fatty acyl chain is covalently tethered to the carrier domain of each subunit as it iteratively cycles to the active sites. The carrier domain of each monomer is connected by a short 10-residue linker to the upper portion of the FAS structure, but the carrier domain and its adjacent flexibly tethered thioesterase domain (Joshi *et al.*, 2005) were not visible in the FAS X-ray structure (Maier *et al.*, 2008; Fig. 9.1C). Mutant complementation and cross-linking studies conclusively established that the acyl-carrier domain shuttles substrates to catalytic domains that, in the FAS crystal structure, are located well beyond reach in the opposite reaction chamber (Smith *et al.*, 2003). Hence, large-scale rearrangements in the high-resolution snapshot captured by the crystal structure of FAS must be required to mediate substrate-transfer reactions evidenced by biochemical studies.

Earlier EM studies suggested that FAS undergoes conformational changes. The first investigation of FAS structure identified several conformations described as "θ," "A," and "H" (Kitamoto *et al.*, 1988). A decade later, Brink and colleagues determined a cryo-EM reconstruction of FAS (Brink *et al.*, 2002), performed normal mode analysis on the resulting structure (Ming *et al.*, 2002), and then initiated 3D refinement with the competing models (Brink *et al.*, 2004). These 3D reconstructions were consistent with conformational flexibility; however, they were based on the prevailing model that the FAS dimer is composed of antiparallel subunits. A subsequent examination of FAS structure by the Asturias and Smith groups revealed that the subunits are parallel and detected changes in FAS conformation that occurs when specific active-site mutants, in the presence of substrates, are arrested at particular catalytic steps (Asturias *et al.*, 2005). Finally, our recent EM analysis of FAS, detailed below, reconciled the functional and X-ray crystallography results by identifying extensive structural rearrangements that enable fatty acid synthesis (Brignole *et al.*, 2009).

3. METHODS AND RATIONALE EMPLOYED IN THE CONFORMATIONAL ANALYSIS OF FAS

3.1. Specimen preparation

Vitrification in a physiologically relevant buffer (Adrian *et al.*, 1984; Taylor and Glaeser, 1974) optimally preserves biological specimens for EM analysis (see Chapter 3 in Volume 481). However, poor contrast and low SNR in EM images of vitrified macromolecules represent nearly insurmountable hurdles when single particles with multiple conformations or compositions must be distinguished during image analysis. Preservation in stain can circumvent this problem by providing images with comparatively higher contrast that facilitate alignment and classification of conformationally heterogeneous particles

(Burgess et al., 2004b; Cheng and Walz, 2009). The use of stain is a calculated compromise, since this preservation method is known to result in dehydration-induced artifacts, notably, flattening of structures against the amorphous carbon support (Boisset et al., 1990, 1993; Cheng et al., 2006; Kellenberger et al., 1982; Radermacher et al., 1994; Ruiz et al., 2001). Modified protocols for preservation in stain such as the addition of trehalose (Harris, 2007) or cryopreservation of stained specimens (Adrian et al., 1998; Golas et al., 2003; Ohi et al., 2004) can ameliorate the flattening that presumably results as the stain surrounding the particle dries (see Chapter 6 in Volume 481).

Previous work from our group established that FAS particles were satisfactorily preserved in stain (Asturias et al., 2005) using a simple protocol previously detailed (Asturias et al., 2004; Ohi et al., 2004) and described below. Due to its oblate shape, FAS has a strong tendency to adsorb in a single orientation such that its surface area contacting the carbon support is maximized. Fortunately, in the preferred orientation, FAS lies flat in the plane of the grid, minimizing artifacts caused by dehydration and collapse in negative stain. By comparison, a minor population of FAS particles adsorbs perpendicular to this orientation and are recognizable in 2D class averages as bottom views but give rise to severely flattened 3D structures (data not shown).

In the approach we employed, FAS was diluted to a concentration of 10–20 ng/μl and 5 μl of the dilute FAS solution was applied to a thin ($\sim$15 nm) amorphous carbon film supported by a 300 mesh Cu/Rh grid. While grids with a finer mesh provide better support for the carbon substrate and contain fewer squares with broken film, wider mesh aids in collection of tilted-pair images as there are fewer grid bars to obstruct the view of the tilted specimen. Immediately before application of the FAS solution (within 5 min), the carbon-coated grids were glow discharged in the presence of amylamine to briefly activate the surface of the carbon film (Aebi and Pollard, 1987) and enhance the deposition of stain and adsorption of particles.

After an $\sim$1-min incubation, the FAS solution was blotted by touching the edge of the grid to filter paper for $\sim$5 s and a 5-μl drop of 1% uranyl acetate was immediately applied. The stain was then blotted and a second drop applied. The wash was repeated, and after applying a fourth drop, the stain was allowed to stand on the grid for 1–2 min. During this final staining period, a thin carbon film was floated from a square of carbon-coated mica onto the surface of a well containing 1% uranyl acetate. Finally, the grid was submerged into the well of stain and lifted through the floating carbon film. Immediately, the grid was blotted by briefly ($\sim$1 s) touching the lower surface of the grid flat against filter paper and any residual stain between arms of the forceps was wicked away. This "carbon sandwich" method of staining ensures that particles are fully embedded in stain between the two layers of carbon (Tischendorf et al., 1974).

3.2. Microscopy and data collection

The general strategy we pursued includes a combination of 2D and 3D analyses that were essential to gain an adequate understanding of FAS conformational changes. Images of untilted specimens were acquired to assess FAS conformation by 2D image alignment and clustering. Determination of 3D structures for each conformation identified through 2D analysis was possible through recording of tilted–untilted image pairs and application of the random conical tilt method (Radermacher et al., 1987).

Micrographs and CCD frames of untilted specimens were collected within 1 μm under focus, so that resolution remained limited to ∼20 Å by the interaction between the stain and protein (Sherman et al., 1981; Steven and Navia, 1980, 1982; Unwin, 1975). As the stained particles clearly contrast with the background, it is unnecessary to collect images farther from focus that would invert and attenuate contrast and require computational correction of the images to partially restore the information (see Chapter 2 in Volume 482). Tilting the specimen produces a gradient of defocus across the micrographs. Therefore, images of the −55°-tilted specimen were acquired with a nominal defocus of ∼1.5-μm and examined on an optical diffractometer to ensure that the entire micrograph was underfocused.

Given the dimensions of the FAS particle (200 Å × 200 Å × 50 Å) and the intended final resolution (∼20 Å), images were recorded at 50,000× magnification on Kodak SO-163 film using an FEI Tecnai F20 microscope. Micrographs were digitized on a Zeiss scanner with a 7-μm step and then binned threefold resulting in a final pixel size of 4.2 Å at the specimen level. This pixel size is severalfold smaller than the best resolution we expected to obtain in stain, and therefore resolution was not limited by digital sampling. Using Leginon (Suloway et al., 2005), other FAS preparations were imaged without tilting as CCD frames at 50,000× with pixel sizes of 2.26 and 1.63 Å.

3.3. Particle selection

Particles were selected automatically from the CCD frames (Rath and Frank, 2004) followed by visual inspection with Web to remove false positives. Until recently, identification of particles in tilted-pair micrographs was accomplished using SPIDER's Web interface (Frank et al., 1996). However, for images of FAS, we used a newly developed program, TiltPicker (Voss et al., 2009), that allows rapid particle selection for random conical tilt. This software includes an integrated automatic particle selection as well as the ability to manually remove false positives and add particle pairs that may have been missed. The program reads micrographs in a variety of formats and writes angles relating the image pairs and the particle coordinates into a single concatenated text document that is easily divided into separate files (comparable to those output by Web) that can be read by SPIDER.

3.4. Particle image preprocessing

For direct comparison with particle images from scanned micrographs, particles windowed from CCD frames were interpolated to 4.2 Å per pixel. Thus, particles from all datasets were contained within 80-pixel boxes, about 50% larger than the diameter of FAS. This box size is appropriate based on the assumption that particle centers were selected within ~25% error and affords an appropriate translation range to bring particles into register. Larger box sizes increase the chance of including nearby particles and unnecessarily increase computation required for operations on particle images. To allow direct comparison of windowed particle images between and within datasets, pixel intensities were similarly scaled according to the density distribution of the background surrounding each particle (SPIDER's CE FIT command; Lambert *et al.*, 1994). Particles were then band-pass Butterworth filtered to retain information between 21 and 330 Å (Fig. 9.2A).

3.5. Preliminary image alignment

EM images contain high-resolution information that can be extracted by registration and averaging of individual particle images that are in the same orientation and conformation (Frank *et al.*, 1978). FAS particle images were aligned without the use of a reference (SPIDER's AP SR command) to avoid bias toward conformations that may not be representative of the data (Penczek *et al.*, 1992). Alignment of FAS images was aided by the fact that, like many other macromolecules, FAS adsorbs in a preferred orientation to the amorphous carbon film used to support stained specimens. Two-dimensional projections of FAS in its preferred orientation have easily recognizable features and, compared to more globular structures of similar mass like RNA polymerase, the protruding domains of FAS facilitate particle image alignment. Therefore, averages from preliminary alignment provide a recognizable view of FAS, albeit complicated primarily by conformational heterogeneity evident as a blurring of features, particularly in the extremities of the lower portion of the structure (Fig. 9.2B, top). In an optimal scenario, alignment brings all particle images in a given conformation or orientation into a common register. In practice, when faced with a heterogeneous set of images, the alignment outcome for any given conformation or orientation is equivocal, but affords a starting point for classification of images into more homogeneous groups.

3.6. Preliminary image classification

Once particle images are aligned and averaged, the variance can be examined for clues about differences among individual particle images, including changes in conformation or orientation. Heterogeneity in the aligned

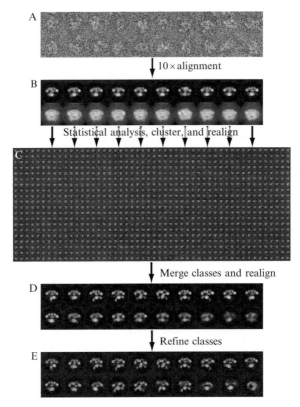

Figure 9.2 (A) Twenty randomly selected negatively stained FAS particles. (B) Averages (top row) and variances (bottom row) of 13,847 particles after 10 independent reference-free alignment attempts. (C) 100 classes were generated from each alignment and then particles were realigned within each class. (D) Classes were merged by alignment and classification of averages. Then particle images were realigned within each merged class. (E) After 20 iterations of supervised classification and reference-free realignment the details in many classes have improved.

particle images appears as pockets of high local variance in the upper and lower portions of the FAS structure that corresponds with blurring of those features in the average (Fig. 9.2B, bottom). Furthermore, correlated variations among individual images can be examined through multivariate statistical analysis to potentially identify patterns of changes in density that can be used to cluster images. While there are several statistical approaches to decompose the variance in a dataset, we typically use correspondence analysis to identify correlated densities in our images (Frank and van Heel, 1982; van Heel and Frank, 1981).

The results of statistical analysis are examined in the form of eigenimages (Fig. 9.3A) that reveal independent (orthogonal) patterns of image variability

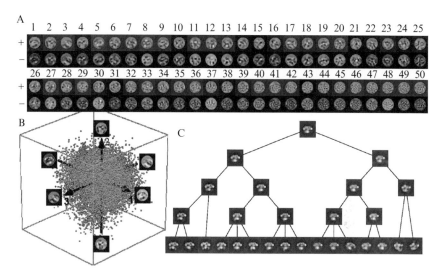

Figure 9.3 Correspondence analysis and hierarchical clustering. (A) Positive and negative eigenimages for the first 50 factors. The first 30 factors capture density variations related to FAS structure, whereas the last 20 factors represent mostly noise. (B) Plot of particle coordinates for the first three factors. Positive and negative eigenimages are shown at the ends of each axis. (C) Dendrogram from hierarchical ascendant clustering of 13,847 particles illustrates relationships between particle clusters and allows an interactive decision about the number of appropriate classes. Several classes of misaligned particles are revealed at lower levels of the dendrogram.

(Bretaudiere and Frank, 1986; Elad *et al.*, 2008). Each particle image can be reconstituted by a linear combination of eigenimages weighted according to that particle's coordinate along each eigenvector. A plot of the images' coordinates reveals the distribution of differences between the particles (Fig. 9.3B). For images of FAS, the particle coordinates are distributed as a continuous cloud, whereas distinct views or conformational states can cause particle images to form discrete groupings in factor space. Once the variation between images is represented in factor space, images are clustered by proximity according to factors that represent meaningful differences. Typically, the first several eigenvectors separate particles according to real differences in structure or orientation, whereas later eigenvectors (typically >20) capture variations in noise and are ignored as a basis for classifying images.

There are several approaches to clustering images including nonlinear mapping (Radermacher and Frank, 1985) and self-organizing maps (Pascual *et al.*, 2000). The method used for analyzing FAS, and perhaps the most conceptually simple, is hierarchical ascendant classification (Frank *et al.*, 1988). This algorithm begins by clustering images that are closest in factor space and then successively combines neighboring clusters (Ward, 1963).

An advantage of this approach is that through SPIDER's Web interface, a dendrogram that relates the classes can be interactively examined to assess the homogeneity of each class (Fig. 9.3C). Therefore, the number of classes needed to adequately describe the data does not need to be decided beforehand as with K-means clustering, another popular and much faster method. Another advantage of hierarchical clustering is reproducibility. Given the same images with the same number of factors, the resulting dendrogram does not change, as opposed to methods that initiate from random seeds and therefore result in variable clustering at each attempt. For practical purposes, because both statistical analysis and clustering are computationally intensive, and because images were sampled well above the achievable resolution, particle images were binned two- to fourfold to reduce the time required for computations, possibly providing a further benefit by improving the signal-to-noise ratio. After clustering, class averages that reflect differences in the particle images were generated from full-sized images to reveal conformational changes of FAS.

3.7. Realignment of particle images within classes

Statistical analysis and clustering of particle images separate similar images from images that represent distinct projections of a structure. Once segregated into different views, alignment parameters can be refined because reference-free alignment performs best when applied to a homogeneous set of images. Therefore, after sorting FAS particles into more homogeneous groups, particle images within each class were realigned (reference-free) to yield averages with improved resolution. As implemented in SPIDER, reference-free alignment of particle images is biased by the image randomly selected to start the alignment process, which exerts a disproportionate influence on the alignment of later images and on the final outcome (Penczek et al., 1992). Because the initial alignment was performed on the entire heterogeneous set of FAS particle images, the entire alignment-classification-realignment scheme was repeated in 10 independent attempts to evaluate multiple possible outcomes, reflecting the stability of the alignment and clustering results (Fig. 9.2C).

3.8. Merging oversampled classes

The 10 independent rounds of alignment–classification–realignment typically produce many similar averages and oversample many outcomes. By aligning and classifying averages, redundant classes were objectively merged while retaining dissimilar classes. After merging, each particle is represented in 10 different classes or, for consistently aligned and classified particles, 10 times in a single class. Particles in each merged class were again realigned (Fig. 9.2D) such that redundant membership weights a particle's influence on the final alignment outcome according to the reproducibility of its match to a particular class. This "weighted" classification and alignment is less sophisticated than the computationally intensive likelihood approaches (see Chapters 10 and 11 in Volume 482) that formally assign weights to each particle over all classes and alignment parameters (Scheres et al., 2005; Sigworth, 1998). However, we have found it to generate reliable results because of its emphasis on testing the stability of alignment and clustering.

3.9. Refinement of the 2D class averages

When dealing with a heterogeneous set of particle images representing a mixture of conformations, compositions, and/or orientations, initial alignment and clustering do not provide an optimal solution. Therefore, the resulting classes were refined by iterating classification and reference-free alignment to separate particles into progressively more homogeneous and distinct groups (Fig. 9.2E). The classification steps were accomplished by multireference alignment that, unlike the unsupervised clustering described above, simultaneously rotates, translates, and mirrors images. After sorting each particle according to its best correlated class average, reference-free alignment was used to realign particles within each class. Twenty iterations of alternating classification and alignment were performed, and after the final iteration, the particle images within each class were realigned to their average.

This iterative classification–alignment procedure has been implemented in a way that facilitates visual comparison of class averages by orienting them at the end of each iteration against a reference image. Additionally, for compositionally heterogeneous samples, classes containing images of smaller subcomplexes may be present that require class-specific radii for alignment. Individual radii are calculated for each class by thresholding the average and 1D projection onto the x- and y-axes. This approach does not require the average to be centered or globular as would a radius calculated from a 1D rotational average. As a useful byproduct, the radius calculations provide the x- and y-shifts to center each average, which is critical during supervised classification because only features of the references within the alignment radius are considered. With these intermediate steps, the approach that we used to address conformational variability in metazoan FAS has been applicable to other macromolecules that exhibit conformational or compositional heterogeneity, such as Mediator (Cai et al., 2010). This procedure for iterative classification and alignment has been incorporated into the Appion pipeline (Lander et al., 2009).

The class averages of FAS that result from the iterative classification–alignment procedure reveal structures that could be related by orientational changes and/or conformational variations. To distinguish between these possibilities, class averages were identified in which detailed features of the upper portion of FAS remain essentially unchanged but those in the lower portion have dramatically moved, clearly indicative of conformation rather than orientation changes. On the other hand, averages of particles in slightly rocked orientations can be easily recognized due to overlap of domains in the upper and lower portions of the structure; however, a wide distribution of views was not observed. Preferential adsorption orientation is an advantage when evaluating conformational heterogeneity by reducing the total number of images that must be collected and analyzed. As a result of adopting a preferred orientation, 2D analysis can separate this modest set of FAS particle images into a manageable number of classes that upon visual inspection predominantly reveal conformational changes.

3.10. Focused alignment and classification

Averaging inadequately aligned images of macromolecules displaying variability in composition, conformation, and/or orientation, results in regions of reduced or absent density. In the case of structural conformers, successful analysis requires separation of particles into a discrete number of stable states. For a flexible macromolecule that exhibits a continuum of conformations (dubbed "fleximers" to distinguish from discrete "conformers"), the goal of alignment and classification is to describe the range and distribution of domain mobility (Burgess *et al.*, 2004b). Examination of the 2D class averages of FAS indicates that the relative angle between the upper and lower portions of the structure flexes over a broad continuum (Fig. 9.4A and B). Such dramatic motions present a difficult problem for alignment of particle images—which portion of the structure should be aligned? In any given class, conformational flexibility requires a compromise with both portions of the structure being slightly out of register.

Domain motions can be defined by fixing in place one portion of the structure and then examining the mobile portion relative to this fixed frame of reference. Once FAS particles are classified into relatively homogeneous groups, the comparatively less variable upper portion of the structure can be isolated and aligned while excluding the more mobile lower portion of the structure. This strategy and rationale was adapted from methods described for studying fleximers of dynein and myosin (Burgess *et al.*, 2003, 2004a,b; Walker *et al.*, 2000).

Focusing alignment on a particular portion of the structure requires a mask to exclude other regions of the images (Fig. 9.4C and D). Masking multiple classes of FAS particles was facilitated by the fact that averages were oriented to a common reference during the preceding iterative classification-alignment (see above). The masked particles (Fig. 9.4E and F) were aligned (reference-free), and the resulting alignment parameters were applied to the original unmasked particles. As observed for images of FAS, focused alignment improves definition of features within the upper portion

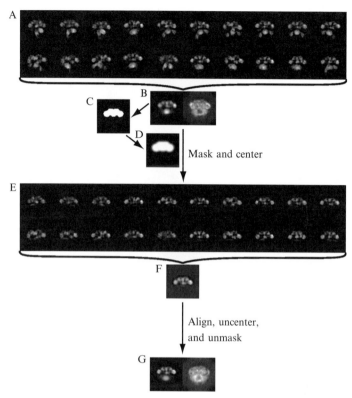

Figure 9.4 Focused alignment. (A) Class averages. (B) Average and variance of particle images from the classes shown in (A). (C) A shape-fitting mask for FAS was constructed by thresholding the average into a binary image (SPIDER's TH M command). The thresholded average was then segmented to isolate the upper portion. (D) Before masking particles the edge of the mask was blurred by box-convolution (SPIDER's BC command) and then dilated by the half-width of the filter by repeating the thresholding and filtration. (E) Each class captures a different conformation and consequently the region to be masked was slightly out of alignment. To remedy this situation, before masking the particles in each class, the mask was applied to class averages, masked averages were aligned, and the alignment parameters for each class average were summed with parameters for particles in the class. The particle images were transformed by the combined parameters, the mask was applied, and finally the masked images were centered. (F) Average of all masked particles. (G) Masked particle images were aligned and, in the absence of a command that restricts rotational search, those that exceed a given rotation ($\sim 20°$) or translation ($\sim 5\%$ of image dimension) were reset to their initial masked position. After reference-free alignment of the masked particles, the alignment parameters were summed with those prior to masking, accounting for and reversing the centering operation. The cumulative transformation was applied to the original unmasked particles. Average and variance images indicate improved alignment of features in the upper portion at the expense of blurring in the lower portion.

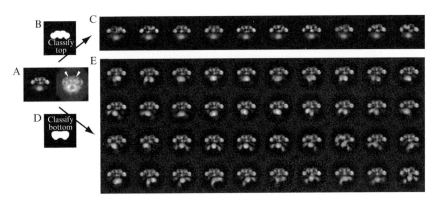

Figure 9.5 Focused classification. (A) Average and variance after alignment of the upper portion. Two regions of high variance in the center of the upper portion are indicated by arrows. (B) Mask that focuses statistical analysis on the upper portion of the structure. (C) Averages from classification of the upper portion reveal a continuum of conformational rearrangement. (D) Mask that focuses statistical analysis on the lower portion of the structure. (E) Compared to classification of the upper portion, almost twice as many factors and four times more classes were necessary to describe the motion of domains in the lower portion of the structure.

of the structure, while more variable features excluded by the mask become blurred (Fig. 9.4G).

Following alignment of the upper portion, two prominent variations were observed within this region (Fig. 9.5A). To better understand this source of conformational variation, statistical analysis and classification was focused on the upper portion (Fig. 9.5B). Eigenimages indicate that the two variations are anticorrelated, and class averages confirm that an asymmetric opening forms in one half of the upper portion of the structure while closing in the other half (Fig. 9.5C). The class averages can easily be arranged according to their coordinate along the first, dominant factor into a sequence that captures motion of domains in the upper portion of the structure. This domain rearrangement reflects not just in-plane motion but suggests out-of-plane domain rotations that cause changes in overlapping densities that are difficult to interpret in averages of 2D projection images.

With the upper portion of the structure aligned, motions of the excluded portion can be analyzed. The average and variance images indicate extensive mobility in the lower portion, and a mask was designed to classify over this entire region (Fig. 9.5D). The resulting class averages (Fig. 9.5E) reveal a dramatic swinging motion as well as a conformational change that obscures the "legs." This latter movement, like the motion delineated by classification of the upper portion, suggests out-of-plane domain rotation. Because rearrangements that result in overlapping densities are difficult to describe from 2D averages, we sought a 3D approach to better comprehend the movement of domains in the upper and lower portions of FAS.

3.11. Random conical tilt: A 3D interpretation of FAS domain movements

Although class averages provide significant insight, a more complete structural understanding of macromolecular conformational variability is afforded by 3D structures of the fleximers identified by 2D analysis. Three-dimensional structures that correspond to conformations apparent in 2D averages were calculated by collecting tilted-pair images and then using the random conical tilt reconstruction method (Radermacher *et al.*, 1987). First, Euler angles were calculated for back-projection of tilted particles by combining the rotational alignment of untilted particle images with the angles of tilt and tilt axes that relate each pair of micrographs as defined during particle selection (see above). Despite its simplicity, caution is advised at this step to avoid producing structures with incorrect or mixed handedness due to the orientation in which the micrographs are digitized. The error in this instance results from the fact that tilt axes are measured between $\pm 90°$ and not over a full $360°$. In our computational routines, ambiguity in the axis angles is resolved by determining the edge of the tilted micrograph that is farthest from focus.

Once projection angles for the tilted particles have been determined, an initial 3D reconstruction for each class was obtained (SPIDER's BP CG command). Since the tilted particle images were not yet centered, six cycles of translational alignment to corresponding projections of the 3D reconstruction were performed. After each cycle of shift refinement, a new 3D structure is calculated (SPIDER's BP 3F command). This procedure is repeated for each class of particles, resulting in a 3D structure that corresponds to each 2D average (Fig. 9.6A). The EM density maps have resolutions of ~ 35 Å ($FSC_{0.5}$) reflecting the limited number of particles (500–1000) that generate the 3D reconstruction for each class, the high defocus of tilted images that lack CTF correction, and compression artifacts from negatively stained specimen preparation. A more useful measure of reliability is comparison with a known high-resolution structure.

The FAS crystal structure (Maier *et al.*, 2008) was fitted into each density map as a single unit. Although the comparatively symmetrical conformation captured by the crystal structure can differ dramatically from the conformation apparent in the EM maps, the relative locations of each domain can be clearly recognized by shape and size, indicating that the 3D reconstructions have sufficient detail to give an accurate impression of domain position. To interpret the conformation captured by each class, after fitting the entire FAS crystal structure into the 3D reconstruction for each class, multidomain structural units (i.e., KR-SD, DH_2, ER_2, $[KS-MAT]_2$) were extracted from

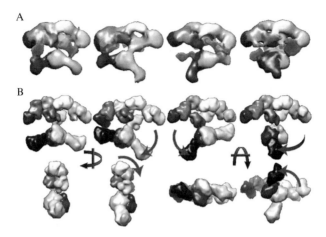

Figure 9.6 (A) Random conical tilt reconstructions of FAS. (B) For a more complete understanding of conformational changes, domains from the crystal structure were fitted into the density maps, and filtered to the approximate resolution of the EM reconstructions. The domains in the lower portion of the structure can be clearly seen to swing over a 50° range and twist by 90° (also shown as bottom views in lower right panels). Domains in the upper portion of the FAS structure can roll by about 30° in either direction (only clockwise rotation is shown as a side view in lower left panels).

the crystal structure and their positions adjusted locally and stored in a single multimodel PDB-format file (Fig. 9.6B). While excellent software exists for fitting high-resolution structures into EM densities (Volkmann and Hanein, 1999; Wriggers *et al.*, 1999), these do not readily consider constraints imposed by flexible linker sequences that tether loosely connected domains. Because conformational changes between the structures can be easily modeled by simple domain movements, structural units were manually adjusted aided by the local "fit-in-map" tool implemented in Chimera (Goddard *et al.*, 2007) so that appropriate distance constraints were maintained given interdomain linkages. To prevent overlap of adjacent domains, occupied densities were removed using the "volume eraser" tool. Chimera also has the capability to use fitted atomic structures to segment density maps that may be used for this purpose.

Once the multidomain structural units were fitted into the EM maps, the angles and axes of domain motions were determined. Previously, this calculation was done manually; however, recent versions of Chimera include the "measure rotation" command to simplify quantitation of domain movements. In the upper portion of the structure, it becomes clear that the DH and ER dimers each roll ±30° to generate an asymmetric opening. Domains in the lower portion of the structure are capable of dramatic swinging motions over a 50° range. The most striking conformation captured by the EM analysis turns domains in the lower portion perpendicular to the upper portion and indicate clearly that the two subunits of the FAS dimer are capable of twisting around one another, likely by a full 180°.

3.12. Quantitative analysis of conformational distribution of various active-site mutants

Given the extensive flexibility of FAS, we sought to understand the catalytic relevance of the observed domain motions. Earlier EM analysis showed that different FAS mutants in the presence of substrates yielded 2D averages that reflect a change in conformation (Asturias *et al.*, 2005). Based on this observation, we reasoned that FAS mutants defective in a particular catalytic activity would, upon addition of substrates, show an altered distribution of conformations related to catalytic arrest.

Initially, we analyzed the conformation distributions independently for several FAS mutants prepared in the presence and absence of substrates. However, comparing the averages from each dataset required subjective decisions due to differences in how particle images clustered. To make a more direct comparison, particle images from all of the FAS preparations were merged into a single large dataset and segregated into the same conformational bins through the alignment and classification scheme, described above (Fig. 9.7A). Then, the number of particles from each FAS preparation assigned to a particular conformation were determined and directly compared (Fig. 9.7B). In the presence of substrates, conformations of FAS with the upper domains rolled into an asymmetric arrangement and lower domains in-plane with the upper portion become dominant, indicating that these conformations are directly related to catalysis. In the absence of substrates, the upper domains are more likely to be symmetrically arranged or the lower domains turned perpendicular to the upper portion, indicating that these conformations represent noncatalytic intermediates.

4. EM AND FAS: A VERSATILE TOOL FOR A FLEXIBLE MACROMOLECULE

The strategy that we employed to study the conformational flexibility of FAS by single-particle EM was based on unsupervised (reference-free) alignment and clustering of particle images and required no *a priori* knowledge of the FAS structure or any assumptions to sort particles according to their conformation. Gradual subdivision of particle images into progressively more classes revealed intermediate domain positions that made possible a description of the continuous flexibility of FAS. The 2D analysis of conformational heterogeneity was extended to 3D structures using tilted

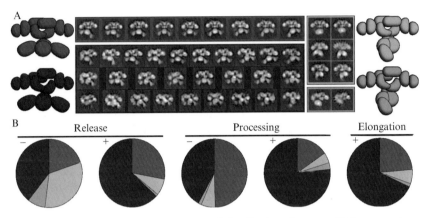

Figure 9.7 Comparison of conformational distributions between FAS preparations. (A) After particles were aligned and classified, class averages were sorted into categories, according to symmetry (red) or asymmetry (blue) in the upper portion of the structure, and according to lower domains in-plane (bright) or perpendicular (faded) with the upper portion. (B) The conformational redistribution of particles for each mutant FAS in the presence (+) or absence (−) of substrates suggests that catalysis is facilitated by conformations with the lower portion in-plane with an asymmetric upper portion (bright blue). The mutations examined resulted in defective acyl-chain release, processing, or elongation. (See Color Insert.)

images and the random conical tilt method (Radermacher *et al.*, 1987). This approach sidesteps reference-based projection matching and common–lines methods and provides a bias-free 3D description of macromolecular conformations. Finally, a molecular interpretation of 2D and 3D EM structures was generated by considering the published X-ray crystallography structure of a static FAS conformation.

4.1. Single-particle EM as a method for studying macromolecular conformation and flexibility

More than a single structure is required to understand how a macromolecular machine functions. Knowledge about structural rearrangements related to association, dissociation, or modification of components is essential to arrive at a complete mechanistic understanding. To facilitate structural analysis, some macromolecular complexes can be induced to populate a single predominant conformation in response to regulators or substrates, as in pH or protease dependent virus maturation (Yu *et al.*, 2008), nucleotide bound states of a group II chaperonin (Zhang *et al.*, 2010), Mediator upon polymerase binding (Cai *et al.*, 2009), and ribosome ratcheting with eEF2/EF-G (Frank and Agrawal, 2000; Taylor *et al.*, 2007). By contrast, attempts to induce FAS to adopt particular conformations by addition of substrates to active-site mutants did not result in population of a single state but only succeeded in perturbing the distribution of particles amongst a conformational continuum (Fig. 9.7B).

Discrete conformations or biochemical states of a macromolecule that adopts only a small number of well-defined conformations become evident when separation of images beyond a certain number of groups results in redundant class averages. In contrast, macromolecular flexibility gives rise to continuous motions, and the goal of EM analysis becomes describing the distribution of particles amongst different fleximers and the range of motions sampled by the complex. In our study of FAS, we based our conclusions on relative changes in conformational distribution that resulted from altering catalytic activities by mutation and addition of substrates. In this way, we identified conformations that become more or less abundant to facilitate particular catalytic events and minimized bias related to perturbations arising from the specimen preparation method (e.g., particle adsorption and stain preservation). Fitting domains from the FAS crystal structure into the low-resolution 3D EM structures and examination of catalytic contacts facilitated by particular structures allowed us to describe how large-scale twisting, swinging, and rolling motions were related to specific catalytic events that occur during fatty acid synthesis (Brignole et al., 2009; Fig. 9.8).

Our approach to analysis of conformational distribution builds upon earlier investigations of dynamic complexes that pioneered the use of EM as a technique for quantifying macromolecular heterogeneity. Notably, Burgess and colleagues used EM to determine the conformational

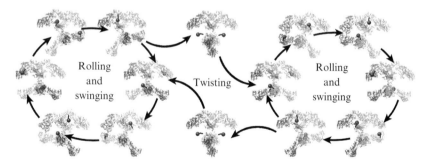

Figure 9.8 The catalytic cycle of FAS is mediated by conformational flexibility: rolling motions of domains in the upper portion, swinging motions of domains in the lower portion, and twisting about the linkers that connect the upper and lower portions. Sequential interactions between each catalytic domain and the carrier domain (spheres) with its covalently bound cargo (black stick) are illustrated. Both reaction chambers are depicted synthesizing fatty acids asynchronously such that the acyl chain is processed in one chamber but elongated in the other. (For interpretation of the references to color in this figure legend, the reader is referred to the Web version of this chapter.)

distribution between structural elements of dynein and myosin, and perturbed this distribution by addition of nucleotide analogs that mimic particular catalytic states (Burgess *et al.*, 2003, 2004a,b; Walker *et al.*, 2000). Heymann *et al.* (2003) have also examined a continuous population of maturing herpes virus capsids. More recently, Southworth and Agard (2008) have identified an equilibrium between open and closed conformers of Hsp90 that can be altered by nucleotide and Cai *et al.* (2010) have used EM to quantify motions in the Mediator head module that can be perturbed by binding of TBP. In a unique example of correlating biochemical kinetics with quantitation of particle conformations, Mulder *et al.* (2010) examined complex mixtures of 30S ribosome assembly intermediates by EM as a time resolved approach.

The power of single-particle EM to provide information about different macromolecular conformations is also demonstrated by new image analysis approaches that make possible the assessment of macromolecular heterogeneity in 3D, ranging from methods to identify structural variations such as local resolution calculation (blocres command in Bsoft; Heymann and Belnap, 2007) or 3D variance analysis (Zhang *et al.*, 2008) to methods that generate several 3D structures from a single dataset by classification of particles within projection groups (Elmlund *et al.*, 2009; Fu *et al.*, 2007) or multimodel maximum likelihood 3D refinement (Scheres *et al.*, 2007). In the past few years, these advances in EM image processing have been used to identify multiple conformational or biochemical states (i.e., subunit occupancy) present in a single specimen preparation, including, for example: SV40 large tumor antigen (Cuesta *et al.*, 2010), ribosome with partial occupancy of EF-G (Gao *et al.*, 2004), RNA polymerase II clamp opening (De Carlo *et al.*, 2003; Kostek *et al.*, 2006), TFIID with partial TBP occupancy (Elmlund *et al.*, 2009; Grob *et al.*, 2006), and RSC chromatin remodeler with and without nucleosome (Chaban *et al.*, 2008).

Finally, continuing improvements in EM image formation and acquisition, in particular, phase plates (Danev *et al.*, 2009; Gamm *et al.*, 2010; see Chapter 14 in Volume 481) and direct detectors (Milazzo *et al.*, 2010), promise to enhance contrast and signal, respectively, so that the approach we employed for visualizing extensive conformational heterogeneity of the relatively small FAS particle in stain might soon become feasible in cryo-EM where macromolecular structure would be preserved in a more physiological state.

ACKNOWLEDGMENTS

The work described here was supported by the US National Institutes of Health through grants R01 DK16073 to S. Smith and F. J. A., and F32 DK080622 to E. J. B.

REFERENCES

Adrian, M., *et al.* (1984). Cryo-electron microscopy of viruses. *Nature* **308**, 32–36.

Adrian, M., *et al.* (1998). Cryo-negative staining. *Micron* **29**, 145–160.

Aebi, U., and Pollard, T. D. (1987). A glow discharge unit to render electron microscope grids and other surfaces hydrophilic. *J. Electron Microsc. Tech.* **7**, 29–33.

Asturias, F. J., *et al.* (2004). Electron microscopic analysis of the RSC chromatin remodeling complex. *Methods Enzymol.* **376**, 48–62.

Asturias, F. J., *et al.* (2005). Structure and molecular organization of mammalian fatty acid synthase. *Nat. Struct. Mol. Biol.* **12**, 225–232.

Boisset, N., *et al.* (1990). Three-dimensional reconstruction of native *Androctonus australis* hemocyanin. *J. Mol. Biol.* **216**, 743–760.

Boisset, N., *et al.* (1993). Three-dimensional immunoelectron microscopy of scorpion hemocyanin labeled with a monoclonal Fab fragment. *J. Struct. Biol.* **111**, 234–244.

Bretaudiere, J. P., and Frank, J. (1986). Reconstitution of molecule images analysed by correspondence analysis: A tool for structural interpretation. *J. Microsc.* **144**, 1–14.

Brignole, E. J., *et al.* (2009). Conformational flexibility of metazoan fatty acid synthase enables catalysis. *Nat. Struct. Mol. Biol.* **16**, 190–197.

Brink, J., *et al.* (2002). Quaternary structure of human fatty acid synthase by electron cryomicroscopy. *Proc. Natl. Acad. Sci. USA* **99**, 138–143.

Brink, J., *et al.* (2004). Experimental verification of conformational variation of human fatty acid synthase as predicted by normal mode analysis. *Structure* **12**, 185–191.

Burgess, S. A., *et al.* (2003). Dynein structure and power stroke. *Nature* **421**, 715–718.

Burgess, S. A., *et al.* (2004a). The structure of dynein-c by negative stain electron microscopy. *J. Struct. Biol.* **146**, 205–216.

Burgess, S. A., *et al.* (2004b). Use of negative stain and single-particle image processing to explore dynamic properties of flexible macromolecules. *J. Struct. Biol.* **147**, 247–258.

Cai, G., *et al.* (2009). Mediator structural conservation and implications for the regulation mechanism. *Structure* **17**, 559–567.

Cai, G., *et al.* (2010). Mediator head module structure and functional interactions. *Nat. Struct. Mol. Biol.* **17**, 273–279.

Chaban, Y., *et al.* (2008). Structure of a RSC–nucleosome complex and insights into chromatin remodeling. *Nat. Struct. Mol. Biol.* **15**, 1272–1277.

Chen, J. Z., *et al.* (2009). Molecular interactions in rotavirus assembly and uncoating seen by high-resolution cryo-EM. *Proc. Natl. Acad. Sci. USA* **106**, 10644–10648.

Cheng, Y., and Walz, T. (2009). The advent of near-atomic resolution in single-particle electron microscopy. *Annu. Rev. Biochem.* **78**, 723–742.

Cheng, Y., *et al.* (2006). Single particle reconstructions of the transferrin–transferrin receptor complex obtained with different specimen preparation techniques. *J. Mol. Biol.* **355**, 1048–1065.

Coleman, R. A., and Lee, D. P. (2004). Enzymes of triacylglycerol synthesis and their regulation. *Prog. Lipid Res.* **43**, 134–176.

Cong, Y., *et al.* (2010). 4.0-A resolution cryo-EM structure of the mammalian chaperonin TRiC/CCT reveals its unique subunit arrangement. *Proc. Natl. Acad. Sci. USA* **107**, 4967–4972.

Cuesta, I., *et al.* (2010). Conformational rearrangements of SV40 large T antigen during early replication events. *J. Mol. Biol.* **397**, 1276–1286.

Danev, R., *et al.* (2009). Practical factors affecting the performance of a thin-film phase plate for transmission electron microscopy. *Ultramicroscopy* **109**, 312–325.

De Carlo, S., *et al.* (2003). Cryo-negative staining reveals conformational flexibility within yeast RNA polymerase I. *J. Mol. Biol.* **329**, 891–902.

Elad, N., *et al.* (2008). Detection and separation of heterogeneity in molecular complexes by statistical analysis of their two-dimensional projections. *J. Struct. Biol.* **162,** 108–120.

Elmlund, H., *et al.* (2009). Cryo-EM reveals promoter DNA binding and conformational flexibility of the general transcription factor TFIID. *Structure* **17,** 1442–1452.

Frank, J., and Agrawal, R. K. (2000). A ratchet-like inter-subunit reorganization of the ribosome during translocation. *Nature* **406,** 318–322.

Frank, J., and van Heel, M. (1982). Correspondence analysis of aligned images of biological particles. *J. Mol. Biol.* **161,** 134–137.

Frank, J., *et al.* (1978). Reconstruction of glutamine synthetase using computer averaging. *Ultramicroscopy* **3,** 283–290.

Frank, J., *et al.* (1988). Classification of images of biomolecular assemblies: A study of ribosomes and ribosomal subunits of *Escherichia coli. J. Microsc.* **150,** 99–115.

Frank, J., *et al.* (1996). SPIDER and WEB: Processing and visualization of images in 3D electron microscopy and related fields. *J. Struct. Biol.* **116,** 190–199.

Fu, J., *et al.* (2007). Unsupervised classification of single particles by cluster tracking in multi-dimensional space. *J. Struct. Biol.* **157,** 226–239.

Gamm, B., *et al.* (2010). Object wave reconstruction by phase-plate transmission electron microscopy. *Ultramicroscopy* **110,** 807–814.

Gao, H., *et al.* (2004). Dynamics of EF-G interaction with the ribosome explored by classification of a heterogeneous cryo-EM dataset. *J. Struct. Biol.* **147,** 283–290.

Goddard, T. D., *et al.* (2007). Visualizing density maps with UCSF Chimera. *J. Struct. Biol.* **157,** 281–287.

Golas, M. M., *et al.* (2003). Molecular architecture of the multiprotein splicing factor SF3b. *Science* **300,** 980–984.

Grob, P., *et al.* (2006). Cryo-electron microscopy studies of human TFIID: Conformational breathing in the integration of gene regulatory cues. *Structure* **14,** 511–520.

Harris, J. R. (2007). Negative staining of thinly spread biological samples. *Methods Mol. Biol.* **369,** 107–142.

Henderson, R., *et al.* (1990). Model for the structure of bacteriorhodopsin based on high-resolution electron cryo-microscopy. *J. Mol. Biol.* **213,** 899–929.

Heymann, J. B., and Belnap, D. M. (2007). Bsoft: Image processing and molecular modeling for electron microscopy. *J. Struct. Biol.* **157,** 3–18.

Heymann, J. B., *et al.* (2003). Dynamics of herpes simplex virus capsid maturation visualized by time-lapse cryo-electron microscopy. *Nat. Struct. Biol.* **10,** 334–341.

Joshi, A. K., *et al.* (2005). Effect of modification of the length and flexibility of the acyl carrier protein-thioesterase interdomain linker on functionality of the animal fatty acid synthase. *Biochemistry* **44,** 4100–4107.

Kellenberger, E., *et al.* (1982). The wrapping phenomenon in air-dried and negatively stained preparations. *Ultramicroscopy* **9,** 139–150.

Kitamoto, T., *et al.* (1988). Structure of fatty acid synthetase from the Harderian gland of guinea pig. Proteolytic dissection and electron microscopic studies. *J. Mol. Biol.* **203,** 183–195.

Kostek, S. A., *et al.* (2006). Molecular architecture and conformational flexibility of human RNA polymerase II. *Structure* **14,** 1691–1700.

Kuhajda, F. P., *et al.* (1994). Fatty acid synthesis: A potential selective target for antineoplastic therapy. *Proc. Natl. Acad. Sci. USA* **91,** 6379–6383.

Kuhlbrandt, W., *et al.* (1994). Atomic model of plant light-harvesting complex by electron crystallography. *Nature* **367,** 614–621.

Lambert, O., *et al.* (1994). Quaternary structure of *Octopus vulgaris* hemocyanin. Three-dimensional reconstruction from frozen-hydrated specimens and intramolecular location of functional units Ove and Ovb. *J. Mol. Biol.* **238,** 75–87.

Lander, G. C., *et al.* (2009). Appion: An integrated, database-driven pipeline to facilitate EM image processing. *J. Struct. Biol.* **166,** 95–102.

Loftus, T. M., *et al.* (2000). Reduced food intake and body weight in mice treated with fatty acid synthase inhibitors. *Science* **288,** 2379–2381.

Maier, T., *et al.* (2006). Architecture of mammalian fatty acid synthase at 4.5 A resolution. *Science* **311,** 1258–1262.

Maier, T., *et al.* (2008). The crystal structure of a mammalian fatty acid synthase. *Science* **321,** 1315–1322.

Menendez, J. A., and Lupu, R. (2007). Fatty acid synthase and the lipogenic phenotype in cancer pathogenesis. *Nat. Rev. Cancer* **7,** 763–777.

Milazzo, A. C., *et al.* (2010). Characterization of a direct detection device imaging camera for transmission electron microscopy. *Ultramicroscopy* **110,** 744–747.

Ming, D., *et al.* (2002). Domain movements in human fatty acid synthase by quantized elastic deformational model. *Proc. Natl. Acad. Sci. USA* **99,** 7895–7899.

Miyazawa, A., *et al.* (2003). Structure and gating mechanism of the acetylcholine receptor pore. *Nature* **423,** 949–955.

Mulder, A. M., *et al.* (2010). Visualizing Ribosome Biogenesis: Time-Resolved Single Particle EM Reveals Parallel Assembly Pathways for 30S Subunit. (in review).

Ohi, M., *et al.* (2004). Negative staining and image classification—Powerful tools in modern electron microscopy. *Biol. Proced. Online* **6,** 23–34.

Pascual, A., *et al.* (2000). Mapping and fuzzy classification of macromolecular images using self-organizing neural networks. *Ultramicroscopy* **84,** 85–99.

Penczek, P., *et al.* (1992). Three-dimensional reconstruction of single particles embedded in ice. *Ultramicroscopy* **40,** 33–53.

Radermacher, M., and Frank, J. (1985). Use of nonlinear mapping in multivariate image analysis of molecule projections. *Ultramicroscopy* **17,** 117–126.

Radermacher, M., *et al.* (1987). Three-dimensional reconstruction from a single-exposure, random conical tilt series applied to the 50S ribosomal subunit of *Escherichia coli. J. Microsc.* **146,** 113–136.

Radermacher, M., *et al.* (1994). Cryo-electron microscopy and three-dimensional reconstruction of the calcium release channel/ryanodine receptor from skeletal muscle. *J. Cell Biol.* **127,** 411–423.

Rath, B. K., and Frank, J. (2004). Fast automatic particle picking from cryo-electron micrographs using a locally normalized cross-correlation function: A case study. *J. Struct. Biol.* **145,** 84–90.

Resh, M. D. (2006). Palmitoylation of ligands, receptors, and intracellular signaling molecules. *Sci. STKE,* re14.

Ruiz, T., *et al.* (2001). The first three-dimensional structure of phosphofructokinase from *Saccharomyces cerevisiae* determined by electron microscopy of single particles. *J. Struct. Biol.* **136,** 167–180.

Scheres, S. H., *et al.* (2005). Maximum-likelihood multi-reference refinement for electron microscopy images. *J. Mol. Biol.* **348,** 139–149.

Scheres, S. H., *et al.* (2007). Disentangling conformational states of macromolecules in 3D-EM through likelihood optimization. *Nat. Methods* **4,** 27–29.

Sherman, M. B., *et al.* (1981). On the negative staining of the protein crystal structure. *Ultramicroscopy* **7,** 131–138.

Sigworth, F. J. (1998). A maximum-likelihood approach to single-particle image refinement. *J. Struct. Biol.* **122,** 328–339.

Smith, S., *et al.* (2003). Structural and functional organization of the animal fatty acid synthase. *Prog. Lipid Res.* **42,** 289–317.

Southworth, D. R., and Agard, D. A. (2008). Species-dependent ensembles of conserved conformational states define the Hsp90 chaperone ATPase cycle. *Mol. Cell* **32,** 631–640.

Steven, A. C., and Navia, M. A. (1980). Fidelity of structure representation in electron micrographs of negatively stained protein molecules. *Proc. Natl. Acad. Sci. USA* **77**, 4721–4725.

Steven, A. C., and Navia, M. A. (1982). Specificity of stain distribution in electron micrographs of protein molecules contrasted with uranyl acetate. *J. Microsc.* **128**, 145–155.

Suloway, C., et al. (2005). Automated molecular microscopy: The new Leginon system. *J. Struct. Biol.* **151**, 41–60.

Swinnen, J. V., et al. (2003). Fatty acid synthase drives the synthesis of phospholipids partitioning into detergent-resistant membrane microdomains. *Biochem. Biophys. Res. Commun.* **302**, 898–903.

Taylor, K. A., and Glaeser, R. M. (1974). Electron diffraction of frozen, hydrated protein crystals. *Science* **186**, 1036–1037.

Taylor, D. J., et al. (2007). Structures of modified eEF2·80S ribosome complexes reveal the role of GTP hydrolysis in translocation. *EMBO J.* **26**, 2421–2431.

Tischendorf, G. W., et al. (1974). Determination of the location of proteins L14, L17, L18, L19, L22, L23 on the surface of the 5oS ribosomal subunit of *Escherichia coli* by immune electron microscopy. *Mol. Gen. Genet.* **134**, 187–208.

Unwin, P. N. (1975). Beef liver catalase structure: Interpretation of electron micrographs. *J. Mol. Biol.* **98**, 235–242.

van Heel, M., and Frank, J. (1981). Use of multivariate statistics in analysing the images of biological macromolecules. *Ultramicroscopy* **6**, 187–194.

Vance, J. E., and Vance, D. E. (2004). Phospholipid biosynthesis in mammalian cells. *Biochem. Cell Biol.* **82**, 113–128.

Volkmann, N., and Hanein, D. (1999). Quantitative fitting of atomic models into observed densities derived by electron microscopy. *J. Struct. Biol.* **125**, 176–184.

Voss, N. R., et al. (2009). DoG Picker and TiltPicker: Software tools to facilitate particle selection in single particle electron microscopy. *J. Struct. Biol.* **166**, 205–213.

Walker, M. L., et al. (2000). Two-headed binding of a processive myosin to F-actin. *Nature* **405**, 804–807.

Ward, J. H. (1963). Hierarchical grouping to optimize an objective function. *J. Am. Stat. Assoc.* **58**, 236–244.

Wriggers, W., et al. (1999). Situs: A package for docking crystal structures into low-resolution maps from electron microscopy. *J. Struct. Biol.* **125**, 185–195.

Yonekura, K., et al. (2003). Complete atomic model of the bacterial flagellar filament by electron cryomicroscopy. *Nature* **424**, 643–650.

Yu, I. M., et al. (2008). Structure of the immature dengue virus at low pH primes proteolytic maturation. *Science* **319**, 1834–1837.

Zhang, Y. M., et al. (2006). Inhibiting bacterial fatty acid synthesis. *J. Biol. Chem.* **281**, 17541–17544.

Zhang, W., et al. (2008). Heterogeneity of large macromolecular complexes revealed by 3D cryo-EM variance analysis. *Structure* **16**, 1770–1776.

Zhang, J., et al. (2010). Mechanism of folding chamber closure in a group II chaperonin. *Nature* **463**, 379–383.

CHAPTER TEN

TOMOGRAPHY OF ACTIN CYTOSKELETAL NETWORKS

Dorit Hanein

Contents

Abstract

Structural biology research is increasingly focusing on unraveling structural variations at the micro-, meso-, and macroscale aiming at interpreting dynamic biological processes and pathways. Toward this goal, high-resolution transmission cryoelectron microscopy (cryo-EM) and cryoelectron tomography (cryo-ET) are indispensable, as these provide the ability to determine 3D structures of large, dynamic macromolecular assemblies in their native, fully hydrated state *in situ*. Underlying such analyses is the implicit assumption that specific structural states yield specific cellular outputs. The dependence on this structure–function paradigm is not unique to studies pertaining a particular pathway or biological process but it sets the foundation for all cell biological analyses of macromolecular assemblies. Yet, the paradigm still awaits formal proof. The field of high-resolution electron microscopy (HREM) is in dire need of establishing approaches and technologies to systematic and quantitative determining structure–function correlates in physiologically relevant environment. Here, using the actin cytoskeletal networks as an example, we will provide snapshots of current advances in defining the structures of these highly dynamic networks *in situ*. We will further detail some of the major stumbling blocks on the way to quantitatively correlate the dynamic state to network morphology in the same window of time and space.

Sanford-Burnham Medical Research Institute, La Jolla, California, USA

Methods in Enzymology, Volume 483
ISSN 0076-6879, DOI: 10.1016/S0076-6879(10)83010-1

1. INTRODUCTION

Actin is one of the most abundant proteins in eukaryotic cells, with a concentration of ~ 100–500 μM in nonmuscle cells. Actin is remarkably conserved across species, with all 375 residues being identical for humans and chickens. Monomeric or G-actin polymerizes to form actin filaments (F-actin) which upon assembly into higher order networks provide cells with mechanical support and driving force for movement (for recent reviews see Dominguez, 2009; Lee and Dominguez, 2010; Pollard and Cooper, 2009; Reisler and Egelman, 2007). Approximately 40 X-ray crystallographic structures of actin monomers are now available; however, the atomic structure of its polymeric, helical F-actin form(s) remains elusive (Kabsch et al., 1990; Oda et al., 2009). Structurally, F-actin consists of actin monomers arranged in two protofilaments that twist around each other, while actin subunits face the same direction rending a polar filament. A large body of data suggests that individual actin filaments exhibit structural polymorphism, with a fairly constant axial rise and a variable rotation between adjacent subunits within the filament (Egelman et al., 1982; Galkin et al., 2008; Schmid et al., 2004; Stokes and DeRosier, 1987).

Eukaryotic actins belong to an ATPase superfamily, defined by shared nucleotide-binding motif consisting of two large domains connected by a hinge to which nucleotide binds (Kabsch and Holmes, 1995). Actin monomers within a filament hydrolyze, through an unidentified structural pathway, a single molecule of ATP to ADP. This process is irreversible in the lifetime of the polymer in vitro (Carlier et al., 1988). It was suggested that hydrolysis is the F-actin "timekeeper" in vivo, demarcating newly polymerized regions from older portions of the filament (Blanchoin and Pollard, 2002; Blanchoin et al., 2000). New ATP-actin monomers preferentially add to the "barbed end" (also known as the fast growing plus end) of the filament and depart the filament primarily from the "pointed end" (minus end) in the form of ADP-actin. The "barbed end" "pointed end" terminology is taken from the appearance of myosin decorated actin filaments imaged in the electron microscope that appear as arrowheads with barbed and pointed ends. The temporal asymmetry of actin monomer behavior in respect to the filaments gives rise to a process known as treadmilling.

In vivo, actin filaments perform their function primarily by assembling into higher order cellular structures. In metazoan cells, at least a dozen distinct supramolecular structures F-actin have been identified. Each of these plays unique F-actin and essential functional roles ranging from cellular motility, organelle transport, and pathogen invasion to tumor progression. To mention but a few: immunological synapse, adherent junction, cytokinetic ring, cortactical spectrin-actin, endocytic pits, phagocytic cups, podosomes, invadapodia, filopodia, microvilli, ruffles, and stress fibers (Chhabra and Higgs, 2007).

Many of the highly diverse biological roles critically rely on dynamic properties of these networks, that is, on their treadmilling. Indeed, *in vivo* transition between G- to F-actin is tightly regulated in a spatiotemporal coordinated manner by a large number of actin-binding and regulatory proteins. Over 100 actin-binding proteins (ABP) have been identified so far. These ABPs are used to initiate polymerization, promote, or restrict elongation of the filaments (severing, capping), maintain a pool of actin monomers (sequestering), promote the assembly of actin networks (cross-linking, bundling, membrane attachment), and regulate the assembly turn-over (Lee and Dominguez, 2010; Pollard and Cooper, 2009). Cancer, various neurological disorders, and cardiomyopathies are few of the diseases associated to abnormalities in the assembly or regulation of the actin cytoskeleton.

While recent advancements in high-resolution light microscopy techniques (HRLM), molecular biology and biophysical assays provided wealth of significant new knowledge on actin cytoskeleton dynamics in these networks, comprehensive molecular detailed structural correlates are critically lacking. Here, we will use the sheet-like lamellipodia/lamella networks as an example to exemplify the major existing barriers toward progress in providing such structure correlates.

2. TESTING THE LAMELLA HYPOTHESIS

Much of current biomedical research and development is guided by the structure–function paradigm, which implies that functional outputs can be predicted from structural information and, in reverse direction, that specific functional outputs are mediated by a specific structural configuration. While the paradigm is established as a means to understanding the function of individual molecules, it has remained a mostly at the level of an assumption for larger macromolecular assemblies that drive more complex cellular outputs.

For example, the dendritic network model of directed cell migration suggests that the leading edge is pushed forward by an array of actin filaments with a characteristic-branched morphology defined by an actin nucleation complex, the Arp2/3 complex (Pollard and Borisy, 2003). Much of this model has been derived from bulk biochemical analyses of actin polymerization *in vitro*, from live imaging of single filament assembly outside the cellular context, and from 2D electron microscopy (EM) images of actin networks in detergent extracted, chemically fixed, dehydrated cells. By extrapolation of this data, it is generally assumed that regions with increased branching activity and filament density would be associated with faster protrusion. Although such structure–function relationship has never been shown directly, the interpretation of experiments in the cell migration literature rests on this inference.

Among the body of work that challenges this view are studies of actin dynamics during epithelial cell protrusion. Using quantitative fluorescent speckle microscopy (qFSM), a live-cell imaging modality delivering maps of the rates of filament turnover and motion with submicron and second scale resolution (Danuser and Waterman-Storer, 2006), it was inferred that cell edge propulsion may be driven by two partially overlapping, yet differentially regulated actin networks (Delorme et al., 2007; Gupton et al., 2005; Ponti et al., 2004). Molecular and functional analyses of the relationship between edge movement, assembly, and contraction forces suggested that forward motion of the cell edge at the onset of a protrusion cycle may be initiated by elongation of the actin assembly, the lamella that is independent of Arp2/3 complex and then a second, Arp2/3-mediated network, the lamellopodia (branched network) reinforces cytoskeleton expansion against pressures from plasma membrane and extracellular environment (Ji et al., 2008). This lamella hypothesis is highly controversial (Danuser, 2009; Vallotton and Small, 2009), similar to the dendritic model that still awaits structural validation.

3. CHALLENGES

In the following sections, we used the lamella hypothesis as an example to explain the challenges of dissecting the ultrastructure of large and transient assemblies and of putting these data into a functional context. This is only one example and similar challenges are encountered in many other structure–function studies.

3.1. First stumbling block, sample preparation

The challenge of structural biology today is to establish approaches that allow capturing the state of a macromolecular assembly at high structural detail during a defined cellular output. It is clear that EM should serve as the gold standard for ultrastructural analyses, nevertheless sample preparation protocols that faithfully preserve these delicate, fragile cytoskeletal networks in situ are still being defined.

3.1.1. Cryo-EM

Recent advances of sample preparation and imaging techniques allow us to more reliable correlate live cells morphology to ultrastructure. One of the key advances was that the development of cryo-EM methods. These methods allow biological samples to be imaged frozen in a near-native, physiological environment, thus bypassing the harsh preparative procedures of detergent extraction, chemical fixation, dehydration, metal shadowing, or critical point drying required by traditional EM and opening the way to HR determination of eukaryotic cytoskeleton ultrastructure (Dubochet *et al.*, 1988; see Volume 483, Chapter 3).

3.1.2. Correlative light and EM

More recently, major advances in relating the structure of cytoskeletal networks to cell function *in situ* were made by correlative light and electron microscopy (LM/EM) approaches. The goal of this approach is to capture the state of a macromolecular assembly at high structural detail preferable at defined cellular output (Lucic *et al.*, 2007, 2008; Muller-Reichert *et al.*, 2007; Sartori *et al.*, 2007; see Volume 481, Chapter 13). This approach aims at combining dynamic information derived from live-cell imaging with high-definition structural characterization of the same region using HR cryo-EM imaging. This approach, in principle, allows us to image a region in the cell with a known dynamic history.

3.1.3. Cryo-ET of eukaryotic cells *in toto*

Electron tomography (cryo-ET) of vitrified whole cells has proven to have great potential for imaging cellular architecture in 3D at a resolution of 4–6 nm of intact cells or cell regions with overall thickness not extending ~ 1 μm. Recently, cryo-ET allows direct proof for the existence of cytoskeletal actin-like networks within bacteria, resolving a long standing controversy for the existence of such assemblies in prokaryotic cells (for recent review see (Li and Jensen, 2009)). The ability to generate 3D volumes of assemblies *in situ* allows filaments or networks to be followed, so branching, cross-linking, or overlapping arrangements that might seem similar in 2D projection of a single planar section, can be faithfully recognized and followed within the volume (Ben-Harush *et al.*, 2010; Medalia *et al.*, 2002b; Urban *et al.*, 2010; see Volume 481, Chapter 12). Lastly, to overcome thickness of the preparations, for example, Salje *et al.* (2009) used cryosectioned frozen *E. coli* cells. Imaging these sections resulted in the first direct *in vivo* proof of an *E. coli* cytoskeletal filament assembly. Frozen-hydrated sections were also used to extract 3D information using tomographic imaging (Al-Amoudi *et al.*, 2007; Gruska *et al.*, 2008; Leis *et al.*, 2009; Pierson *et al.*, 2010; see Volume 481 Chapter 8). An alternative approach for thinning of the frozen-hydrated specimen is just being introduced via focused ion beam (FIB) technology (Marko *et al.*, 2006, 2007; Rigort *et al.*, 2010). In this new approach, correlative cryofluorescence microscopy allows to navigate the large cellular volumes and to localize specific cellular targets, followed by FIB thinning, and cryo-ET.

Despite these advances, sample preparation protocols optimized to promote normal cell growth, that are compatible with HR fluorescence imaging, sustain cryogenic plunging, and are amenable for cellular cryo-ET data acquisition, still need to be identified. Furthermore, to ascertain that these cell culture preparations reproducibly preserve the protein content, 3D architecture, and antigenicity of the macromolecular machineries involved in motility, still remain a challenge. The fidelity of the structural studies is critically dependent on the reproducibility in preservation of the characteristic "composition signature" of these sites. At least 20–40 ABPs have been "colocalized" within the lamellopidia or lamella during their formation and disassembly. Their presence or absence within the preparations contribute to the controversy in proving (disproving) the lamella hypothesis.

The elegant work of Medalia *et al.* (2002a) first demonstrated that cryo-ET of intact *Dictyostelium discoideum* cells could reveal the connections of the F-actin network with the plasma membrane as well as possible F-actin branching. Similar views of branched networks were obtained when cells of higher eukaryotes were studied (Ben-Harush *et al.*, 2010; Delorme *et al.*, 2007; Gupton *et al.*, 2005; Medalia *et al.*, 2007). In contrast, cryo-EM and cryo-ET imaging of lamellipodia suggested that F-actins are almost exclusively unbranched (Urban *et al.*, 2010). These studies bring to light the second stumbling block Small *et al.* (2008).

3.2. Second stumbling block, image analysis

The putative spatial overlap of distinct actin networks and the dimensions of the filaments (10 nm) in the volume of the leading edge of a cell will probably be recognizable only by the application of HR image processing tools to the 3D tomographic volumes, supported by sophisticated image segmentation and topology classification algorithms. Visual inspection of projection images is insufficient to test the lamella hypothesis by cryo-ET (see, e.g., Fig. 10.1). Due to the inability to tilt above 70° in all available electron microscopes, reconstruction artifacts such as the missing wedge produced by standard tomographic data collection schemes can significantly hamper the interpretability of the resulting reconstruction. To minimize these artifacts, collecting dual-axis tilt data, thus filling in some of the missing data are essential (Kremer *et al.*, 1996; Penczek *et al.*, 1995). Although challenging on the sample, instrument performance, and volume reconstruction, dual-axis tomography significantly increases the fidelity of filament network determination. We and others have shown the feasibility of this approach to vitrified samples with reconstituted actin networks (Rouiller *et al.*, 2008), and in *Pyrodictium* (Nickell *et al.*, 2003). High-end TEM instruments that provide functionality to rotate the grid by 90° around the beam axis facilitating dual-axis data acquisition schemes are available.

Owing to the extremely low signal-to-noise ratio (~ 0.01), the low contrast, relatively low resolution (> 4 nm), and the crowded nature of the specimen, the electron densities are difficult to interpret. Indeed, identification of key macromolecular motifs is the most time-consuming step in the tomography pipeline. Segmentation of the density into more manageable subvolumes, in conjunction with automatic template matching approaches and

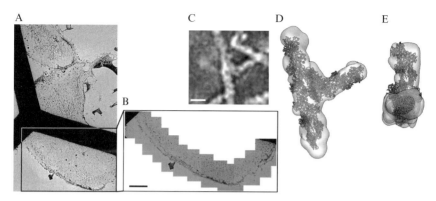

Figure 10.1   Testing the lamella hypothesis. (A) Two-dimensional projection of the leading edge of an eukaryotic cell. The thickness of the cell periphery is ∼300 nm. Grid bar width: 25 μm. (B) Twelve consecutive dual-axis tilt series will be needed to capture the entire cell edge. A far from trivial task both from imaging and image analysis perspectives. Bar 10 μm. (C) A slice through the volume of one of the movies provided as supplementary information in Urban *et al.*, 2010; ncb2044-s5. This movie was used by the authors for demonstrating the absence of Arp2/3 mediated branch junctions in the captured region of the cell. The authors ascertain the absence of branches in the tomogram by visual inspection of the slices. Bar = 10 nm. (D) We show the segmented density of the central feature in (C) overlay with the fitted 3D branch model determined in (Rouiller *et al.*, 2008). The molecular model was fitted into the extracted density with no modifications, reflecting the good fit between the *in vitro* derived molecular model and the *in vivo* imaged density. (E) The extracted branch junction seen in (D) viewed along the mother filament axis. It is unclear why this branch junction was not detected by visual inspection (Urban *et al.*, 2010), however this discrepancy is used to exemplify the need for employing more robust, objective, quantitative image analysis tools rather than by eye inspection.

subvolume averaging should be used to extract structural motifs from the 3D volumes. A major effort in the HR cryo-ET community is to enhance tomographic reconstruction and to devise such fully automated segmentation, and template matching software suites (Volume 482, Chapter 13, Volume 483, Chapters 2, 3). Here, we argue that much of the reported differences in the detection and distinction of filament assemblies with different topology, for example, branched versus straight, and their macromolecular interaction partners is related to the lack of implementation of enabling protocols for unbiased image analysis that allows employment (and scrutiny) of statistical and mathematical tools to ascertain meaningful structure–function correlation (Fig. 10.1C-E).

3.3. Third stumbling block, transience

Interactions between the lamellipodium and lamella are transient in space and time. The short-term transience of these systems, the state of motion of the leading edge changes between protrusion and retraction in cycles as short as

60–100 s on the length scale of 2–3 μm (Machacek and Danuser, 2006). It can be assumed that the underlying cytoskeleton morphology undergoes a complete restructuring at the same time scale. Thus, the life time of a functional state is probably in the order of 15–20 s, assuming up to five distinct states per cycle. The challenge for EM analyses is the capture of well-defined functional states of cytoskeleton structures at the relevant time-scale and resolution. However, very limited precautions are being taken to guarantee that the structural analyses occurred at the time and length scales of seconds and microns over which cell morphogenic processes are regulated. Therefore, to establish a quantitative structure–function relationship between actin network dynamics and cell motility the states of cell edge and cytoskeleton must be frozen within seconds (< 10 s) from the last frame taken by HRLM and noninvasive HR EM imaging and image analysis protocols need to be developed to deconvolve potentially different network architectures in a volume of 2×2 μm.

High pressure (HP) freezing apparatus (see Volume 481, Chapters 3, 8) has been optimized to allow dual LM/EM correlation and at the end of the LM observation, HP freezing is achieved in less than 10 s using the RTS (rapid transfer apparatus). However, the maximum cooling speed that can be reached at the surface of the sample is much smaller than what can be reached by direct cooling (plunge-freezing). Thus cryoprotection agent and pressure ($\sim$2000 atm) are required. Commonly used cryoprotecting agents (such sugar $> 20\%$, v/v) prevent ice formation to certain extent, however, causes osmotic dehydration of samples. High molecular weight sugar polymers (like dextran) were found to minimize this effect, however, at the same time increase crowding due to their protein-like nature. Furthermore, sugars exponentially decrease sample resistance to beam-induced damage, thus impinging of HR structural definition (see Volume 482, Chapter 15). Plunge-freezing is the only technique to achieve vitrification of pure water or physiological buffers in a reproducible way (Cavalier et al., 2009), however, no plunge-freezing vitrification techniques available today can address the need of correlation of live-cell imaging and cryo-EM of vitrified eukaryotic cells with the time-shifts less than 10 s. Vitrification robots are available to warrant controlled, reproducible vitrification. However, these systems do not have the capability of interfacing with a HRLM using oil immersion optics. Technology and protocols need to be developed to allow accommodating vitrification in the relevant time window, while overcoming the obstacles of the geometry and configuration required for HRLM oil immersion optics.

3.4. Lastly, fourth stumbling block, heterogeneity

The transience of cell and cytoskeleton states is expected to generate structural heterogeneity. Meaningful statistical evaluation of such structural distributions requires the combination of higher throughput EM imaging with HR live-cell light microscopy. Transient processes produce a vast spatial heterogeneity of states. Seeing a few snapshots of a structure at different length scales and associating them with a coarse definition of cellular outputs produce incomplete and arbitrary models of structure–function relationships—we suspect a major source of the current disputes in the cell motility field. Different labs may merely see different instances of the same process. To tackle the state heterogeneity, approaches for HR cryo-ET with high-throughput need to be developed (see Volume 481, Chapter 12 and Volume 483, Chapter 11). This will transform HR cryo-ET data acquisition from a laborious, weakly reproducible imaging method to a quantitative technique that affords employment of state-of-the art statistical and machine learning tools to ascertain significant correlations between cell motility behavior and actin cytoskeleton structure.

Driven by the advances of computer-controlled microscopes, digital cameras, and aberration-reducing electron-optics, cellular cryo-ET have become a fast developing imaging technique (Volume 481, Chapter 24). However, to date it is still limited to imaging anecdotal regions of a single eukaryotic cell. Tomographic reconstruction of the full structural variation of the leading edge of a single cell (and subsequent analysis of structurally homogeneous volumes) is a 2-year project for a highly trained and perservering postdoctoral fellow. Establishing structure–function relationships between cytoskeleton and cell morphodynamics by unbiased machine learning of the structural and functional heterogeneity rather requires cryo-ET of tens ($\sim$30) of cells; and understanding the structural bases of the migration deficiencies requires structure–function relationships from $\sim$10 experimental conditions. Clearly, at present, a meaningful structural analysis of a disease condition is outside reach of cryo-ET. Furthermore, the available imaging devices for cryo-ET image acquisition limit the spatial span of image acquisition. Between 8 and 16 consecutive tomograms are required for a full data set of one leading edge of eukaryotic cell (Fig. 10.1A-B). Neither the sample nor the commonly available TEMs would be able to perform for this extent of time (vacuum deterioration, buildup of ice contamination on the sample, etc.). Hence, the development of new instrumentation is necessary to bring cryo-ET to the level of analyzing the structural underpinnings of transient processes in macromolecular assemblies.

Structural biology research is increasingly focusing on unraveling biological and biochemical processes and pathways at the micro-, meso-, and macroscale aiming to correlate space and time. Toward this goal, HR transmission electron cryomicroscopy (cryo-TEM) is indispensable, as it provides ability to determine 3D structures of large, dynamic macromolecular assemblies in the native, fully hydrated state *in situ*. Nevertheless, this technology has not yet been harnessed to quantitatively correlate dynamic states to morphology of fully hydrated sample, in the same window of time, and using rigorous statistical sampling and mathematical tools to ascertain meaningful structure–function correlation in those situations. To harness the power of quantitation of structure–cell function relationships, we need to combine advances in technology development with adequate specimen preparation and correlative mathematical methods that allow quantitative link between HR live-cell imaging and cryo-ET.

ACKNOWLEDGMENTS

I thank Drs Danuser and Volkmann for advice and assistance with the manuscript, and Drs Ochoa and Volkmann for providing the data included in Fig 10.1. The funding source for Dr Dorit Hanein for this study is the National Institutes of Health Cell Migration Consortium; Grant Number: U54 GM064346 and NIGMS Grant Number P01 GM066311.

REFERENCES

Al-Amoudi, A., Diez, D. C., Betts, M. J., and Frangakis, A. S. (2007). The molecular architecture of cadherins in native epidermal desmosomes. *Nature* **450,** 832–837.
Ben-Harush, K., Maimon, T., Patla, I., Villa, E., and Medalia, O. (2010). Visualizing cellular processes at the molecular level by cryo-electron tomography. *J. Cell Sci.* **123,** 7–12.
Blanchoin, L., and Pollard, T. D. (2002). Hydrolysis of ATP by polymerized actin depends on the bound divalent cation but not profilin. *Biochemistry* **41,** 597–602.
Blanchoin, L., Pollard, T. D., and Mullins, R. D. (2000). Interactions of ADF/cofilin, Arp2/3 complex, capping protein and profilin in remodeling of branched actin filament networks. *Curr. Biol.* **10,** 1273–1282.
Carlier, M. F., Pantaloni, D., Evans, J. A., Lambooy, P. K., Korn, E. D., and Webb, M. R. (1988). The hydrolysis of ATP that accompanies actin polymerization is essentially irreversible. *FEBS Lett.* **235,** 211–214.
Cavalier, A., Spehner, D., and Humbel, B. M. (2009). Handbook of cryo-preparation methods for electron microscopy. *In* "Methods in Visualization Series," (G. Morel, ed.). CRC Press, Boca Raton.
Chhabra, E. S., and Higgs, H. N. (2007). The many faces of actin: Matching assembly factors with cellular structures. *Nat. Cell Biol.* **9,** 1110–1121.
Danuser, G. (2009). Testing the lamella hypothesis: The next steps on the agenda. *J. Cell Sci.* **122,** 1959–1962.
Danuser, G., and Waterman-Storer, C. M. (2006). Quantitative fluorescent speckle microscopy of cytoskeleton dynamics. *Annu. Rev. Biophys. Biomol. Struct.* **35,** 361–387.
Delorme, V., Machacek, M., DerMardirossian, C., Anderson, K. L., Wittmann, T., Hanein, D., Waterman-Storer, C., Danuser, G., and Bokoch, G. M. (2007). Cofilin activity downstream of Pak1 regulates cell protrusion efficiency by organizing lamellipodium and lamella actin networks. *Dev. Cell* **13,** 646–662.
Dominguez, R. (2009). Actin filament nucleation and elongation factors—structure-function relationships. *Crit. Rev. Biochem. Mol. Biol.* **44,** 351–366.
Dubochet, J., Adrian, M., Chang, J.-J., Homo, J.-C., Lepault, J., McDowall, A. W., and Schultz, P. (1988). Cryo-electron microscopy of vitrified specimens. *Q. Rev. Biophys.* **21,** 129–228.
Egelman, E. H., Francis, N., and DeRosier, D. J. (1982). F-actin is a helix with a random variable twist. *Nature* **298,** 131–135.

Galkin, V. E., Orlova, A., Cherepanova, O., Lebart, M. C., and Egelman, E. H. (2008). High-resolution cryo-EM structure of the F-actin-fimbrin/plastin ABD2 complex. *Proc. Natl. Acad. Sci. USA* **105,** 1494–1498.

Gruska, M., Medalia, O., Baumeister, W., and Leis, A. (2008). Electron tomography of vitreous sections from cultured mammalian cells. *J. Struct. Biol.* **161,** 384–392.

Gupton, S. L., Anderson, K. L., Kole, T. P., Fischer, R. S., Ponti, A., Hitchcock-DeGregori, S. E., Danuser, G., Fowler, V. M., Wirtz, D., Hanein, D., and Waterman-Storer, C. M. (2005). Cell migration without a lamellipodium: Translation of actin dynamics into cell movement mediated by tropomyosin. *J. Cell Biol.* **168,** 619–631.

Ji, L., Lim, J., and Danuser, G. (2008). Fluctuations of intracellular forces during cell protrusion. *Nat. Cell Biol.* **10,** 1393–1400.

Kabsch, W., and Holmes, K. C. (1995). The actin fold. *FASEB J.* **9,** 167–174.

Kabsch, W., Mannherz, H. G., Suck, D., Pai, E. F., and Holmes, K. C. (1990). Atomic structure of the actin:DNase I complex. *Nature* **347,** 37–44.

Kremer, J. R., Mastronarde, D. N., and McIntosh, J. R. (1996). Computer visualization of three-dimensional image data using IMOD. *J. Struct. Biol.* **116,** 71–76.

Lee, S. H., and Dominguez, R. (2010). Regulation of actin cytoskeleton dynamics in cells. *Mol. Cells* **29,** 311–325.

Leis, A., Rockel, B., Andrees, L., and Baumeister, W. (2009). Visualizing cells at the nanoscale. *Trends Biochem. Sci.* **34,** 60–70.

Li, Z., and Jensen, G. J. (2009). Electron cryotomography: A new view into microbial ultrastructure. *Curr. Opin. Microbiol.* **12,** 333–340.

Lucic, V., Kossel, A. H., Yang, T., Bonhoeffer, T., Baumeister, W., and Sartori, A. (2007). Multiscale imaging of neurons grown in culture: From light microscopy to cryo-electron tomography. *J. Struct. Biol.* **160,** 146–156.

Lucic, V., Leis, A., and Baumeister, W. (2008). Cryo-electron tomography of cells: Connecting structure and function. *Histochem. Cell Biol.* **130,** 185–196.

Machacek, M., and Danuser, G. (2006). Morphodynamic profiling of protrusion phenotypes. *Biophys. J.* **90,** 1439–1452.

Marko, M., Hsieh, C., Moberlychan, W., Mannella, C. A., and Frank, J. (2006). Focused ion beam milling of vitreous water: Prospects for an alternative to cryo-ultramicrotomy of frozen-hydrated biological samples. *J. Microsc.* **222,** 42–47.

Marko, M., Hsieh, C., Schalek, R., Frank, J., and Mannella, C. (2007). Focused-ion-beam thinning of frozen-hydrated biological specimens for cryo-electron microscopy. *Nat. Methods* **4,** 215–217.

Medalia, O., Typke, D., Hegerl, R., Angenitzki, M., Sperling, J., and Sperling, R. (2002a). Cryoelectron microscopy and cryoelectron tomography of the nuclear pre-mRNA processing machine. *J. Struct. Biol.* **138,** 74–84.

Medalia, O., Weber, I., Frangakis, A. S., Nicastro, D., Gerisch, G., and Baumeister, W. (2002b). Macromolecular architecture in eukaryotic cells visualized by cryoelectron tomography. *Science* **298,** 1209–1213.

Medalia, O., Beck, M., Ecke, M., Weber, I., Neujahr, R., Baumeister, W., and Gerisch, G. (2007). Organization of actin networks in intact filopodia. *Curr. Biol.* **17,** 79–84.

Muller-Reichert, T., Srayko, M., Hyman, A., O'Toole, E. T., and McDonald, K. (2007). Correlative light and electron microscopy of early *Caenorhabditis elegans* embryos in mitosis. *Methods Cell Biol.* **79,** 101–119.

Nickell, S., Hegerl, R., Baumeister, W., and Rachel, R. (2003). Pyrodictium cannulae enter the periplasmic space but do not enter the cytoplasm, as revealed by cryo-electron tomography. *J. Struct. Biol.* **141,** 34–42.

Oda, T., Iwasa, M., Aihara, T., Maeda, Y., and Narita, A. (2009). The nature of the globular- to fibrous-actin transition. *Nature* **457,** 441–445.

Penczek, P., Marko, M., Buttle, K., and Frank, J. (1995). Double-tilt electron tomography. *Ultramicroscopy* **60,** 393–410.

Pierson, J., Fernandez, J. J., Bos, E., Amini, S., Gnaegi, H., Vos, M., Bel, B., Adolfsen, F., Carrascosa, J. L., and Peters, P. J. (2010). Improving the technique of vitreous cryo-sectioning for cryo-electron tomography: Electrostatic charging for section attachment and implementation of an anti-contamination glove box. *J. Struct. Biol.* **169,** 219–225.

Pollard, T. D., and Borisy, G. G. (2003). Cellular motility driven by assembly and disassembly of actin filaments. *Cell* **112,** 453–465.

Pollard, T. D., and Cooper, J. A. (2009). Actin, a central player in cell shape and movement. *Science* **326,** 1208–1212.

Ponti, A., Machacek, M., Gupton, S. L., Waterman-Storer, C. M., and Danuser, G. (2004). Two distinct actin networks drive the protrusion of migrating cells. *Science* **305,** 1782–1786.

Reisler, E., and Egelman, E. H. (2007). Actin structure and function: What we still do not understand. *J. Biol. Chem.* **282,** 36133–36137.

Rigort, A., Bauerlein, F. J., Leis, A., Gruska, M., Hoffmann, C., Laugks, T., Bohm, U., Eibauer, M., Gnaegi, H., Baumeister, W., and Plitzko, J. M. (2010). Micromachining tools and correlative approaches for cellular cryo-electron tomography. *J. Struct. Biol.* (in press) Feb 21 Epub ahead of publication.

Rouiller, I., Xu, X. P., Amann, K. J., Egile, C., Nickell, S., Nicastro, D., Li, R., Pollard, T. D., Volkmann, N., and Hanein, D. (2008). The structural basis of actin filament branching by the Arp2/3 complex. *J. Cell Biol.* **180,** 887–895.

Salje, J., Zuber, B., and Lowe, J. (2009). Electron cryomicroscopy of *E. coli* reveals filament bundles involved in plasmid DNA segregation. *Science* **323,** 509–512.

Sartori, A., Gatz, R., Beck, F., Rigort, A., Baumeister, W., and Plitzko, J. M. (2007). Correlative microscopy: Bridging the gap between fluorescence light microscopy and cryo-electron tomography. *J. Struct. Biol.* **160,** 135–145.

Schmid, M. F., Sherman, M. B., Matsudaira, P., and Chiu, W. (2004). Structure of the acrosomal bundle. *Nature* **431,** 104–107.

Small, J. V., Auinger, S., Nemethova, M., Koestler, S., Goldie, K. N., Hoenger, A., and Resch, G. P. (2008). Unravelling the structure of the lamellipodium. *J. Microsc.* **231,** 479–485.

Stokes, D. L., and DeRosier, D. J. (1987). The variable twist of actin and its modulation by actin-binding proteins. *J. Cell Biol.* **104,** 1005–1017.

Urban, E., Jacob, S., Nemethova, M., Resch, G. P., and Small, J. V. (2010). Electron tomography reveals unbranched networks of actin filaments in lamellipodia. *Nat. Cell Biol.* **12,** 429–435.

Vallotton, P., and Small, J. V. (2009). Shifting views on the leading role of the lamellipodium in cell migration: Speckle tracking revisited. *J. Cell Sci.* **122,** 1955–1958.

CHAPTER ELEVEN

Visual Proteomics

Friedrich Förster,* Bong-Gyoon Han,[†] *and* Martin Beck[‡]

Contents

Abstract

Visual proteomics attempts to generate molecular atlases by providing the position and angular orientation of protein complexes inside of cells. This is accomplished by template matching (pattern recognition), a cross-correlation-based process that matches the structure of a specific protein complex to the

* Max Planck Institute of Biochemistry, Martinsried, Germany
† Life Sciences Division, Lawrence Berkeley National Laboratory, University of California, Berkeley, California, USA
‡ European Molecular Biology Laboratory, Heidelberg, Germany

Methods in Enzymology, Volume 483
ISSN 0076-6879, DOI: 10.1016/S0076-6879(10)83011-3

densities of the whole volume or subvolume of a cell, that is typically acquired by cryoelectron tomography. Thereby, a search is performed that scans the entire volume for structural templates contained in a database. In this chapter, we primarily describe the practical experiences gained with visual proteomics during the *Leptospira interrogans* proteome project [Beck *et al.* (2009). Visual proteomics of the human pathogen *Leptospira interrogans*. *Nat. Methods* **6,** 817.]. We give a practical guide how to implement the method and review critical experimental and computational aspects in detail. Based on a survey that has been undertaken for protein complexes from *Desulfovibrio vulgaris*, we review the difficulty of generating reference structures in detail. Finally, we discuss the high yield targets for technical improvements.

1. Introduction

Proteomics approaches often rely on mass spectrometric (MS) measurements and are carried out on the combined lysates of multiple cells. Thereby, the proteins contained in a sample are identified, but any spatial information about them is lost and cell specific properties are averaged out over the population of lysed cells. The visual proteomics concept promises to overcome these limitations and to identify individual protein complexes in intact cells (Nickell *et al.*, 2006). This method consists of three processing steps that attempt to localize individual structures within cryoelectron tomograms. At first, a library is assembled that contains the reference structures of the targeted protein complexes resampled to the relevant electron optical conditions. Subsequently, the local cross-correlation coefficient between each reference structure and tomogram is calculated for all possible positions and orientations and stored in a cross-correlation volume. Finally, the distribution of cross-correlation values within such volumes is translated into a position list by peak extraction and statistical methods. The combined structural signatures of multiple protein complexes, detected in frozen-hydrated specimens, have the potential to describe the spatial proteome of a specific cell as a molecular atlas (Fig. 11.1). Such information is invaluable to biologists in the age of systems biology, where awareness has grown that individual cells have specific properties that contribute to the behavior of the entire population and that protein interactions are more dynamic than anticipated a decade ago.

In 2000, Bohm *et al.* (2000) provided a proof-of-concept of the approach using simulations as well as a first application to cryoelectron tomograms. However, the unambiguous identification of protein complexes in cryoelectron tomograms of intact cells remains a considerable challenge today. The major obstacles that currently prevent a straightforward implementation are: (i) the strong interaction of electrons with matter limits the application of cryoelectron tomography (CET) to

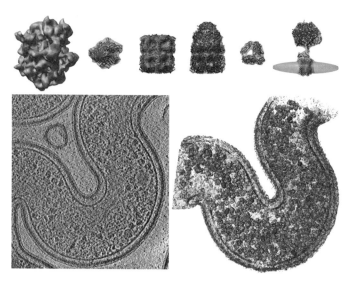

Figure 11.1 Visual proteomics of the human pathogen *Leptospira interrogans*. Reference structures are shown in the top panel. The ribosome, RNA polymerase II, GroEL, GroEL-ES, Hsp, and ATP synthase were template matched in the tomogram shown as centered slice through the reconstruction on the left. The templates assigned into this volume are shown surface rendered on the right (membrane in blue, cell wall in brown). (See Color Insert.)

relatively thin specimens. Therefore, visual proteomics of whole cells is currently effectively limited to prokaryotes; (ii) the availability of template structures for these species limits the number of targets; (iii) at the currently achievable resolution only large protein complexes have a chance of being identified; and (iv) the low signal-to-noise ratio (SNR) peculiar to cryo-electron tomograms hampers a robust and reliable detection. Technical improvements in specimen thinning, better detectors for electron microscopy, the introduction of phase plates, development of C_S correctors, as well as growing structural libraries hold great promise to overcome at least some of these limitations in the future.

2. DATA ACQUISITION

2.1. Required experimental setup

The acquisition of cryoelectron tomograms of highest quality in terms of SNR is desirable for the method outlined here. To achieve this goal, the thickness of the frozen-hydrated specimen should not exceed 300 nm in order to avoid extensive multiple electron scattering. Sample preparation is described in greater detail in Chapter 3, Vol. 481. The use of a transmission electron microscope (TEM) equipped with liquid nitrogen cooled stage, field emission gun, and image filter is essential. The acceleration voltage (*av*) may range from 200 to 300 kV. While working at 300 kV enables the use of slightly thicker specimen, the use of 200 kV pays off for thinner specimens with an increase in contrast. Any standard tomography acquisition software may be used, however, a good control over the dosage spent on the specimen during the acquisition is advantageous (see Section 2.3).

For most of the computational postprocessing steps a standard work station running under any operating system can be used. The template matching itself is computationally expensive and access to a Linux cluster is highly desirable. The protocol outlined in this chapter requires the following software: Matlab version 6.5 or higher, the TOM toolbox for tomography (Nickell *et al.*, 2005), MolMatch for the actual template matching (Förster, 2005), and a number of scripts and example files that are available online (see also Chapter 15, Vol. 482).

2.2. Choosing acquisition parameters

In CET, the experimental setup for the tomographic data collection should be chosen according to the task. One of the factors limiting the resolution in a tomogram is the angular increment $\Delta \alpha$ of the tilt series. The Crowther criterion provides an estimate for the maximal resolution r_{max} of an object with a diameter (*d*; Crowther *et al.*, 1970):

$$r_{max} = \frac{1}{d \Delta \alpha}. \tag{11.1}$$

In Eq. (11.1) the angular increment is measured in radians. This equation is based on the fact that the Fourier transformation of a projection corresponds to a central slice of the Fourier transformation of the 3D reconstruction (Chapter 1, Vol. 482). The thickness of each slice is reciprocally proportional to *d*. The Crowther criterion defines to what resolution the different slices overlap in Fourier space; interpolation can be used to retrieve the information to that resolution. For example, when an object with a diameter of 25 nm is sampled with an angular increment of 9° r_{max} will not exceed $(4 \text{ nm})^{-1}$. However, it is good practice in image processing to oversample data to minimize the loss of information due to interpolation artifacts, which are in the core of every 3D reconstruction algorithm. Typically, data is oversampled at least by a factor of 2; that is, if the targeted object with a diameter of 25 nm is to be imaged with r_{max} of $(4 \text{ nm})^{-1}$, it should be sampled with an angular increment of no more than 4.5°.

In CET, the resolution is not isotropic: due to the limited tilt-angle of typically -60 to $+60°$, a wedge shaped area in Fourier space remains unsampled. Accordingly, the information is unresolved in the respective sections, which leads to a typical elongation of objects along the z-axis in real space ("missing wedge effect").

In addition to the sampling, we must consider the imaging conditions for data acquisition. The imaging of the TEM can be described as a linear system: in this approximation, an electron micrograph is the projection of the electrostatic potential of the object convoluted with the "contrast transfer function" (CTF). The CTF is an oscillating function of the applied defocus value. Since the contrast largely vanishes in focus, micrographs are taken at a specific defocus value. Due to the oscillation of the CTF, the contrast is reversed in certain resolution ranges. Since it is difficult to deconvolute electron micrographs ("CTF correction") in CET, we typically low-pass filter tomograms at the first zero-crossing of the estimated CTF. Thus, the chosen defocus value effectively sets an upper resolution limit to the tomographic data. For example, the 1st CTF zero of a TEM operated at 300 kV and a defocus of 8 μm is at $(4 \text{ nm})^{-1}$. Nevertheless, CTF correction of tomograms is a very active field of research (Chapter 1, Vol. 482) and CTF correction methods for tomograms will certainly be applied more commonly in the future, which would make it possible to use data beyond the 1st CTF zero (Fernandez et al., 2006; Winkler, 2007; Zanetti et al., 2009).

Furthermore, the chosen magnification is also vitally important for the targeted resolution. The resolution cannot exceed the Nyquist frequency:

$$Ny = \frac{1}{2d_{\text{pix}}}. \tag{11.2}$$

In Eq. (11.2) d_{pix} is the pixel size (ps) at the specimen level. When choosing the ps, it must furthermore be considered that the signal transfer of CCD cameras decays substantially as a function of frequency at that image data should generally be oversampled to avoid extensive information loss during image processing. As a consequence, we recommend choosing Ny at least two to three times the targeted resolution.

We emphasize that the above considerations only determine the maximal resolution. The precision of the alignment of projections to a common 3D coordinate system can also limit the maximum resolution (Lawrence, 1992; Chapter 13, Vol. 482). However, in most cases the errors from projection alignment are negligible compared to the limitation due to the applicable electron dose. The signal is typically not significant to r_{max}, that is, the signal is not distinguishable from the noise. When significance criteria from signal-particle analysis (typically Fourier ring correlation) are applied, the signal of cryoelectron tomograms is typically not significant beyond $(5\text{--}10 \text{ nm})^{-1}$. The theoretically attainable resolution, that is, if signal loss due to detector imperfections and interpolation could be excluded, is approximated to be $(2 \text{ nm})^{-1}$ (Henderson, 2004). Nevertheless, considerable information is contained in cryoelectron tomograms beyond the resolution limit set by the electron dose, but it is buried in the noise. Template matching strategies, also coined "matched filters," can help to retrieve this information. The idea is to use prior knowledge to filter the data accordingly (see below).

After summarizing the different factors on signal content in CET, we would recommend choosing the imaging parameters such that the maximum resolution is in the range of $(3\text{--}4 \text{ nm})^{-1}$ when no CTF correction is applied. For a typical experimental setup, we recommend a cumulative electron dose of ~ 100 e$^-$/Å^2, a defocus of 6–10 μm, a *ps* of 5–8 Å at the specimen level and an angular increment $\Delta\alpha$ of 1.5–4°.

2.3. Acquisition and reconstruction of tomograms

The individual cells of a population will contain different amounts of the target protein complexes in varying subcellular locations. The strength of the visual proteomics approach is that it can reveal such properties. However, in order to compare tomograms taken from different cells or different subvolumes of the same cell, a comparable SNR throughout all data sets is required, because the SNR directly contributes to the true-positive discovery rate (see below). This is a difficult task: The specimen thickness will vary across the EM grid and objects close to grid bars will not be accessible at high tilt-angles. Common acquisition software does not necessarily support a satisfactory dosage control. It is therefore essential to use identical condenser beam settings and exposure times throughout data acquisition. Data sets of similar specimen thickness should be subselected for further analysis following the general principle that the thinner, the higher the SNR. The specimen thickness can be approximated from the electron count of a filtered versus a nonfiltered image acquired at 0° tilt-angle.

Three-dimensional reconstructions should be carried out by weighted back projection or using iterative algorithms, for example, the simultaneous iterative reconstruction technique (Leis *et al.*, 2006). A low-pass filter at the first zero of the CTF should be applied to the weighted projections when no CTF correction is attempted. A further comparative assessment between data sets, particularly of the resolution, can be obtained by Fourier ring correlation comparisons between an original projection and the corresponding reprojection of the tomogram calculated from all the other projections (Cardone *et al.*, 2005). When a 2k × 2k camera is used, the tomograms are typically reconstructed with a binning factor of 2 and then visually inspected in order to select the desired region that is afterward reconstructed with a binning factor of 1. Finally, a region of the final reconstruction might be selected by segmentation in order to reduce the computational expense for some of the subsequent steps, in particular the peak extraction (see Section 4.2).

3. TEMPLATES

3.1. Selecting template structures

There are a number of different aspects to be taken into account when reference structures are selected to build a library of templates. Is the reference structure available in high resolution in the protein data bank (PDB file) or as an electron optical density map? PDBs are converted into templates by summing up and resampling the electron density of all the atoms (see Section 3.3). This approach is advantageous because the resulting template maps represent the true electron density and can be directly related to each other. In contrast, EM maps are not on an absolute scale; the gray values are usually normalized to a mean value of zero. However, to simulate particle identification performance for different templates it is crucial to ensure that the signals are on the same scale. Scaling the density of EM-derived maps to other templates derived from EM maps or PDBs is not a trivial task because average densities differ among proteins and the densities of other components, such as nucleic acids vary from proteins (see Section 5.1). As an approximation, the mean value and standard deviation of the density distribution of EM maps can be set to the average mean value and standard deviation observed in all X-ray structures investigated within the same study.

Does the targeted protein complex form different oligomeric states or transient structures in the imaged system? To date, visual proteomics studies have targeted highly stable protein complexes that exist primarily in the fully assembled form. If the rigidity assumption is not supported by biochemical data—or even worse: data indicate a high degree of structural variation—an investigation by visual proteomics might be biased.

What is the abundance of the targeted protein complex in the cell and to what extent does the template library account for all large protein complexes that exist in the cell? A study of the quantitative proteome of *Leptospira interrogans* (Malmstrom *et al.*, 2009) has revealed a dynamic range of protein concentrations covering more than 3 orders of magnitude. If, for example, the desired protein complex exists on average in 10 copies per cell, and a single tomogram covers approximately 10% of the cell volume, it will in average contain one target particle. In such a scenario, a statistically significant discrimination of true- from false-positive template matches, as described below (Section 5), is simply impossible. Moreover, other protein complexes might exhibit similar structural signatures that are difficult to distinguish at the given resolution. The protein family of

AAA-ATPases, for example, has numerous members fulfilling diverse functions in the cell (Hanson and Whiteheart, 2005). AAA-ATPases adopt the same fold and they typically assemble to hexameric rings (Vale, 2000). At the typical resolution of cryoelectron tomograms, it will be impossible to distinguish the different members of this diverse protein family. Nevertheless, the *Leptospira* study has also shown that one specific paralogue is usually by far the most abundant complex of one family. As a consequence, the chances of "overlooking" a large protein complex of high abundance are limited.

Can reference structures of homologs account for the desired target protein complex? Thus far, visual proteomics studies have either targeted highly conserved protein complexes using reference structures from other species, or, as in case of *Mycoplasma pneumonia*, templates were obtained from the same organism using single-particle cryoelectron microscopy (Kuhner *et al.*, 2009). Recently, it was shown that in some cases even subtle changes in the primary structures of homologs may result in substantial changes on the quaternary structure level (Section 3.2).

3.2. Quaternary structure conservation of protein complexes in *Desulfovibrio vulgaris*

A structural-proteomic survey has been performed for the largest and most abundant protein complexes in *D. vulgaris* Hildenborough (*Dv*H; Han *et al.*, 2009). One of the goals of this survey was to establish the extent to which the previously known structures of homologous protein complexes could be used as templates in the visual proteomics characterization of a new microorganism of interest. Except for the limiting conditions of stability, abundance, and high MW that ultimately determine if a protein complex can be purified, the choice of protein complexes studied was free from other assumptions that would bias the survey. This survey included 15 protein complexes (Table 11.1) purified by a tagless technique (Dong *et al.*, 2008), together with 70S ribosomes purified by a separate protocol. Out of these 16 macromolecular complexes, the quaternary structures of eight were successfully analyzed by single-particle EM structural study at resolutions better than 30 Å. In addition, the subunit compositions and stoichiometries of all the protein complexes were analyzed by biochemical methods and mass spectroscopy. The biochemical identity of 13 complexes could be established with confidence based on sequence homology to other proteins with known functions. For three complexes, however, the sequence homology to other proteins was too low to assign any function.

Out of the 13 complexes with reliable functional assignment that were included in this study, only three complexes (the 70S ribosome, GroEL, and the phosphoenolpyruvate synthase complex) showed full conservation of subunit stoichiometry and quaternary structures, while the remaining 10 complexes showed variation between different bacteria, sometimes even

Table 11.1 Biochemical identity, abundance, and stoichiometry of large macromolecular complexes from DvH

Database annotation (gene)[a]	Particle weight estimated by SEC (from EM when known) (kDa)	Number of particles per cell; stoichiometry
Pyruvate:ferredoxin oxidoreductase[b] (DVU3025)	1000 (1052)	4000; $[\alpha\beta\delta\gamma]_8$
Lumazine synthase (riboflavin synthase β-subunit; DVU1198)	600 (996)[c]	300; $\alpha_2\beta_{60}$
Riboflavin synthase α-subunit (DVU1200)		
RNA polymerase β-subunit (DVU1329)		
RNA polymerase β'-subunit (DVU2928)		
RNA polymerase α-subunit (DVU2929)	1100 (885)	500; $[\beta\beta'\alpha_2\omega\mathrm{NusA}]_2$
RNA polymerase ω-subunit (DVU3242)		
NusA (DVU0510)		
60 kDa chaperonin (GroEL; DVU1976)	530 (409 and 818)	700[d]; α_7 (C7) and $[\alpha_7]_2$
Putative protein (DVU0671)	440 (473)	700; α_8
Inosine-5'-monophosphate dehydrogenase (DVU1044)	440 (418)	800; α_8
Phosphoenolpyruvate synthase[e] (DVU1833)	370 (265)	1200; α_2
Hemolysin-type calcium-binding repeat protein (DVU1012)	800	1400; α_{2-3}
Carbohydrate phosphorylase (DVU2349)	670 ($\geq$584)	700; α_{6-7}
Putative protein (DVU0631)	600	100; α_{10-14}
Predicted phospho-2-dehydro-3-deoxyheptonate aldolase (DVU0460)	530	200; α_{16-20}
Alcohol dehydrogenase (DVU2405)	370	12,000; α_{9-10}
Ketol-acid reductoisomerase (DVU1378)	370	600; α_{8-12}
Pyruvate carboxylase (DVU1834)	340	800; $[\alpha\beta]_2$ or $[\alpha\beta]_4$[f]
Proline dehydrogenase/delta-1-pyrroline-5-carboxylate dehydrogenase (DVU3319)	300	1100; α_3

This table is adapted from Table 1 in Han *et al.* (2009).

[a] Entries in italic font indicate protein complexes for which 3D reconstructions were obtained by single-particle electron microscopy (EM).

[b] Homologs of pyruvate ferredoxin oxidoreductase are sometimes fused and sometimes split into multiple chains.

[c] Contribution of the riboflavin synthase α-subunit to the particle weight is not included.

[d] Particle copy number estimated on the assumption that the protein is present in the cell as a D7 14-mer rather than as the C7 heptamer.

[e] EM result indicates either a dimer or tetramer. Size-exclusion chromatography cannot distinguish between these possibilities.

[f] Pyruvate carboxylase is present in some bacteria as a single polypeptide chain and in other bacteria as two chains. We use $\alpha\beta$ to represent the single-chain form.

within the same bacterial genus. The type of variation that was observed can be illustrated by four examples taken from the study of Han *et al.* (2009). First, RNA polymerase II was purified mainly as an unusual dimeric complex of the core enzyme and transcription elongation factor NusA. This dimeric stoichiometry was not observed previously in other bacteria. Second, *Dv*H pyruvate:ferredoxin oxidoreductase exists as octomeric complex while the same protein exists as a dimer even in another species of the same genus, *Desulfovibrio africanus* (Chabrière *et al.*, 2001). Third, lumazine synthase (also known as riboflavin synthase β-subunit) shows various species-dependent multimeric states that include pentamers (Morgunova *et al.*, 2005), decamers (Klinke *et al.*, 2005), and icosahedra. Even though *Dv*H lumazine synthase has an icosahedral form, as it does in *B. subtilis* (Ritsert *et al.*, 1995) and *Aquifex aeolicus* (Zhang *et al.*, 2001), the pentameric subunits are rotated about 30° relative to the fivefold symmetry axis (Fig. 11.2), resulting in a cage structure with a bigger diameter than previous X-ray model structures reported. Fourth, a *Dv*H homolog of carbohydrate phosphorylase shows a ring like structure in the EM images and eluted as a complex large enough to be at least a hexamer in size-exclusion

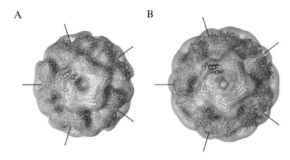

Figure 11.2 Comparison of two types of icosahedral structures of lumazine synthase (riboflavin synthase beta subunit) formed by the proteins from *Aquifex aeolicus* (A) and *D. vulgaris* Hildenborough (B). The positions and directions of some of the fivefold axes are indicated with red lines to help the comparison of two structures. A single pentameric ring subunit is shown as a yellow ribbon representation with a single monomer shown in red. Note that the vertices of the pentamers are rotated by different amounts in the two complexes, resulting in two structures with different diameters. (A) Semitransparent isosurface representation of the complex from *A. aeolicus*, computed at the same resolution as that estimated for the structure obtained by electron microscopy for the complex from *Dv*H. A ribbon diagram of the atomic model of the complex (PDB: 1HQK) is embedded in the low-resolution isosurface. (B) Semitransparent isosurface representation of the complex, obtained by electron microscopy, is shown together with the ribbon diagram of a *Dv*H homology model. The pentameric ring is rotated by ≈30° around the icosahedral fivefold axis to produce a good fit within the EM density map. The homology model was obtained by using the MODBASE (Pieper *et al.*, 2006) server located at http://modbase.compbio.ucsf.edu/modbase-cgi/index.cgi. This figure is adapted from Fig. 2 in Han *et al.* (2009). (See Color Insert.)

chromatography (SEC). The previously known structures of enzymes in this family of proteins were either monomers or dimers. In addition to these unexpected differences in structure, the $D\upsilon$H GroEL particles were initially eluted as a heptameric, single-ring structure with C7 symmetry. However, after the addition of ATP and Mg^{+2}, most of the GroEL particles assembled into the more conventional double-ring form with D7 symmetry, suggesting that the double-ring GroEL is probably the major form under the physiological conditions within $D\upsilon$H. At the same time, it is interesting to note that a single-ring form was also purified from mitochondria (Dubaquie et al., 1998) and a few other bacteria (Ferrer et al., 2004; Ishii et al., 1995; Pannekoek et al., 1992).

The high structural diversity observed in the study of Han et al. (2009) suggests that variation is much more likely to occur than conservation as regards the size, shape, and subunit composition of large prokaryotic complexes. This observation warns us that the use of previously determined quaternary structures from homologous proteins as templates for interpreting tomograms should be practiced with caution. In the specific case of $D\upsilon$H, it was essential to set up $D\upsilon$H-specific templates for 10 complexes out of 13 protein complexes with identified functions.

3.3. Generation of templates from atomic maps

In order to generate template structures that best account for the signal observed in CET, the imaging process has to be simulated as realistically as possible. First, the electron density needs to be calculated from the coordinates and identities of the atoms specified in the PDB. Subsequently the density is convoluted with the CTF, which describes the imaging in the TEM in linear approximation (Fig. 11.3A).

For biological materials, the electron optical density is proportional to the electrostatic potential of the macromolecule. To create a template from an existing atomic model, the coordinates of the individual atoms contained in a PDB file are translated into the 3D electron density. The following code executed in Matlab together with the TOM toolbox will at first define the dimensions of the array and ps; then load the information from the PDB file; transform it into an electron density map at the defined ps; center the density according to the center of mass, and finally display the result:

```
cube = 64; ps = 12.6/2;
GroELS = tom_pdbread('1AON_groE.pdb');
GroELSem = tom_pdb2em(GroELS, ps, cube);
shift = tom_cm(GroELSem);
GroELSems = tom_shift(GroELSem,[(cube/2)+1 (cube/2)+1 (cube/2)+1]-shift);
tom_dspcub(GroELSems,2);
```

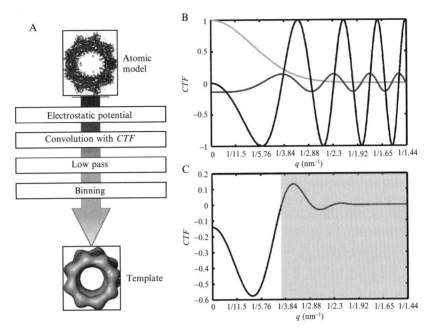

Figure 11.3 Generation of templates from atomic maps. (A) The electrostatic potential of the target structure is calculated from the atomic model. The potential is then convolved with the CTF, low-pass filtered at the first zero of the CTF and binned. (B) The CTF consists of two additive contributions, phase contrast (blue) and amplitude contrast (red), which are damped by the MTF (green). (C) The resulting CTF is low-pass filtered at the first zero. (For interpretation of the references to color in this figure legend, the reader is referred to the web version of this chapter.)

The procedure places the density into the center of the specified cube, which has to be chosen large enough to accommodate the entire template. All operations are carried out at half the *ps* prior to binning in order to avoid sampling artifacts. Next, the template needs to be convoluted with the relevant CTF (Fig. 11.3B). The following code defines *ps*, defocus, *av*, and Nyquist frequency; calculates the CTF; convolutes the template, and displays the result:

```
ps = 1.26 / 2; defocus = − 6.5; av = 200;
Ny = 1/(2*ps);
ctf = tom_create_ctf(defocus, ones(64,64,64), ps, av);
GroELSconv = real(tom_ifourier(ifftshift(fftshift(tom_fourier(GroELSems)).
    *ctf)));
tom_dspcub(GroELSconv, 2);
```

The *ps* needs to be redefined because it is to be given in Å for tom_pdb2em and in nm for tom_ctf. In the example above, any negative contrast values beyond the 1st zero of the CTF are eliminated. If the modulation transfer function (MTF) of the detector is known, the template may be convoluted with the MTF as shown above. Finally, the template is filtered to the relevant r_{max} of the tomograms (Fig. 11.3C; Section 2.3); the volume is binned to the relevant ps; the mean value and standard deviation are normalized to 0 and 1, respectively; the results are displayed and the volume is stored:

Resolution = 5;
Shell = (cube/2) * (1/(Resolution * Ny));
GroELSconv = tom_bandpass(GroELSems,0,shell,shell/10);
GroELSconv = tom_bin(GroELSconv);
[mean max min std] = tom_dev(GroELSconv,'noinfo');
GroELS = GroELSconv-mean;
GroELS = GroELS./std;
tom_dspcub(GroELS, 2);
tom_emwrite('GroELS_12.6A_conv.em',GroELS);

Finally, a spherical mask needs to be generated that defines the area around the template, to which the template matching is constrained. To achieve optimal performance a mask tightly enclosing the reference structure is preferable because densities, which are adjacent to the putative particles in the tomograms, but not considered in the template will otherwise decrease correlation values. The following code generates a spherical mask with a radius of 10 pixels and 2 pixels Gaussian smoothing at the edges, displays the masked template for visual inspection, and stores the mask:

mask = tom_spheremask(ones(32,32,32),10,2);
tom_dspcub(GroELS.*mask,2);
tom_emwrite('GroELS_mask10.em',mask);

A template and mask generated and stored in this way can be directly used with MolMatch for template matching (Section 4). All of the described steps will have to be repeated for all the desired templates and mask radii should be adjusted based on visual inspection.

4. TEMPLATE MATCHING

4.1. Handling MolMatch

MolMatch is software to calculate a measure for the local similarity of a tomogram and a template (Förster, 2005) and is freely available online (www. biochem.mpg.de/foerster/Content_Software/MOLMATCH). Probably the most common measure for similarity of two sets of N-dimensional data is cross-correlation. In essence, the cross-correlation corresponds to the normalized scalar product of the two data sets (interpreted as an N-dimensional vector).

The normalization comprises subtraction of the respective mean-values and subsequent division by their standard division. The advantage is that no absolute scale for the data sets is required. For detection of patterns in tomograms, the concept has to be extended in two respects: (i) only small local features within the tomograms are searched, that is, the dimensions of tomogram and template differ vastly. (ii) Electron tomograms are incompletely sampled leading to typical distortions of the volume in the beam direction ("missing wedge effect"). To account for both, we implemented a local, constrained correlation function (*CCF*) to correlate a volume (*V*), and a template (*T*; Förster, 2005). The *CCF* is constrained to the experimentally sampled fraction of the Fourier space by convoluting *T* with a point-spread function (*PSF*) that causes the same "smearing" along the beam direction as *V*:

$$
CCF(\mathbf{r}, \varphi, \vartheta, \psi) =
$$

$$
\frac{\sum_{\Delta=(1,1,1)}^{(N_x,N_y,N_z)} V_{\mathbf{r}+\Delta} \cdot \left(T_{\Delta,\varphi,\vartheta,\psi} \otimes PSF_\Delta \right) \Theta_\Delta}{\sqrt{\sum_{\Delta=(1,1,1)}^{(N_x,N_y,N_z)} \left(\left(T_{\Delta,\varphi,\vartheta,\psi} \otimes PSF_\Delta \right) \Theta_\Delta \right)^2} \cdot \sqrt{\sum_{\Delta=(1,1,1)}^{(N_x,N_y,N_z)} \Theta_\Delta \cdot \left(V_{\mathbf{r}+\Delta} - \bar{V}_{\mathbf{r}} \right)^2}}
$$

$$(11.3)$$

In Eq. (11.3) the *CCF* measures the local similarity of *V* and the mean-free *T* at every tomogram voxel *r*, given a mask *Θ* specifying the local environment and a *PSF*. The *CCF* is further a function of the template orientation, specified by three angles (φ, ϑ, ψ) (subsequent rotations around the *z*-, *x*-, and *z*-axis). The three volumes *T*, *Θ*, and *PSF* are of much smaller dimensions than the tomogram *V*; they are pasted into the tomogram such that $r = (1,1,1)$ corresponds to the center of all three volumes of the same dimensions. All expressions of the *CCF* can be calculated very efficiently using Fourier transformations (Förster, 2005; Ortiz *et al.*, 2006; Roseman, 2004).

MolMatch is a C-based program to calculate the above expression. More precisely, the maximum value of the *CCF* with respect to the orientation (φ, ϑ, ψ) and the corresponding orientation are stored because the amount of data would be immense otherwise.

All input volumes need to be prepared externally, for example, in MATLAB. The mask has been prepared above and the *PSF* for a volume of a specified dimension (*dim*), minimum- and maximum-tilt-angles (*mintilt* and *maxtilt*) can be generated with the following script:

```
dim = 32;mintilt =−60;maxtilt = 60;
psf = zeros(dim,dim,dim); psf(1,1,1) = 1;
wedge = av3_wedge(psf, mintilt, maxtilt); psf = real(tom_ifourier(tom_fourier
      (psf).*fftshift(wedge)));
tom_emwrite('psf_−60+60.em',psf);
```

After preparation of the required input volumes MolMatch can be invoked in a Unix shell. For the given example of GroEL, MolMatch could be run the following way:

mpirun -np 36 molmatch.exe Lepto_phantom_cell.em_SNR_0.5_BPWei-Proj_0.5.em GroELS_12.6A_conv.em Out 0 50 10 0 50 10 0 90 10 psf_-60+60.em GroELS_mask10.em 256

Here, MolMatch is preceded by the specification of the number of used processors for the MPI library (36 processors in this case). The first three arguments following MolMatch are the tomogram file (*Lepto_phantom_cell. em_SNR_0.5_BPWeiProj_0.5.em*), the template (*GroELS_12.6A_conv.em*), and the output filename (*Out*). The next nine numbers specify the angular sampling: for φ, ψ, and ϑ the minimum, maximum, and the increment are specified. Since GroEL exhibits a D7 point symmetry, φ and ψ can be restricted to 0–50° and ϑ to 0–90° at the chosen increment of 10° for all angles. Asymmetric templates will generally require the full range of 0–360° for φ and ψ and 0–180° for ϑ. The angular increment should be chosen according to the targeted resolution. For example, if the template volume spans 20 nm in one dimension and the targeted resolution is 4 nm^{-1}, the increment should be at least a sin(4.2/20) $\approx$ 20°, but better only 10° as oversampling is generally beneficial. The angular range is followed by the PSF (*psf_−60+60.em*) and the mask (*GroELS_mask10.em*). The last parameter is of entirely technical nature: it specifies how the volumes are subdivided into smaller cubes to facilitate computation of large volumes that exceed the available memory in the used computers. For typical applications, it should be chosen equal to the smallest dimension of the tomogram, typically the third dimension (z).

4.2. Creating motif lists

The MolMatch output is used to generate "motif lists," that is, lists containing coordinates and orientations of putative targets. MolMatch produces two output volumes, from which the motif lists are generated: *Out.ccf.norm* (maximum of the local constrained correlation with respect to the orientation), and *Out.ang* (indices of orientations corresponding to maximum correlation). The correlation function should ideally exhibit distinct maxima at putative particles, whereas the orientation indices are only interpretable when the angular range used as input is known. The motif list is generated sequentially: First, *Out.ccf.norm* is screened for its maximum value. The value and the corresponding coordinates and orientation (three Euler angles calculated from the index in *Out.ang* and the rotational parameters for MOLMATCH) are stored in the motif list. *Out.ccf.norm* is then set to a small value (-1) in a sphere with defined *radius* centered at the peak to prevent recurrent detection. The procedure is repeated *nparts* times to detect nparts putative particles.

Optionally, particles that are too close to the border of the tomogram (specified by *tomoedge*) are rejected from the motif list.

corr = *tom_emread('Out.ccf.norm'); angfile* = *'Out.ang';*
nparts = *1000; radius* = *10; tomoedge* = *15;*
maxphi = *50; maxpsi* = *50; maxthe* = *90; minphi* = *0; minpsi* = *0; minthe*
= *0; incr* = *10;*
motlfile = *'motl_GroEL.em';*
motiflist = *av3_createmotl(corr, angfile, nparts, radius, tomoedge, maxphi, maxpsi,*
maxthe, incr, minphi, minpsi, minthe);
tom_emwrite(motlfile, motiflist);

The motif lists then need to be assessed to discard false-positive detections (Section 5). In addition, motif lists can be used to refine and classify 3D averages of the detected particles using the AV3 package.

4.3. Scoring

A scoring function provides a measure for the probability of an event. For example, the *CCF* described in Eq. (11.3) is a score for the probability that a feature in a tomogram corresponds a specific macromolecular complex. To assess the performance of a scoring function, it has to be calibrated using a defined test data set (also called training data). Thereby, the likelihood of false-positive discoveries is calculated as a function of the score threshold that discriminates true from false hits. Such concepts for assessment of performance are widely used in other areas of biology, for example, for assigning peptides to tandem MS spectra (Keller *et al.*, 2002) or for assessing the statistical significance of sequence alignments (Henikoff, 1996; Rost, 2002). How the performance of visual proteomics can be assessed is described in detail in Section 5.

Scoring functions often rely on more than one single readout but combine several readouts as linear combinations of subscores. Thereby, the weight of such subscores is optimized in training data sets using a linear discriminate analysis of true and false positives. Once optimized, the combined discrimination power of all subscores is likely superior to single readouts. A scoring function *SF* for visual proteomics (Beck *et al.*, 2009) that relies on three different knowledge-based, empirical readouts has been used during the Leptospira proteome project:

$$SF = A * CCF_{\text{Par}} + B * \frac{CCF_{\text{Par}}}{CCC_{\text{TopComp}}} + C * \frac{CCF_{\text{Par}}}{CCF_{\text{TopDecoy}}} \quad (11.4)$$

In Eq. (11.4), A, B, and C are weighting factors for the subscores, CCF_{Par} is the *CCF* of the targeted template as described in Eq. (11.3), CCF_{TopComp} is the highest *CCF* value of any other (competing) template within the same position of the tomogram, and $CCF_{TopDecoy}$ is the highest of any nontargeted, decoy template, which can be random structures or shapes. The rational of this SF is to reward a penalty to hits in positions where competing or decoy templates have a high CCF. In principle, the list of subscores could be extended by quantities that are not derived from cross-correlations (e.g., occupied volume of templates or the local SNR of the tomogram).

To apply the scoring function to the $CCFs$ calculated by MolMatch for all targets and decoy templates used in a study, the individual motif lists have to be merged as layers into a single 3D motif list file. This can be done automatically and for the entire template library at once using *motlgen*, a wrapper script for *av3_createmotl*:

load Templates
load Decoys
help motlgen
fileroot = 'PhantomCell_SNR_0.5_BPWeiProj_0.5';
motlgen (fileroot, Templates, Decoys);

Thereby, *Templates* is an array that contains the library annotation, a table containing names, sizes, and other definitions of the template library. The command "*help motlgen*" will display the specifics of the definitions contained in the annotation table in addition to the help text on the *motlgen* script. *Decoys* is the same as *Templates* but for additional nontargeted templates that are used for scoring. The *motlgen* wrapper executes the peak extraction for multiple templates and stores the retrieved information in motif lists that are structured as usual but contain competing assignments from other templates as layers along the third dimension. Afterward, the score can be calculated:

score_motl = score_matches('motl_',Templates,Decoys);

Thereby, the number of matches to be assigned positive (class 1) based on the score threshold defined by *Templates*. This number strongly influences the performance and must be carefully chosen (see Section 5 for detail). Finally, the resulting hits can be assessed for double assignments (coordinates that are assigned to more than target) in space:

fetch_double_hits(score_motl, Templates);

This script identifies double hits, displays statistics, and reassigns the competing hits that have a lower relative score within their score distribution as false prior to visualization (see below).

4.4. Visualization of molecular atlases

For visualization of the detection results the template can be rotated and positioned at the determined coordinates and the corresponding angles. These particles can be displayed in the context of the typically manually segmented tomogram. In most cases, an isosurface representation is chosen because it is much easier to capture the content of a tomogram in this representation compared to common volume-rendered approaches.

A convenient, albeit costly, program for 3D visualization is Amira. It is also possible to develop plug-ins for Amira. The Frangakis laboratory has developed such a plug-in to display templates in a tomogram at positions and orientations specified in a motif list (Pruggnaller *et al.*, 2008). The plug-in is freely available (www.biophys.uni-frankfurt.de/frangakis/Amiratools.htm).

An alternative, professional visualization software is 3D Studio Max. The software uses realistic ray-tracing technology and scripting is possible as well. This software has been used to visualize ribosomal morphologies in different studies (Brandt *et al.*, 2009; Ortiz *et al.*, 2006).

5. Assessment of Performance

A major challenge in the visual proteomics approach is to discriminate true-positive from false-positive detections. Due to the moderate SNR of the original data, the observed distribution of cross-correlation values comprises an overlay of the distributions of true- and false-positive hits. In order to assess the performance, the amount of overlap of both distributions has to be estimated. Such an approach is commonly used in MS-based proteomics to determine false-positive discovery rates for peptides assigned to tandem MS spectra (Keller *et al.*, 2002). It does not attempt to associate individual hits with the true- or false-positive distribution, but builds a statistical model that allows calculating the likelihood for all matches in the observed distribution to be false-positive hits.

The performance of visual proteomics depends on a variety of tomogram-specific parameters that affect the SNR and consequently all the targeted protein complexes in a similar way. These are primarily acquisition settings, specimen thickness, and molecular crowding. However, also target-specific parameters, such as molecular weights, cellular abundance of the target, as well as cellular abundance of protein complexes competing for assignments play a critical role. Therefore, the performance cannot be determined in general but needs to be estimated separately for each individual template contained in the library.

5.1. Estimating true-positive discovery rates from artificial tomograms

One way of estimating the performance of visual proteomics is to simulate the entire process *in silico*. Thereby, artificial tomograms are generated as realistically as possible and subjected to template matching. Since the position and orientation of all templates in the artificial tomogram is known, the distribution of true- and false-positive hits can be calculated separately and false as well as true-positive discovery rates can be determined. This approach does not only take the image formation mechanism into account, but also *a priori* information about template abundances, and, if procurable, the abundances of protein complexes competing for assignments as well as the total protein concentration inside the cell. This information can be approximated using quantitative proteomics approaches (Malmstrom *et al.*, 2009). As an additional benefit, the performance for the desired template can be assessed in conjunction with acquisition parameters before the actual data collection.

The first step to *in silico* assessment is to build an artificial tomogram (phantom cell) that contains the correct copy number of the template structures to account for their concentration within the cell. The following code makes the necessary definitions, generates an empty phantom cell and a "book keeping" volume that is required to avoid multiple assignments:

```
nTemp = 10; radius = 8; offset = 10;
volFile = 'phantom_cell.em '; bkFile = 'book_keeping_volume.em ';
bkVol = ones (128,128,128); tom_emwrite(bkFile, bkVol);
vol = zeros (128,128,128); tom_emwrite(volFile, vol);
```

Thereby, *nTemp* defines the number of molecules to paste into the phantom cell, *radius* defines the local proximity around the center of mass occupied by the template, *offset* defines the size of an empty edge of the phantom cell, *volFile* the file name of the phantom cell, and *bkFile* the name of the "book keeping" volume.

Next, the template structure is loaded and pasted 10 times into the phantom cell at random positions and orientations. These positions are bookmarked in the "book keeping" volume by setting the density in their proximity to zero. The pasting procedure will not place templates into previously bookmarked positions. Finally, the modified phantom cell and book keeping volume as well as the respective motif list (*motl*) are stored.

```
GroELS = tom_emread('GroELS_12.6A.em ');
[motl bkVol vol] = paste_template(nTemp, GroELS.Value, radius, bkFile,
    volFile, offset);
tom_emwrite('GroELS_motl.em', motl);
tom_emwrite(bkFile, bkVol);
tom_emwrite(volFile, vol);
```

This procedure can be repeated multiple times to account for multiple templates. In addition, decoy density can be added analogously to simulate molecular crowding.

Next, adequate noise has to be added to simulate the image formation process. This can be done by adding a contribution where the CTF- and

MTF-convoluted signal is multiplied with noise ("quantum noise") and a pure MTF-convoluted noise contribution ("background noise"; Forster *et al.*, 2008). To simulate the tomogram, noisy projections are calculated and then reconstructed in 3D. The following code makes definitions as described above and simulates the imaging of a given phantom cell at a given *ps*, *defocus*, *av*, angular increment scheme (*angles*), and *SNR*:

ps = 1.26; *defocus* = − 6.5; *av* = 200;
angles = [− 63:1.5:− 40 − 38:2:38 40:1.5:63]; SNR = 0.5;
tomogen('Lepto_phantom_cell.em', SNR, angles, defocus, ps, av);

The resulting volume can be directly used for template matching as described in Section 4. The noise level should be estimated from real data sets. This can, for example, be done in the following way: Artificial tomograms are generated at different noise levels and subjected to template matching as well as real data sets. Thereby, an easily detectable template of sufficient size and abundance, for example the ribosome, should be targeted. To select the adequate noise level, the distributions of cross-correlation values of candidate hits from real data sets and phantom cells are compared. Once the noise level is calibrated, all reference structures contained in the library template matched to the phantom data using MolMatch as described in Section 4. Afterward, motif lists are generated for all templates using the following code (see Section 4.4 for detail):

load Templates
fileroot = 'PhantomCell_SNR_0.5_BPWeiProj_0.5';
motlgen (fileroot, Templates);

Finally, the performance is assessed by comparing two sets of motif lists: the one containing positions of templates originally pasted into the phantom cell and the other containing matched positions from the same cell. The following code loads the template library annotation, defines the file locations of both motif lists and launches the performance calculation:

load Templates
MatchedMotlRoot = 'MatchMotl_'; *PastedMotlRoot* = 'PasteMotl_';
assess_performance(MatchedMotlRoot, PastedMotlRoot, Templates);

This procedure produces a number of plots. The observed distribution of cross-correlation values is overlaid with the corresponding distributions of the true- and false-positive hits. The specificity *Sp* and the sensitivity *Se* are calculated and plotted against the cross-correlation threshold, whereby specificity is defined in Eq. (11.5):

$$Sp = \frac{n_{\text{TruePos}}}{n_{\text{TruePos}} + n_{\text{FalsePos}}} \qquad (11.5)$$

with $n_{TruePos}$ being the number of true-positive hits and $n_{FalsePos}$ being the number of false-positive hits at a given cross-correlation threshold; and sensitivity is defined in Eq. (11.6),

$$Se = \frac{n_{TruePos}}{n_{Templates}} \qquad (11.6)$$

with $n_{Templates}$ being the number of templates accounting for the given target protein complex that are contained in the tomogram. The output allows an assessment if the desired protein complex can be detected under the given cellular and imaging conditions and it reveals the dependency of the performance on the CCC threshold (Fig. 11.4).

5.2. Performance assessment in real data sets

Alternatively, the performance of visual proteomics can be assessed in real data sets based on *a priori* knowledge about the template structures or their spatial distribution. This approach, although preferable over simulations, is not generally applicable to all template structures. Matching with a mirrored template of "nonnative" handedness can provide complementary information about the distribution of cross-correlation values. When the

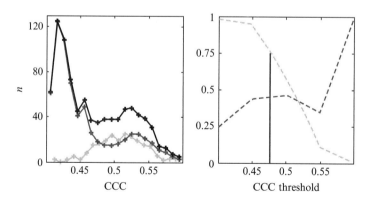

Figure 11.4 Performance analysis of template matching in artificial tomograms. RNA polymerase, a high abundant protein complex of only moderate size was chosen as an example. The observed distribution of cross-correlation coefficients (blue, left) is overlaid with the corresponding distributions of the true- (green) and false-positive hits (red). The latter two do overlap to a large extend, therefore the performance is imperfect. Specificity (red) and sensitivity (green) are shown on the right. At an arbitrarily chosen CCC threshold of ~0.48 about 75% of the discovered hits are true positive, however, less than 50% of all targets are discovered. (For interpretation of the references to color in this figure legend, the reader is referred to the web version of this chapter.)

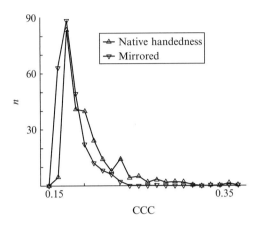

Figure 11.5 Distribution of the cross-correlation coefficients (CCCs) extracted obtained from a tomogram of a *Leptospira* cell after template matching with the ribosome as template (black) and the mirrored ribosome as decoy template (grey).

distribution of cross-correlation values observed using the 70S ribosome as a template is compared to the corresponding distribution observed for the mirrored ribosome, a shift is apparent (Fig. 11.5). The number of single particles to the right of the cross-over of both curves roughly matches the cellular abundance of the ribosome indicating that at this point both distributions decline into noise correlation.

Another promising approach is to match template structures of different subunits of a protein complex separately, for example, the small and the large ribosomal subunit, to check if they reasonably colocalize in translating ribosomes. The implementation of this control for tightly associated subunits is, however, quite challenging: The local cross-correlation has to be tightly constrained in order to mask out the adjacent subunits, barely including the surrounding solvent. However, the contrast arises primarily from the difference in density between the target and the surrounding solvent and hence template matching performance decreases. Only if resolution and SNR are sufficient to recognize intrinsic features within the different subunits, this approach might be employed, which is not realistic with the current experimental setup, even for very large protein complexes. Alternatively, loosely associated structures provide a good target for similar controls. The colocalization of ribosomes engaged in polysomal arrays (Beck *et al.*, 2009; Brandt *et al.*, 2009) or even RNA polymerases associated with translating ribosomes can account as examples for such cases. Furthermore, *a priori* knowledge of the cellular localization or orientation of protein complexes can provide clues about the performance. For example, membrane associated protein complexes exhibit a specific positioning and orientation relative to the membrane (Ortiz *et al.*, 2006) and RNA polymerases cluster to the genomic DNA. Finally, cellular treatments such as drug or stress stimuli might induce a considerable up- or downregulation of individual target protein complexes (Beck *et al.*, 2009; Malmstrom *et al.*, 2009). Such perturbations should change the observed distribution of cross-correlation values; however, impose the challenge of comparing data obtained from different tomograms and cells (Section 2.3). Subtle changes in the SNR and stochastic variations of the expression level within the cell population complicate the data evaluation in this case.

6. The Spatial Proteome of *L. interrogans*

L. interrogans is a Spirochete with a strongly elongated and helically coiled cell shape and a pathogen that causes Leptosirosis in a wide range of species. Since the diameter of a cross-section of a typical cell is no more than 100–180 nm, *L. interrogans* is an ideal specimen for cryo-ET allowing for excellent electron beam penetration and the acquisition of tomograms with relatively good SNR (Beck *et al.*, 2009). At first, the proteome was measured by tandem MS in shotgun mode. Gene products accounting for about 62% of the ~3700 open reading frames were detected. A query for the expressed protein complexes in structural databases retrieved 26 candidate protein complexes of a certain minimal size (>250 kDa) and a certain sequence conservation that might be suitable for template matching (Table 11.2). Targeted proteomics was used to determine the cellular protein concentration on a proteome-wide scale (Malmstrom *et al.*, 2009). Protein abundance was found to range over 3.5 orders of magnitude. The most abundant protein detected in this study was LipL32, an integral part of the peptido glycan layer, with ~40,000 copies per cell. The most abundant candidate protein complex was the ribosome with about 4500 copies per cell, occupying ~20% of the cytoplasmic volume. Another 10 protein complexes had abundances of at least 100 copies per cell, all others were of low abundance.

Next, a subset of protein complexes was selected to estimate the performance of visual proteomics *in silico* (Fig. 11.6A). For this purpose, phantom cells were generated that contained the templates at their cytoplasmic concentration in a membrane enclosed space and the image formation process was simulated to generate artificial tomograms and subjected to template matching (Fig. 11.6B). Ideally, the distribution of cross-correlation values obtained by template matching for a specific protein complex can be described by an overlay of two Gaussian functions, one accounting for the false-positive and the other for the true-positive distribution. The false-positive distribution is, however, likely composed of signals from several different species. Therefore, the data interpretation based curve shape *per se*

Table 11.2 Structures of protein complex candidates for template matching

Template	Reference	Oligomeric state	Molecular weight (kDa)	Complexes per cell
Ribosome	pdb_2AW7_2AWB	Multicomponent protomer	2200	4500
RNA polymerase	pdb_2GHO	Multicomponent protomer	340	3000
ATP synthase	pdb_1QO1	6-mer	450	1500
GroEL/ES	pdb_1aon	21-mer	870	1100
GroEL	pdb_1KP8	14-mer	810	
Glutamine synthase	pdb_2gls	12-mer	620	320
Citrate synthase	pdb_2h12	6-mer	300	220
Transaldolase	pdb_1vpx	20-mer	520	150
clpP	pdb_1YG6	14-mer	304	140
Carbamoyl phosphatase	pdb_1bxr	8-mer	650	120
Enolase	pdb_1w6t	8-mer	390	100
Asparryl-tRNA-synthase	pdb_1eqr	3-mer	200	100
Cytosolic amino-peptidase	pdb_1gyt	12-mer	660	90
GTP-cyclohydrolase	pdb_1GTP	20-mer	500	75
clpB	emd_1243	6-mer	600	70
Ornithine carbamoyltransferase	pdb_1a1s	12-mer	420	65
Phoshoribosyl-pyrophosphatase	pdb_1dkr	6-mer	210	55
Acetyl-CoA-carboxylase	pdb_1vrg	6-mer	350	40
Hsp15	pdb_2BYU	12-mer	220	40
Dihydrolipoamide acetyl-transferase	pdb_1dpb	24-mer	640	30
HslU-V	pdb_1G3I	24-mer	830	20
LIC11615_UbiD	pdb_2idb	6-mer	350	17
Aspartate carbamoyltransferase	pdb_1f1b	12-mer	300	10
Acetolactate synthase	pdb_1OZF	4-mer	250	10
Bacterioferretin	pdb_1BFR	24-mer	450	4
Lumazine synthase	pdb_1NQU	60-mer	1000	4

A passing number, oligomeric states, molecular weights, and cellular abundances are listed.

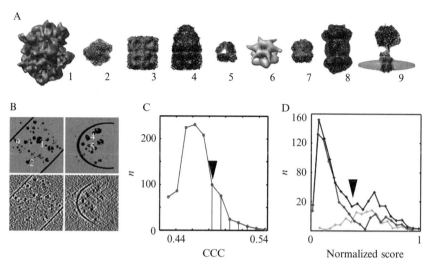

Figure 11.6   *In silico* estimation of the performance of visual proteomics. (A) Subset of protein complexes selected for performance analysis is shown surface rendered with chain traces in black (if applicable): 1. 70S ribosome, 2. RNA polymerase II, 3. GroEL, 4. GroEL-ES, 5. Hsp, 6. Clp B, 7. Clp P, 8. HslU-V, 9. ATP-synthase. (B) Centered slices through a *Leptospira* phantom cell shown perpendicular (left) and parallel (right) to the electron optical axis, as well as before (top) and after (bottom) simulating the image formation process. (C) Distribution of cross-correlation coefficients obtained from real data sets by template matching. The area marked under the curve corresponds to the expected abundance of the template; the arrowhead marks the intuitive threshold for assignments as indicated by the curve shape. (D) Score distributions obtained from phantom cells of the observed (blue), true- (green) and false-positive hits (red). The arrowhead marks the threshold that accounts for a specificity of 40%. (For interpretation of the references to color in this figure legend, the reader is referred to the Web version of this chapter.)

will not allow the deduction of false-positive discovery rates or specificities (Fig. 11.6C). Since the position and orientation of all protein complexes in the artificial tomograms are known, the distribution of true- and false-positive hits can be calculated in order to determine specificities.

To improve the discrimination of true- from false-positive hits, a scoring function was developed that not only incorporates the cross-correlation value of the target but also the cross-correlation values of competing templates and a number of decoy templates within the same position of the tomogram (Fig. 11.6D, Section 4.3). This scoring scheme outperformed the classical workflow in some cases, particularly when protein complexes of similar structural signature were competing for assignments.

A set of 18 different phantom cells was evaluated multiple times to investigate the performance of visual proteomics under the influence of noise, missing wedge direction, protein abundance, and molecular crowding. In the course of these simulations, the following factors were found to have a profound impact: (i) background noise may reduce the performance depending on the MTF of the particular CCD camera used in a study; (ii) the missing wedge can introduce angular bias to the discovery rate of templates with anisotropic signal content, such as the elongated ATP-synthase; (iii) the specificity increases with the template abundance; and (iv) decreases with the degree of molecular crowding. The last two effects are generally more pronounced for smaller than for larger templates (Fig. 11.7). While the specificity achieved for high abundant megadalton complexes is satisfactory, true-positive discovery rates higher than 50% are difficult to achieve when protein complexes of smaller molecular weights are targeted. For protein complexes of very low cellular abundance, it is quite a challenge to obtain robust statistical models.

Tomograms covering subvolumes of 37 different *L. interrogans* cells were acquired. A subset of 12 tomograms of similar SNR was selected and subjected to template matching and scoring (Fig. 11.1). The local concentration of the targeted protein complexes varied within and across data sets: the cells displayed an average ribosome concentration of $\sim$20 μM ($\sim$40 mg/ml) in the cytoplasm, but the local concentration ranged from 5 to 30 μM ($\sim$10–65 mg/ml). The local fluctuations in case of total GroEL together with (GroEL-ES) were larger and ranged from $\sim$8 to 100 μM ($\sim$0.5–6.5 mg/ml). In most, but not all tomograms, the ratio of ribosomes to RNA polymerases was locally maintained. Stress-induced changes in the GroEL to GroEL-ES ratio and Hsp abundance were apparent, but could not

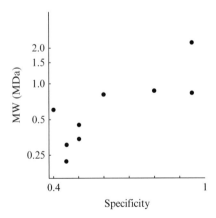

Figure 11.7 The performance of visual proteomics depends on the size of the targeted protein complexes. The specificity for nine different protein complexes with molecular weights ranging from 0.2 to 2.5 MDa was determined *in silico* at a pixel size of 1.26 nm, SNR of 0.5 with an underfocus of 6.5 μm (first zero of CTF at 4 nm). The specificity refers to the fraction of true-positive hits at a discovery rate of 0.5.

be substantiated by statistical significance testing. Some templates provided auxiliary information for the orthogonal validation of the data obtained to assess the performance in real data sets as described above (Section 5.2).

7. OUTLOOK

A number of limitations to visual proteomics might be overcome by further technical developments. Currently, the most critical limitation is the moderate SNR of CET. The development of alternative detection concepts is probably most important to capture the entire signal emanated from the sample. In addition, the introduction of phase plates in conjunction with C_S correctors may optimize the contrast transfer. Specimen thinning techniques will be pivotal to image cell types, which exceed 500 nm or more; thus, application of visual proteomics to eukaryotic cells depends on progress in this field.

In-depth characterization of the vast majority of large protein complexes in systems under study, by biochemical as well as structural techniques, has become feasible and will largely contribute to template library reliability and completeness. The further development of classifiers and scoring functions for pattern recognition holds great potential to increase true-positive discovery rates. To detect low abundant protein complexes with confidence, the throughput has to be increased on the data acquisition and postprocessing side. The implementation of a better dosage control during the data acquisition will be necessary to compare large number of data sets. The signal in visual proteomics primarily arises from the contrast given by the difference in density between the targeted protein complex and the surrounding solvent. Improved contrast, for example through phase plates, is therefore likely to increase the performance in the future.

One major limitation is given by nature and will remain: The dosage of electrons that can be applied to a specimen ultimately limits the attainable resolution of CET. As a consequence, visual proteomics in conjunction with CET is not likely to become broadly applicable. The biological questions that can be answered will be limited to protein complexes of a certain minimal molecular weight that can be targeted by template matching. Such relatively large protein complexes, however, function in central cell biological processes, such as transcription, translation, protein folding, and degradation.

ACKNOWLEDGMENTS

B. G. H. was supported by the U.S. Department of Energy Contract DE-AC02-05CH11231. F. F. is grateful to a Career Development Award from the Human Frontier Science Program.

REFERENCES

Beck, M., et al. (2009). Visual proteomics of the human pathogen Leptospira interrogans. Nat. Methods 6, 817.

Bohm, J., et al. (2000). Toward detecting and identifying macromolecules in a cellular context: Template matching applied to electron tomograms. Proc. Natl. Acad. Sci. USA 97, 142–145.

Brandt, F., et al. (2009). The native 3D organization of bacterial polysomes. Cell 136, 261.

Cardone, G., et al. (2005). A resolution criterion for electron tomography based on cross-validation. J. Struct. Biol. 151, 117.

Chabrière, E., et al. (2001). Crystal structure of the free radical intermediate of pyruvate: Ferredoxin oxidoreductase. Science 294, 2559.

Crowther, R. A., et al. (1970). Three dimensional reconstructions of spherical viruses by Fourier synthesis from electron micrographs. Nature 226, 421.

Dong, M., et al. (2008). A "tagless" strategy for identification of stable protein complexes genome-wide by multidimensional orthogonal chromatographic separation and iTRAQ reagent tracking. J. Proteome Res. 7, 1836.

Dubaquie, Y., et al. (1998). Purification of yeast mitochondrial chaperonin 60 and co-chaperonin 10. Methods Enzymol. 290, 193.

Fernandez, J. J., et al. (2006). CTF determination and correction in electron cryotomography. Ultramicroscopy 106, 587.

Ferrer, M., et al. (2004). Functional consequences of single:double ring transitions in chaperonins: Life in the cold. Mol. Microbiol. 53, 167.

Förster, F. (2005). Quantitative Analysis of macromolecules in cryoelectron tomograms using correlation methods. Dissertation. Technical University of Munich, Munich.

Förster, F., et al. (2008). Classification of cryo-electron sub-tomograms using constrained correlation. J. Struct. Biol. 161, 276.

Han, B. G., et al. (2009). Survey of large protein complexes in D. vulgaris reveals great structural diversity. Proc. Natl. Acad. Sci. USA 106, 16580.

Hanson, P. I., and Whiteheart, S. W. (2005). AAA+ proteins: Have engine, will work. Nat. Rev. 6, 519.

Henderson, R. (2004). Realizing the potential of electron cryo-microscopy. Q. Rev. Biophys. 37, 3.

Henikoff, S. (1996). Scores for sequence searches and alignments. Curr. Opin. Struct. Biol. 6, 353.

Ishii, H., et al. (1995). Equatorial split of holo-chaperonin from Thermus thermophilus by ATP and K^+. FEBS Lett. 362, 121.

Keller, A., et al. (2002). Empirical statistical model to estimate the accuracy of peptide identifications made by MS/MS and database search. Anal. Chem. 74, 5383.

Klinke, S., et al. (2005). Crystallographic studies on decameric Brucella spp. Lumazine synthase: A novel quaternary arrangement evolved for a new function? J. Mol. Biol. 353, 124.

Kuhner, S., et al. (2009). Proteome organization in a genome-reduced bacterium. Science 326, 1235.

Lawrence, M. C. (1992). Least-squares method of alignment using markers. In "Electron Tomography," (J. Frank, ed.), p. 197. Plenum Press, New York.

Leis, A. P., et al. (2006). Cryo-electron tomography of biological specimens. IEEE Signal. Process. Mag. 23, 95.

Malmstrom, J., et al. (2009). Proteome-wide cellular protein concentrations of the human pathogen Leptospira interrogans. Nature 460, 762.

Morgunova, E., *et al.* (2005). Crystal structure of lumazine synthase from *Mycobacterium tuberculosis* as a target for rational drug design: Binding mode of a new class of purinetrione inhibitors. *Biochemistry* **44**, 2746.

Nickell, S., *et al.* (2005). TOM software toolbox: Acquisition and analysis for electron tomography. *J. Struct. Biol.* **149**, 227.

Nickell, S., *et al.* (2006). A visual approach to proteomics. *Nat. Rev.* **7**, 225.

Ortiz, J. O., *et al.* (2006). Mapping 70S ribosomes in intact cells by cryoelectron tomography and pattern recognition. *J. Struct. Biol.* **156**, 334.

Pannekoek, Y., *et al.* (1992). Identification and molecular analysis of a 63-kilodalton stress protein from *Neisseria gonorrhoeae*. *J. Bacteriol.* **174**, 6928.

Pieper, U., *et al.* (2006). MODBASE: A database of annotated comparative protein structure models and associated resources. *Nucleic Acids Res.* **34**, D291.

Pruggnaller, S., *et al.* (2008). A visualization and segmentation toolbox for electron microscopy. *J. Struct. Biol.* **164**, 161.

Ritsert, K., *et al.* (1995). Studies on the lumazine synthase/riboflavin synthase complex of *Bacillus subtilis*: Crystal structure analysis of reconstituted, icosahedral beta-subunit capsids with bound substrate analogue inhibitor at 2.4 Å resolution. *J. Mol. Biol.* **253**, 151.

Roseman, A. M. (2004). FindEM—A fast, efficient program for automatic selection of particles from electron micrographs. *J. Struct. Biol.* **145**, 91.

Rost, B. (2002). Enzyme function less conserved than anticipated. *J. Mol. Biol.* **318**, 595.

Vale, R. D. (2000). AAA proteins. Lords of the ring. *J. Cell Biol.* **150**, F13.

Winkler, H. (2007). 3D reconstruction and processing of volumetric data in cryo-electron tomography. *J. Struct. Biol.* **157**, 126.

Zanetti, G., *et al.* (2009). Contrast transfer function correction applied to cryo-electron tomography and sub-tomogram averaging. *J. Struct. Biol.* **168**, 305.

Zhang, X., *et al.* (2001). X-ray structure analysis and crystallographic refinement of lumazine synthase from the hyperthermophile *Aquifex aeolicus* at 1.6 A resolution: Determinants of thermostability revealed from structural comparisons. *J. Mol. Biol.* **306**, 1099.

CHAPTER TWELVE

CRYOELECTRON TOMOGRAPHY OF EUKARYOTIC CELLS

Asaf Mader,*,† Nadav Elad,*,† *and* Ohad Medalia*,†

Contents

Abstract

Biological processes involve a high degree of protein dynamics resulting in a constant remodeling of the cellular landscape at the molecular level. Orchestrated changes lead to significant rearrangement of the eukaryotic cytoskeleton and nuclear structures. Visualization of the cellular landscape in the unperturbed state is essential for understanding these processes. The development of cryoelectron tomography (cryo-ET) and its application to eukaryotic cells has provided a major step forward toward better realizing these processes. In conjunction with rapid freezing techniques, that is, vitrification by plunge-freezing and high-pressure freezing, cryo-ET is most suitable for investigating cellular ultrastructures in a close-to-life state. Here, we review the application of cryo-ET to the study of eukaryotic cells, with special emphasis on sample preparation, cytoskeleton organization, and macromolecular structures observed at a resolution of 4–6 nm.

* Department of Life Sciences, Ben Gurion University of the Negev, Beer-Sheva, Israel
† The National Institute for Biotechnology in the Negev, Ben Gurion University of the Negev, Beer-Sheva, Israel

Methods in Enzymology, Volume 483
ISSN 0076-6879, DOI: 10.1016/S0076-6879(10)83012-5

1. INTRODUCTION

Nearly every major process that occurs in a cell is conducted by multiprotein complexes, often referred to as molecular machines. These complexes, composed of proteins and nucleic acids, form functional units capable of performing specific cellular tasks. Detailed structural studies could shed light on understanding individual cellular processes. Such information is needed to comprehend the complexity of the cell and to understand how multiple protein assemblies participate in tightly regulated processes within the crowded cellular environment (Campbell, 2008).

Since its inception, electron microscopy (EM) has been central for the visualization of cellular structures with high spatial resolution. Indeed, our basic knowledge of cells, their various organelles, membranous compartments, cytoskeleton filaments, and even small cytoplasmatic particles, such as ribosomes, have been provided by EM imaging. Conventional sample preparation techniques, however, require physical and chemical perturbations, such as chemical fixation, dehydration, plastic-embedding, sectioning, and heavy metal staining.

Technical advents in microscopic imaging techniques, such as cryoelectron microscopy (cryo-EM), have led to unprecedented capabilities for examining unperturbed cellular structures. Cryo-EM, in conjunction with automated electron tomography (ET; Dierksen *et al.*, 1992, 1993), now allows three-dimensional (3D) visualization of polymorphic cellular structures through the recording of projection series obtained at varying angles around one or two axes, followed by reconstruction of the 3D volume (Koning and Koster, 2009; Lucic *et al.*, 2005). Thus, cryoelectron tomography (cryo-ET) enables 3D investigation of prokaryotes (Gan *et al.*, 2008; Iancu *et al.*, 2010; Konorty *et al.*, 2008; Li and Jensen, 2009; Milne and Subramaniam, 2009), eukaryotes (Henderson *et al.*, 2007; Kurner *et al.*, 2004; Medalia *et al.*, 2002), and purified molecular complexes (Saibil, 2000; Sali *et al.*, 2003) embedded in amorphous, vitrified buffers. As such, cryo-ET bridges structural studies with molecular and cell biology approaches, providing complimentary information toward unraveling the functional networks found within cells (Ben-Harush *et al.*, 2010).

Vitrification by rapid freezing (10^{-5} s) arrests cellular processes instantly (McDowall *et al.*, 2010), circumventing molecular changes owing to chemical fixation, heavy metal staining, or dehydration and, therefore, ensures close-to-life representation of the proteome. Still, many eukaryotic cells are too thick to be imaged directly by cryo-ET, due to the inelastic scattering of electrons that occurs while penetrating a sample thicker than a micron. When sample thickness exceeds the mean free path of the electrons, multiple inelastic scattering occurs, degrading the quality of the image, even when medium voltages (300–400 keV) and energy filters are used (Grimm *et al.*, 1996). Dealing with thicker cellular regions and tissues requires physically processing the samples using high-pressure freezing (Moor, 1987) followed by cryosectioning (Al-Amoudi *et al.*, 2004; Ladinski *et al.*, 2010), cryosectioning with a focused-ion-beam (FIB; Marko *et al.*, 2007) or thinning freeze-plunged cells (Rigort *et al.*, 2010). Hence, cryo-ET has been more successfully applied to prokaryotes and viruses, to provide novel structural information. Nevertheless, cryo-ET of eukaryotic cells is possible but is mostly limited to flat thin cells or cell peripheries. Exceptionally, thin eukaryotic cells can be studied as a whole, such as *Ostreococcus tauri* (Henderson *et al.*, 2007), where all parts of the cells can be reconstructed within a single tomogram (Fig. 12.1).

Traditionally, cellular components and events have been tracked and captured using fluorescent-light microscopy. In contrast, cryo-ET only allows study of an area of $\sim 2\ \mu m^2$, representing only 1.5% of the peripheral areas of a typical eukaryotic cell. Therefore, it is difficult to localize specific biological processes or complexes of interest using low-dose cryo-EM conditions. Accordingly, correlating cryo-EM with fluorescent-light microscopy imaging (i.e., correlative light-electron microscopy—CLEM) was recently introduced to unambiguously identify specific cellular components and processes (Chen and Briegel, 2010; Lucic *et al.*, 2008; Plitzko *et al.*, 2009; Robinson *et al.*, 2001; Vicidomini *et al.*, 2010; Weston *et al.*, 2010). Thus, fluorescent microscopy assists by locating and mapping a desired cellular feature, which can later be found under the electron beam

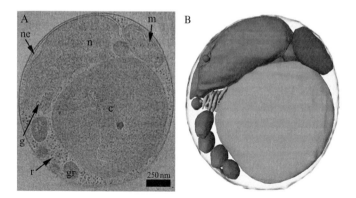

Figure 12.1   **Cryo-ET of entire eukaryotic cell.** (A) Single *x-y* slice through the middle of a single *O. tauri* cell. In these experiments, the cells were found to be mainly non-dividing, 21.6-nm. Shown are nuclei (n), nuclear envelope (ne), chloroplasts (c), mitochondria (m), Golgi bodies (g), granules (gr), and ribosome-like particles (r). (B) 3-D rendered view of *O. tauri* cell, 36-nm. Adapted from Henderson *et al.* (2007). (See Color Insert.)

and studied in 3D at higher magnification by cryo-ET. Hence, combining these two methodologies allows for a resolving of the molecular architecture of cells in terms of specific functional processes and states.

In this review, we address the application of cryo-ET to the analysis of the cytoplasm and nuclear periphery of eukaryotic cells. In particular, we will focus on cytoskeleton-based structures, macromolecular complexes such as the nuclear pore complexes (NPCs), and different methods allowing for localization of cellular processes.

2. SPECIMEN PREPARATION

Cryo-ET provides a unique opportunity to study macromolecules *in situ*. However, this technique is strictly limited by specimen thickness, as described above. Therefore, when growing cells on EM grids, conditions should be optimized to allow cells to spread maximally.

Unprecedented information has been deduced by studying whole thin prokaryotic cells by cryo-ET, without the need for sectioning (Komeili *et al.*, 2006; Kürner and Baumeister, 2006; Seybert *et al.*, 2006; Zhang *et al.*, 2007). On the other hand, sample preparation of eukaryotic cells for cryo-ET is more challenging, since most eukaryotic cells exceed the sample thickness limitation (1000 nm thickness for 300 keV). Hence, to directly visualize unaltered eukaryotic cells, we are restricted to sufficiently flat cells or to cell peripheries. Dealing with thicker specimens, such as concentrated cell suspensions or tissue samples, requires high-pressure freezing (Moor, 1987), a technique that enables the vitrification of an about 200 μm thick biological sample through the use of a $\sim$2000 bar jet of liquid nitrogen (Studer *et al.*, 2008). The frozen-hydrated specimen is then cryosectioned into 50–200 nm thick sections, by cryoultramicrotomy (Al-Amoudi *et al.*, 2004; Hsieh *et al.*, 2006; McDowall *et al.*, 1983), or thinned by an FIB (Marko *et al.*, 2007). FIB thinning usually utilizes a gallium ion beam for gentle removal of a few nanometers of material, while keeping the specimen in the vitrified state, thereby minimizing artifacts, such as roughness, knife marks, or deformation attributed to cryosectioning. Cryoplaning can also facilitate thinning of large areas of vitreous ice prior to cryofluorescence, FIB thinning, and cryo-ET. Through the use of a customized 35° diamond knife with a clearance angle of 6°, planing the ice surface can be carried out using a nominal "feed" (increment) of 6 nm and a "block" speed of 5 mm/s at -150 °C. An antistatic device was reported to assist in clearing debris from the knife edge. Thus, cryoplaning was introduced as a new concept for mechanical prethinning samples for cryo-ET (Rigort *et al.*, 2010).

The first application of cryo-ET for the study of an intact eukaryotic cell was documented by Medalia *et al.* (2002). In this study, the authors addressed the soil-living amoeba, *Dictyostelium discoideum*, resolving macro-molecular complexes, namely ribosomes and 26S proteasomes, as well as actin filaments. These cells remained viable and were able to adhere and spread on the electron microscope grid, a prerequisite for studying intact eukaryotic cells. However, it is noteworthy that *D. discoideum* are fast motile cells that spread and adhere to surfaces within minutes.

The unique eukaryotic cell, *O. tauri*, can also be studied as a whole. This unicellular green alga does not need to be attached to the EM grid for enabling the application of cryo-ET to resolve its tightly packed organelles and ultrastructure (Henderson *et al.*, 2007). Figure 12.1 shows a cross-section and 3D segmentation of a cell harvested at the dark-to-light transition. Since its organelles were not visibly dividing, this example was classified as "nondividing" cell.

So, while most eukaryotic cells are too thick to be captured in the intact state, their interactions with the extracellular matrix (ECM) and the external surfaces of their peripheral regions, such as adhesion sites, lamelleapodia, and other membrane protrusions, can be studied. For example, the actin network arrangement of intact *D. discoideum* filopodia was described in detail (Medalia *et al.*, 2007), as were luminal particles within mammalian cellular microtubules (MT; Garvalov *et al.*, 2006).

A beautiful example of the use of cryo-ET to structurally characterize a eukaryotic structure came from the work of Nicastro *et al.* (2005, 2006). Their reconstruction of a specific molecular complex, namely the MT-based scaffold structure of the flagella and axonemal dynein of *Chlamydomonas* and sea urchin sperm, exemplify the possibility for studying native eukaryotic structure at medium resolution. The eukaryotic flagellum is thin enough to allow for structural analysis addressing the molecular mechanism underlying flagellar beating. Since this organelle is highly complex and its motor protein, dynein, is large enough for tomographic reconstruction, this study showed the power of studying a given process at a close-to-life state. Figure 12.2A shows a transmission electron micrograph of an intact quiescent, frozen-hydrated sea urchin sperm flagellum. Due to the superposition of elements along the direction of the beam, fine details are not attainable However, after reconstruction, a detailed 3D view was achieved. The structure of the dynein motor, that is, the motor that moves each doublet MT relative to its neighbor, was identified and shown to include links connecting dynein arms with one another along the axoneme, as well as structures that connect inner and outer dynein arms (Fig. 12.2B). Comparing axoneme from wild-type *Chlamydomonas* and from the pf9-3 mutant (Myster *et al.*, 1997) which fails to assemble the I1 inner arm complex, a two-headed dynein isoform composed of two dynein heavy chains (1 alpha and 1 beta) and three intermediate chains, revealed differences in structure (Fig. 12.2C).

Adherent eukaryotic cells can be cultured on EM grids. Therefore, growth conditions for each cell line should be adjusted to allow optimal

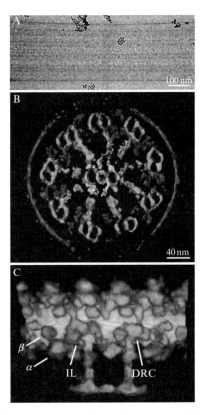

Figure 12.2 Cryo-ET of sperm flagellum from sea urchin and *Chlamydomonas*. (A) Transmission electron micrograph of an intact quiescent, frozen-hydrated sea urchin sperm flagellum. (B) A cross-section through a surface-rendered view of an axoneme. The structures were translationally averaged using 96-nm axial periodicity. Shown are the plasma membrane (brown), microtubules (gray), outer dynein arms (ODAs) and inner dynein arms (IDAs; red), radial spokes (orange), central pair protrusions (yellow), and bipartite bridges (pink). Scale bar: 40 nm. (C) Volume rendering presents the organization of the wild-type *Chlamydomonas* dynein arm viewed from the B tubule of the adjacent doublet, with the proximal end on the left. The 1-alpha and 1-beta motor domains of the I1complex, the IL which is the DIC/DLC-tail complex (dynein intermediate chains /dynein light chains) and the dynein regulatory complex (DRC) that regulates phosphatase and kinase activities on the doublet MTs are indicated. Adapted from Nicastro *et al.* (2005, 2006). (See Color Insert.)

cell growth and maximal attachment. Parameters which are often adjusted include grid geometry, continuous carbon versus holey grids, and protein coating of the carbon-coated grids.

Carbon-coated gold and platinum EM grids are typically favored over the widely used copper grids, due to the toxicity of copper ions to cells. To circumvent the toxic effect of copper ions, *D. discoideum* cells were cultured on copper grids coated with carbon on both sides (Medalia *et al.*, 2002), such that the surface of interaction between the buffer and the metal was minimized.

Carbon-coated EM grids are the most commonly used support for EM studies. Traditionally, cryo-EM is conducted using perforated carbon films, as the microscopy is performed on macromolecular complexes which are not absorbed to the support. However, cells can only spread when adhering to a support. Although a perforated support allows for removal of excessive solution without direct contact with cells, not all eukaryotes adhere to these EM grids. An alternative support which may be of some advantage is silicon membrane-coated grids. These commercially available EM grids (QUAN-TIFOIL, Jena) present different surface properties that allow certain cells to spread better, while allowing cryo-ET to be performed. The grid surface can be coated with proteins or other polymers by modifying the general protocol used for EM grid coating (Grassucci *et al.*, 2007). Since most eukaryotic cells interact with ECM proteins, coating the EM grids with such proteins prior to cell seeding can often assist the cells to better spread on the coated grid. Such coating of grids can be achieved by incubating glow-discharged grids on a drop of an ECM protein solution before placing the grids into a culture dish. For example, overlying the grids on 50 μg/ml fibronectin (Calbiochem, Darmstadt, Germany) for 45 min with the carbon side facing toward the solution.

Cells cultured on grids should be carefully examined to permit normal adherence and spreading to achieve an appearance indistinguishable from cells grown in a conventional tissue culture dish. The density of cells should be kept at the subconfluence level to achieve optimal results, since high cell density will eventually lead to thicker cellular structure that cannot be studied by cryo-ET. Consequently, we suggest the following steps for successful investigation:

1. Place the grids inside a culture dish containing growth medium. The grid carbon film should face up.
2. Seed the culture gently, by pipetting the adherent cells into the culture dish.
3. Leave the plate for 10–15 min to allow for cell attachment. Monitor cell density before incubating at 37 °C.
4. Incubate the grid in an incubator (37 °C, 5% CO_2) until the cells spread.

Finally, for cryo-ET, fiducial markers are often used to align projection images. Therefore, colloidal gold (10–15 nm in diameter) is routinely added before cell vitrification (Dubochet *et al.*, 1982). However, fiducial markers should be resuspended in physiological buffer when studying eukaryotic cells. For this purpose, the colloidal gold should be synthesized and pro-tected by physisorption of a protein, such as BSA, as previously described (Slot and Geuze, 1985).

3. Relying on Correlative Light and Electron Microscope for Cellular Structural Study of Eukaryotic Cells

The immense efforts devoted to uncovering protein–protein interactions and characterizing the interactome (Rual *et al.*, 2005) has also aimed at positioning macromolecular complexes within their native cellular environment. Approaches such as visual proteomics enable characterizing macromolecular complexes within their native cellular environment *in situ*, allowing for a full view of functional protein interactions (Beck *et al.*, 2010; Nickell *et al.*, 2006).

Cryo-ET provides the means to visualize single macromolecular assemblies in their native cellular context. Still, identifying specific macromolecules in the cellular context or in a cellular process remains challenging. The limited field of view and lack of genetically expressed markers associated with EM hampers the unambiguous identification of cellular process under cryoconditions. Correlating fluorescent light with cryo-EM thus offers a route to circumvent this obstacle, in some cases.

Fluorescence microscopy has revolutionized cell biology (Yuste, 2005). With the discovery of the green fluorescent protein (GFP) and its derivates, specific organelles, cellular events, and macromolecular complexes can be tracked by live cell microscopy (Chalfie *et al.*, 1994; Heim *et al.*, 1995; Shimomura *et al.*, 1962). Integrating cryo-ET with fluorescent-light microscopy can thus provide invaluable information relating the functional state of cellular processes and the location of specific organelles.

Fluorescence microscopy was previously correlated with EM studies using detergent-extracted, chemically fixed cells (Nemethova *et al.*, 2008). The design of a cryochamber for light and fluorescent microscopy (Sartori *et al.*, 2007; Schwartz *et al.*, 2007; van Driel *et al.*, 2009) now allows one to scan vitrified samples under the fluorescent microscope and use the same sample for cryo-ET. Cells, grown on grids, can be vitally labeled with one or more fluorescent dyes, such as fluorescent protein, vitrified, and directly examined. When considering a long process, the nonfixed biological specimen can undergo live fluorescent inspection and only then be vitrified for cryo-ET.

The correlative approach combining light (i.e., fluorescence) and EM (i.e., cryo-ET) is illustrated in Fig. 12.3. Cells cultured on an EM grid were mapped under the fluorescent microscope, with regions of interest, for example, focal adhesion sites, identified, and imaged prior to vitrification of the EM grid, without the use of fixative (Fig. 12.3A). Next, the exact region of interest was located by cryo-EM and a tilt series was collected and reconstructed (Fig. 12.3B).

A technical difficulty in CLEM is to pinpoint the exact fluorescent spot observed in the fluorescent microscope in the electron microscope under

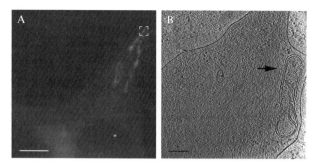

Figure 12.3   Correlative fluorescent and cryoelectron microscopy. (A) Cells expressing GFP-paxilin were examined under a fluorescent microscope. One cell is shown with the fluorescent region under scrutiny. (B) The cells were vitrified and cryo-ET was conducted on the protrusion shown in (A). The mitochondrion is clearly seen (black arrow). The actin cytoskeleton, as well as macromolecular complexes, can also be observed in the tomographic section, 10 nm in thickness. Scale bars: (A) 5 μm; (B) 200 nm.

cryoconditions. Grids containing markers, that is finder grids, are most suitable for this purpose. A major development came with the introduction of holders that can convert the coordinates from fluorescent microscope to cryo-EM, allowing easy localization of the region of interest in a fully automatic, computerized fashion (Lucic *et al.*, 2007; Sartori *et al.*, 2007), thereby minimizing the time needed to localize a given cellular structure under the electron beam.

A different strategy utilizes noninvasive genetic manipulations to generate covalent fusions with a protein, such as metallothionein (i.e., a protein that can bind gold atoms) designed to serve as an electron-dense GFP analogue for EM (Mercogliano and DeRosier, 2006). In this report, aurothiomalate, an anti-arthritic gold compound, reacted with metallothionein *in vitro* to form metallothionein–gold complexes, containing 17 gold atoms. This is a promising approach but needs to be further developed so as to attach larger quantities of gold atom to allow for an unambiguous visualization of heavy metal clusters *in situ*. Another promising approach uses quantum dots (QD), namely inorganic nanocrystals that fluoresce at distinct wavelengths, depending on their size (2–12 nm) and shape (Liu *et al.*, 2005). The core, typically a CdSe or CdTe crystal, is electron-dense, enabling discrimination of the distinct QDs at the EM level (Giepmans *et al.*, 2005). Since QDs possess high fluorescence, they allow single molecule detection. Moreover, the electron-dense crystal is directly visible by EM, and because of the different sizes and shapes, three different QDs can potentially be simultaneously used to label distinct proteins at the EM level (Deerinck *et al.*, 2007; Nisman *et al.*, 2004). QDs can be conjugated to streptavidin-tagged antibodies target to specific biotinylated-receptor proteins (Howarth *et al.*, 2008). This approach can potentially be applied for cryo-CLEM, allowing for the visualization of specific cellular processes. A variety of techniques have been used to incorporate QDs into cells and label cellular proteins, including mechanical delivery, receptor-mediated internalization, chemical transfection, and passive uptake. Conjugation of cell-penetrating peptides to QDs is in progress (Akerman *et al.*, 2002; Xue *et al.*, 2007), although some successful studies have already been reported (Hoshino *et al.*, 2004). Still, QDs exhibit relative low contrast in the electron microscope and, therefore, are not easily identified.

Although the optical resolution of an ordinary fluorescent microscope is about 200 nm, new light imaging modalities recently introduced new super spatial resolution capabilities. These include photoactivated localization microscopy (PALM; Betzig *et al.*, 2006), stimulated emission depletion microscopy (STED; Hell, 2003), stochastic optical reconstruction microscopy (STORM; Rust *et al.*, 2006), and structured illumination (Schermelleh *et al.*, 2008), with >40 nm resolution (as achieved with 3DSI, STED, PALM, STORM) providing the most suitable platform for CLEM. However, applying these methods without the use of chemical fixation is challenging and still needs additional development. Since most of the techniques that can break Abbe's law of resolution rely on multiple image collection of the sample and switching fluorophores between two states, the sample is typically fixed to achieve maximal resolution.

4. Cryoelectron Tomography of Cytoskeleton-Driven Processes

Eukaryotic cells move by an intricate mechanism of extension and retraction of their filamentous actin (F-actin) cytoskeleton network. During locomotion, cells extend thin dynamic protrusions at their leading edge, such as lamellipodia and filopodia. The actin cytoskeleton in eukaryotic cell is also the infrastructure for many fundamental processes and a major constituent of the cell shape regulation, adhesion, division, and motility machineries.

EM of actin cytoskeletons was traditionally performed on sectioned chemically fixed, detergent-extracted (Brown *et al.*, 1976; Svitkina *et al.*, 1997) or ventral membranes of cells (Heuser and Kirschner, 1980). While analysis of such preparations provided insight into the architecture of actin networks (Small *et al.*, 1994), the spatial resolution of the structures revealed was limited. For instance, interactions between filamentous cytoplasmic structures were difficult to resolve after metal decoration or replica formation. Moreover, artificial alteration of the cytoskeleton by detergent treatment prohibited the visualization of F-actin–plasma membrane interactions. In any case, bundling and cross-linking proteins, elements crucial for maintaining actin network architecture, may dissociate from the filaments and redistribute within a specimen during the course of traditional EM preparation procedures.

EM of vitrified yet unaltered cells has proven to be a key technique for the 3D reconstruction of actin architecture (Medalia *et al.*, 2002, 2007). As shown in Fig. 12.4, cryo-ET of intact *D. discoideum* cells reveals the intricate actin network, membranes, and cytoplasmatic macromolecular complexes (Fig. 12.4A) without the distortion or artifacts associated with detergent treatment. Using cryo-ET also permits viewing and study of the unaltered

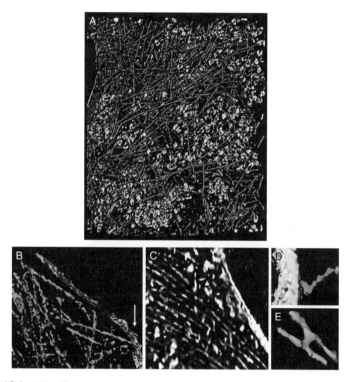

Figure 12.4   Visualizing the cytoplasm of a *Dictyostelium discoideum* cell. (A) Surface-rendered view of a volume of 815 nm × 870 nm × 97 nm. Colors were subjectively attributed to linear elements to mark the actin filaments (reddish); other macromolecular complexes, mostly ribosomes (green), and membranes (blue) are shown. (B, D) Actin anchorage to the membrane is illustrated in a surface-rendering of a cortical region demonstrating the plasma membrane associated with actin filaments ((B) is a rendered volume of 500 nm × 300 nm × 20 nm). The high magnification in (D) shows a kink-like structure close to the filament membrane-associating site, creating a filament-membrane bridge at a nonperpendicular angle. (C, E) Actin filament bundles in the cell cortex, and rendered volumes showing actin lateral connections to the membrane (C) or lateral bridges between two filaments (E), (extracted from (C)). Adapted from Medalia *et al.* (2002). (See Color Insert.)

anchorage of the actin filament network with the plasma membrane (Fig. 12.4B and 4D), as well as massive actin filament bundling at the cell cortex (Fig. 12.4C and E). Thus, cryo-ET is becoming a major structural tool to study cytoskeleton network architecture (Urban *et al.*, 2010), comprising actin and MTs, using the identical experimental setups routinely used in the field of cell biology, namely intact cells.

5. CRYOTOMOGRAPHY OF MIDBODIES

Another organelle we have recently characterized by cryo-ET is the midbody, a tightly packed MT-based bridge which transiently connects the two daughter cells at the end of cytokinesis. Assembled from condensed, central spindle-MTs and numerous associated proteins, the midbody gradually narrows until daughter cell partitioning occurs at this site (Barr and Gruneberg, 2007; Eggert *et al.*, 2006; Glotzer, 2005). Unlike the flagellum, the midbody features several hundreds of MTs arranged in parallel when assembled, together with several tens of MTs at it narrowest point, where abscission occurs. Furthermore, the midbody is extremely populated and highly electron-dense, properties which make it a challenging sample for cryo-ET, in particular, when one attempts to track individual MTs and detect macromolecular complexes.

Fluorescence microscopy and proteomic studies localize an ever-growing number of proteins to the midbody, including MT-associated and actin-associated proteins, as well as membrane trafficking and regulatory proteins (Echard *et al.*, 2004; Otegui *et al.*, 2005; Skop *et al.*, 2004; Steigemann and Gerlich, 2009). Some of these show distinctive sublocalizations when tagged fluorescently, for example PRC1, the MT-bundling protein that localizes to the overlap region (stem body; Mollinari *et al.*, 2002), the MT-severing protein, spastin, that is localized to two narrow segments on either side of the overlap region (Connell *et al.*, 2009), and centriolin, which is involved in membrane vesicle targeting and fusion and localized to a ring structure around the midbody (Gromley *et al.*, 2005). Nevertheless, the detailed macromolecular composition and MT organization that occur during the abscission process cannot be determined by current light microscopy techniques due to the compactness of the structure. Such high-resolution details require EM.

Midbody morphology was viewed early on by EM using sectioning and heavy metal staining (Byers and Abramson, 1968; McDonald *et al.*, 1979; McIntosh *et al.*, 1975; Mullins and Biesele, 1977; Paweletz, 1967). These studies and others (e.g., Euteneuer and McIntosh, 1980; McDonald *et al.*, 1979; Tippit *et al.*, 1980) determined that interdigitating antiparallel MTs, ending at the center of the telophase spindle (termed "polar MTs"), create the thicker overlap region. Tracking midbody MTs through serial sections confirmed this observation, and revealed that the midbody featured an additional population of MTs which cross the overlap region (termed "continuous" or "free" MTs) (Mastronarde *et al.*, 1993; McIntosh *et al.*, 1975).

Cryo-ET of midbodies enabled in-depth 3D analysis, providing a novel view of MT organization within this organelle and insight into its components. Since the investigated cells are typically round during mitosis, their midbody was obscured by thick ice or by the flanking daughter cells. Cryo-ET was, therefore, not unfeasible for intact midbodies, and midbody isolation was carried out according to established protocols (Mullins and McIntosh, 1982; Sellitto and Kuriyama, 1988). Isolated midbodies were easily identified under the electron beam and were suitable for cryo-ET, although their dense overlap region and its thickness prevented the gaining of detailed information in most tomograms. Nevertheless, tracking of MTs was possible in several tomograms and provided interesting results. At late telophase, the midbody was shown to be dominated by a core-bundle of MTs that transverse the electron-dense region, with their plus ends found outside the overlap region. The polar MTs that terminate in the overlap region surround this continuous bundle in an outer shell. A marked change in architecture was, however, observed in the outer shell of late-stage midbodies, where the polar MTs lost their interdigitation and retracted from the overlap region. These observations suggest that the midbody, having acquired a distinct MT architecture, as compared to the preceding central spindle, actively facilitates the final stage of cytokinesis. The spindle network, initially designed for separating the duplicated chromosome and later, for pushing the poles of the mitotic cell apart, is apparently redesigned upon formation of the midbody and adapts to its new role in daughter cell partitioning.

This example indicates that cytoskeleton elements can be detected even in an extremely dense environment and when sample is as thick as a micron. In some cases, as shown above, the resolution of the 3D map is sufficient to permit biological meaningful experiments and conclusions. Such detailed MT architecture could only be investigated when the whole mount midbodies are considered in their hydrated state, emphasizing the ability of cryo-EM to unravel detailed structural organization.

6. STRUCTURAL ANALYSIS OF THE NUCLEAR PORE COMPLEX BY CRYO-ET

The nuclear envelope (NE) is perforated by NPCs, which fuse the outer nuclear membrane (ONM) and the inner nuclear membrane (INM) to form aqueous translocation channels. These multiprotein assemblies allow passive diffusion of small molecules, as well as receptor-dependent, mediated translocation of large proteins and ribonucleoproteins. Macromolecular cargo usually harbor specific nuclear localization signals (NLSs) or nuclear export signals (NESs) that are recognized by transport receptors, mediating cargo passage through the NPC. Such receptors, referred to as karyopherins, chaperone cargo during transport across the NPC by means of hydrophobic interactions with phenylalanine-glycine-rich nucleoporin repeat domains (FG-repeats; Stewart, 2007; Suntharalingam and Wente, 2003).

NPCs are composed of ~ 30 different proteins termed nucleoporins (Nups) that are arranged as multimers containing multiple copies (Cronshaw *et al.*, 2002; Rout *et al.*, 2000; Terry *et al.*, 2007). As such, the NPC is one of the largest molecular machines in the cell, exhibiting a molecular weight of over 100 MDa in vertebrates (Beck and Medalia, 2008; Lim *et al.*, 2008). The structural organization of the NPC is largely conserved from yeast to man, comprising a pseudoeightfold symmetric central framework termed the spoke complex, a central pore of about 50 nm diameter, and filamentous structures on the cytoplasmic and nuclear sides of the complex (Elad *et al.*, 2009; Fig. 12.5). At its nuclear face, the NPC is found in close interaction with the

Figure 12.5 Structural analysis of the nuclear pore complex by Cryo-ET. (A) A 50-nm thick tomographic slice through an intact nucleus of *Dictyostelium discoideum*. The *D. discoideum* NPC was observed in cryotomograms of intact nuclei. (B) A reconstruction of the NPC from native spread NE of *Xenopus* oocytes. Luminal and cytoplasmic faces of the NPCs are shown in the upper and lower panels, respectively. Adapted from Beck *et al.* (2007), Elad *et al.* (2009), and Frenkiel-Krispin *et al.* (2010).

nuclear lamina, a meshwork of lamin filamentous structures (Ben-Harush *et al.*, 2009), and other associated proteins.

Structural examination of NPCs began several decades ago (Callan and Tomlin, 1950). However, the development of cryo-EM (Adrian *et al.*, 1984) has permitted detailed and artifact-free analysis of NPCs embedded in physiological buffers (Akey and Radermacher, 1993; Yang *et al.*, 1998). A landmark achievement in the field came with the application of cryo-ET to the intact NE and nuclei, enabling enhanced structural preservation with minimal purification steps (Beck *et al.*, 2004; Stoffler *et al.*, 2003). The nucleus had been considered as a dense and a thick specimen, beyond the limits of cryo-ET analysis. However, as will be discussed, cryo-ET of intact nuclei revealed the structure of the NPC, one of the largest macromolecular complexes in cells, in unprecedented detail.

While structural analysis of active NPCs has obvious advantages, the integrity of the NE is crucial for retaining pore activity (Gorlich and Kutay, 1999). Additionally, biochemical procedures which are traditionally used for the purification of NPCs can lead to a loss of components of the transport machinery and the cargo. Nuclei can be readily prepared from *D. discoideum* by filtering the cells through a polycarbonate filter. These nuclei retain their nuclear transport activity and can be instantly vitrified on EM grids (Beck *et al.*, 2004). Nuclei from this lower eukaryote are small (~ 2 μm) and are connected to the bulk of the ER by thin tubular connections (Muller-Taubenberger *et al.*, 2001). Figure 12.5A shows a slice through the reconstructed volume of such a nucleus, with discontinuity of the NE being reflected in a side view of NPC.

Using a 3D alignment algorithm (Forster *et al.*, 2005), the structure of the *D. discoideum* NPC was described without imposing eightfold symmetry. Instead, the structure was analyzed as eight protomers *in silico* (Beck *et al.*, 2007). Such analysis resulted in an improvement in resolution of the structure of the central framework to about 6 nm. Due to its inherent flexibility, such analysis did not, however, consider the nuclear basket. The estimated orientation of the NPC was estimated manually, since it is embedded on an ellipsoid surface, the nuclear membrane, reducing the need for large angular scan during the averaging procedure.

Luminal and cytoplasmic aspects of the *D. discoideum* NPC are presented in Fig. 12.5B. The central framework is ~ 60 nm in height and has an outer diameter of ~ 120 nm. The diameter of the main channel is about 50 nm. The structure of the metazoan NPC, as reconstructed from spread NEs of *Xenopus* oocytes, resembles that of *D. discoideum*. In both cases, the outer diameter of the structure and the opening of the main channel are very similar in dimension, namely 120 and 50 nm, respectively, although the *Xenopus* NPC seems to be taller (~ 100 nm, Fig. 12.5C; Stoffler *et al.*, 2003). Moreover, the distribution of densities appears to vary between the two species.

7. Concluding Remarks

Cryo-ET of eukaryotic cells has the potential to become a major approach in answering outstanding questions in cellular structural biology. The ability to structurally characterize systems which were routinely studied by means of cell biology techniques, for example, fluorescently based microscopy of wild-type and mutant cell lines, now using structural tools is intriguing. However, a major challenge still facing this technique concerns identifying macromolecular complexes. The variety of orientations macromolecules can assume and the current resolution of cellular tomograms, that is 4–6 nm, prohibit unbiased identification of many complexes, although some successful template-matching approaches have been introduced (Frangakis *et al.*, 2002; Seybert *et al.*, 2006). An alternative approach, based on electron-dense labeling, should be developed, focused on the design of a GFP analogue for cryo-EM. Such a strategy would provide a general solution for identifying complexes whose structure is not yet determined or which transiently interact to form large assemblies in cells.

In conclusion, to make cryo-ET applicable not only to cellular protrusions and thin regions of the eukaryotic cell but also to allow a structural view of organisms, development of a reliable freeze-hydrated, artifact-free sectioning technique that can be applied to tissues and cells is called for. Alternative microdissection techniques based on ion beam milling are currently being developed (Marko *et al.*, 2007; Rigort *et al.*, 2010). Realization of this technology would open a window for the entry of cryo-ET into other branches of biology and provide, for instance, a bridge between structural and developmental biology. Thus, we are currently at the first stages of applying cryo-ET to eukaryotic cells, entering biological territories which are not yet charted.

ACKNOWLEDGMENTS

This study was supported grants from the Israeli Science Foundation and the German–Israeli Cooperation Project (DIP) (H.2.2) and by an ERC Starting Grant to O. M.

REFERENCES

Adrian, M., *et al.* (1984). Cryo-electron microscopy of viruses. *Nature* **308,** 32–36.
Akerman, M. E., *et al.* (2002). Nanocrystal targeting in vivo. *Proc. Natl. Acad. Sci. USA* **99,** 12617–12621.
Akey, C. W., and Radermacher, M. (1993). Architecture of the *Xenopus* nuclear pore complex revealed by three-dimensional cryo-electron microscopy. *J. Cell Biol.* **122,** 1–19.

Al-Amoudi, A., *et al.* (2004). Cryo-electron microscopy of vitreous sections. *EMBO J.* **23,** 3583–3588.

Barr, F. A., and Gruneberg, U. (2007). Cytokinesis: Placing and making the final cut. *Cell* **131,** 847–860.

Beck, M., and Medalia, O. (2008). Structural and functional insights into nucleocytoplasmic transport. *Histol. Histopathol.* **23,** 1025–1033.

Beck, M., *et al.* (2004). Nuclear pore complex structure and dynamics revealed by cryoelectron tomography. *Science* **306,** 1387–1390.

Beck, M., *et al.* (2007). Snapshots of nuclear pore complexes in action captured by cryoelectron tomography. *Nature* **449,** 611–615.

Beck, M., *et al.* (2010). Visual proteomics in *Leptospira interrogans. Methods Enzymol.* **1.** Cryo-EM. Part 2 Lessons learned from specific examples chapter.

Ben-Harush, K., *et al.* (2009). The supramolecular organization of the *C. elegans* nuclear lamin filament. *J. Mol. Biol.* **386,** 1392–1402.

Ben-Harush, K., *et al.* (2010). Visualizing cellular processes at the molecular level by cryoelectron tomography. *J. Cell Sci.* **123,** 7–12.

Betzig, E., *et al.* (2006). Imaging intracellular fluorescent proteins at nanometer resolution. *Science* **313,** 1642–1645.

Brown, S., *et al.* (1976). Cytoskeletal elements of chick embryo fibroblasts revealed by detergent extraction. *J. Supramol. Struct.* **5,** 119–130.

Byers, B., and Abramson, D. H. (1968). Cytokinesis in HeLa: Post-telophase delay and microtubule-associated motility. *Protoplasma* **66,** 413–435.

Callan, H. G., and Tomlin, S. G. (1950). Experimental studies on amphibian oocyte nuclei. 1. Investigation of the structure of the nuclear membrane by means of the electron microscope. *Proc. R. Soc. Lond. B Biol. Sci.* **137,** 367–378.

Campbell, I. D. (2008). The Croonian lecture 2006. Structure of the living cell. *Philos. Trans. R. Soc. Lond. B Biol. Sci.* **363,** 2379–2391.

Chalfie, M., *et al.* (1994). Green fluorescent protein as a marker for gene expression. *Science* **263,** 802–805.

Chen, S., and Briegel, A. (2010). Correlative cryo-light and –electron microscopy. *Methods Enzymol.* **1.** Cryo-EM. Part 1 Data collection chapter.

Connell, J. W., *et al.* (2009). Spastin couples microtubule severing to membrane traffic in completion of cytokinesis and secretion. *Traffic* **10,** 42–56.

Cronshaw, J. M., *et al.* (2002). Proteomic analysis of the mammalian nuclear pore complex. *J. Cell Biol.* **158,** 915–927.

Deerinck, T. J., *et al.* (2007). Light and electron microscopic localization of multiple proteins using quantum dots. *Methods Mol. Biol.* **374,** 43–53.

Dierksen, K., *et al.* (1992). Towards automatic electron tomography. *Ultramicroscopy* **40,** 71–87.

Dierksen, K., *et al.* (1993). Towards automatic electron tomography. 2. Implementation of autofocus and low-dose procedures. *Ultramicroscopy* **49,** 109–120.

Dubochet, J., *et al.* (1982). Electron-microscopy of frozen water and aqueous-solutions. *J. Microsc. Oxf.* **128,** 219–237.

Echard, A., *et al.* (2004). Terminal cytokinesis events uncovered after an RNAi screen. *Curr. Biol.* **14,** 1685–1693.

Eggert, U. S., *et al.* (2006). Animal cytokinesis: From parts list to mechanisms. *Annu. Rev. Biochem.* **75,** 543–566.

Elad, N., *et al.* (2009). Structural analysis of the nuclear pore complex by integrated approaches. *Curr. Opin. Struct. Biol.* **19,** 226–232.

Euteneuer, U., and McIntosh, J. R. (1980). Polarity of midbody and phragmoplast microtubules. *J. Cell Biol.* **87,** 509–515.

Forster, F., *et al.* (2005). Retrovirus envelope protein complex structure in situ studied by cryo-electron tomography. *Proc. Natl. Acad. Sci. USA* **102,** 4729–4734.

Frangakis, A. S., *et al.* (2002). Identification of macromolecular complexes in cryoelectron tomograms of phantom cells. *Proc. Natl. Acad. Sci. USA* **99,** 14153–14158.

Frenkiel-Krispin, D., *et al.* (2010). Structural analysis of a metazoan nuclear pore complex reveals a fused concentric ring architecture. *J. Mol. Biol.* **395,** 578–586.

Gan, L., *et al.* (2008). Molecular organization of Gram-negative peptidoglycan. *Proc. Natl. Acad. Sci. USA* **105,** 18953–18957.

Garvalov, B. K., *et al.* (2006). Luminal particles within cellular microtubules. *J. Cell Biol.* **174,** 759–765.

Giepmans, B. N., *et al.* (2005). Correlated light and electron microscopic imaging of multiple endogenous proteins using Quantum dots. *Nat. Methods* **2,** 743–749.

Glotzer, M. (2005). The molecular requirements for cytokinesis. *Science* **307,** 1735–1739.

Gorlich, D., and Kutay, U. (1999). Transport between the cell nucleus and the cytoplasm. *Annu. Rev. Cell Dev. Biol.* **15,** 607–660.

Grassucci, R. A., *et al.* (2007). Preparation of macromolecular complexes for cryo-electron microscopy. *Nat. Protoc.* **2,** 3239–3246.

Grimm, R., *et al.* (1996). Determination of the inelastic mean free path in ice by examination of tilted vesicles and automated most probable loss imaging. *Ultramicroscopy* **63,** 169–179.

Gromley, A., *et al.* (2005). Centriolin anchoring of exocyst and SNARE complexes at the midbody is required for secretory-vesicle-mediated abscission. *Cell* **123,** 75–87.

Heim, R., *et al.* (1995). Improved green fluorescence. *Nature* **373,** 663–664.

Hell, S. W. (2003). Toward fluorescence nanoscopy. *Nat. Biotechnol.* **21,** 1347–1355.

Henderson, G. P., *et al.* (2007). 3-D ultrastructure of *O. tauri*: Electron cryotomography of an entire eukaryotic cell. *PLoS ONE* **2,** e749.

Heuser, J. E., and Kirschner, M. W. (1980). Filament organization revealed in platinum replicas of freeze-dried cytoskeletons. *J. Cell Biol.* **86,** 212–234.

Hoshino, A., *et al.* (2004). Quantum dots targeted to the assigned organelle in living cells. *Microbiol. Immunol.* **48,** 985–994.

Howarth, M., *et al.* (2008). Monovalent, reduced-size quantum dots for imaging receptors on living cells. *Nat. Methods* **5,** 397–399.

Hsieh, C. E., *et al.* (2006). Towards high-resolution three-dimensional imaging of native mammalian tissue: Electron tomography of frozen-hydrated rat liver sections. *J. Struct. Biol.* **153,** 1–13.

Iancu, C. V., *et al.* (2010). Organization, structure, and assembly of alpha-carboxysomes determined by electron cryotomography of intact cells. *J. Mol. Biol.* **396,** 105–117.

Komeili, A., *et al.* (2006). Magnetosomes are cell membrane invaginations organized by the actin-like protein MamK. *Science* **311,** 242–245.

Koning, R. I., and Koster, A. J. (2009). Cryo-electron tomography in biology and medicine. *Ann. Anat.* **191,** 427–445.

Konorty, M., *et al.* (2008). Structural analysis of photosynthetic membranes by cryo-electron tomography of intact *Rhodopseudomonas viridis* cells. *J. Struct. Biol.* **161,** 393–400.

Kürner, J., and Baumeister, W. (2006). Cryo-Electron Tomography Reveals the Architecture of a Bacterial Cytoskeleton, in Complex Intracellular Structures in Prokaryotes, Springer-Verlag, Berlin-Heidelberg, pp. 313–318.

Kurner, J., *et al.* (2004). New insights into the structural organization of eukaryotic and prokaryotic cytoskeletons using cryo-electron tomography. *Exp. Cell Res.* **301,** 38–42.

Ladinski, M., *et al.* (2010). High-pressure freezing and cryosectioning thick samples. *Methods Enzymol.* **1.** Cryo-EM. Part 1 Sample preparation chapter.

Li, Z., and Jensen, G. J. (2009). Electron cryotomography: A new view into microbial ultrastructure. *Curr. Opin. Microbiol.* **12,** 333–340.

Lim, R. Y., *et al.* (2008). Towards reconciling structure and function in the nuclear pore complex. *Histochem. Cell Biol.* **129**, 105–116.

Liu, T., *et al.* (2005). The fluorescence bioassay platforms on quantum dots nanoparticles. *J. Fluoresc.* **15**, 729–733.

Lucic, V., *et al.* (2005). Structural studies by electron tomography: From cells to molecules. *Annu. Rev. Biochem.* **74**, 833–865.

Lucic, V., *et al.* (2007). Multiscale imaging of neurons grown in culture: From light microscopy to cryo-electron tomography. *J. Struct. Biol.* **160**, 146–156.

Lucic, V., *et al.* (2008). Cryo-electron tomography of cells: Connecting structure and function. *Histochem. Cell Biol.* **130**, 185–196.

Marko, M., *et al.* (2007). Focused-ion-beam thinning of frozen-hydrated biological specimens for cryo-electron microscopy. *Nat. Methods* **4**, 215–217.

Mastronarde, D. N., *et al.* (1993). Interpolar spindle microtubules in PTK cells. *J. Cell Biol.* **123**, 1475–1489.

McDonald, K. L., *et al.* (1979). Cross-sectional structure of the central mitotic spindle of *Diatoma vulgare*. Evidence for specific interactions between antiparallel microtubules. *J. Cell Biol.* **83**, 443–461.

McDowall, A. W., *et al.* (1983). Electron microscopy of frozen hydrated sections of vitreous ice and vitrified biological samples. *J. Microsc.* **131**, 1–9.

McDowall, A. W., *et al.* (2010). Plunge-freezing isolated macomolecular complexes, viruses, and small cells. *Methods Enzymol.* **1**. Cryo-EM. Part 1 Sample preparation chapter.

McIntosh, J. R., *et al.* (1975). Structure and physiology of the mammalian mitotic spindle. *Soc. Gen. Physiol. Ser.* **30**, 31–76.

Medalia, O., *et al.* (2002). Macromolecular architecture in eukaryotic cells visualized by cryoelectron tomography. *Science* **298**, 1209–1213.

Medalia, O., *et al.* (2007). Organization of actin networks in intact filopodia. *Curr. Biol.* **17**, 79–84.

Mercogliano, C. P., and DeRosier, D. J. (2006). Gold nanocluster formation using metallothionein: Mass spectrometry and electron microscopy. *J. Mol. Biol.* **355**, 211–223.

Milne, J. L., and Subramaniam, S. (2009). Cryo-electron tomography of bacteria: Progress, challenges and future prospects. *Nat. Rev. Microbiol.* **7**, 666–675.

Mollinari, C., *et al.* (2002). PRC1 is a microtubule binding and bundling protein essential to maintain the mitotic spindle midzone. *J. Cell Biol.* **157**, 1175–1186.

Moor, H. (1987). Theory and practice of high-pressure freezing. In "Cryotechniques in Biological Electron Microscopy," (R. A. Steinbrecht and K. Zierold, eds.), pp. 175–191. Springer-Verlag, Berlin.

Muller-Taubenberger, A., *et al.* (2001). Calreticulin and calnexin in the endoplasmic reticulum are important for phagocytosis. *EMBO J.* **20**, 6772–6782.

Mullins, J. M., and Biesele, J. J. (1977). Terminal phase of cytokinesis in D-98s cells. *J. Cell Biol.* **73**, 672–684.

Mullins, J. M., and McIntosh, J. R. (1982). Isolation and initial characterization of the mammalian midbody. *J. Cell Biol.* **94**, 654–661.

Myster, S. H., *et al.* (1997). The *Chlamydomonas* Dhc1 gene encodes a dynein heavy chain subunit required for assembly of the I1 inner arm complex. *Mol. Biol. Cell* **8**, 607–620.

Nemethova, M., *et al.* (2008). Building the actin cytoskeleton: Filopodia contribute to the construction of contractile bundles in the lamella. *J. Cell Biol.* **180**, 1233–1244.

Nicastro, D., *et al.* (2005). 3D structure of eukaryotic flagella in a quiescent state revealed by cryo-electron tomography. *Proc. Natl. Acad. Sci. USA* **102**, 15889–15894.

Nicastro, D., *et al.* (2006). The molecular architecture of axonemes revealed by cryoelectron tomography. *Science* **313**, 944–948.

Nickell, S., *et al.* (2006). A visual approach to proteomics. *Nat. Rev. Mol. Cell Biol.* **7**, 225–230.

Nisman, R., *et al.* (2004). Application of quantum dots as probes for correlative fluorescence, conventional, and energy-filtered transmission electron microscopy. *J. Histochem. Cytochem.* **52**, 13–18.

Otegui, M. S., *et al.* (2005). Midbodies and phragmoplasts: Analogous structures involved in cytokinesis. *Trends Cell Biol.* **15**, 404–413.

Paweletz, N. (1967). On the function of the "Flemming body" during division of animal cells. *Naturwissenschaften* **54**, 533–535.

Plitzko, J. M., *et al.* (2009). Correlative cryo-light microscopy and cryo-electron tomography: From cellular territories to molecular landscapes. *Curr. Opin. Biotechnol.* **20**, 83–89.

Rigort, A., *et al.* (2010). Micromachining tools and correlative approaches for cellular cryo-electron tomography. *J. Struct. Biol.* (In press)

Robinson, J. M., *et al.* (2001). Correlative fluorescence and electron microscopy on ultrathin cryosections: Bridging the resolution gap. *J. Histochem. Cytochem.* **49**, 803–808.

Rout, M. P., *et al.* (2000). The yeast nuclear pore complex: Composition, architecture, and transport mechanism. *J. Cell Biol.* **148**, 635–651.

Rual, J. F., *et al.* (2005). Towards a proteome-scale map of the human protein–protein interaction network. *Nature* **437**, 1173–1178.

Rust, M. J., *et al.* (2006). Sub-diffraction-limit imaging by stochastic optical reconstruction microscopy (STORM). *Nat. Methods* **3**, 793–795.

Saibil, H. R. (2000). Conformational changes studied by cryo-electron microscopy. *Nat. Struct. Biol.* **7**, 711–714.

Sali, A., *et al.* (2003). From words to literature in structural proteomics. *Nature* **422**, 216–225.

Sartori, A., *et al.* (2007). Correlative microscopy: Bridging the gap between fluorescence light microscopy and cryo-electron tomography. *J. Struct. Biol.* **160**, 135–145.

Schermelleh, L., *et al.* (2008). Subdiffraction multicolor imaging of the nuclear periphery with 3D structured illumination microscopy. *Science* **320**, 1332–1336.

Schwartz, C. L., *et al.* (2007). Cryo-fluorescence microscopy facilitates correlations between light and cryo-electron microscopy and reduces the rate of photobleaching. *J. Microsc.* **227**, 98–109.

Sellitto, C., and Kuriyama, R. (1988). Distribution of a matrix component of the midbody during the cell cycle in Chinese hamster ovary cells. *J. Cell Biol.* **106**, 431–439.

Seybert, A., *et al.* (2006). Structural analysis of Mycoplasma pneumoniae by cryo-electron tomography. *J. Struct. Biol.* **156**, 342–354.

Shimomura, O., *et al.* (1962). Extraction, purification and properties of aequorin, a bioluminescent protein from the luminous hydromedusan, Aequorea. *J. Cell. Comp. Physiol.* **59**, 223–239.

Skop, A. R., *et al.* (2004). Dissection of the mammalian midbody proteome reveals conserved cytokinesis mechanisms. *Science* **305**, 61–66.

Slot, J. W., and Geuze, H. J. (1985). A new method of preparing gold probes for multiple-labeling cytochemistry. *Eur. J. Cell Biol.* **38**, 87–93.

Small, J. V., *et al.* (1994). Visualization of actin filaments in keratocyte lamellipodia: Negative staining compared with freeze-drying. *J. Struct. Biol.* **113**, 135–141.

Steigemann, P., and Gerlich, D. W. (2009). Cytokinetic abscission: Cellular dynamics at the midbody. *Trends Cell Biol.* **19**, 606–616.

Stewart, M. (2007). Molecular mechanism of the nuclear protein import cycle. *Nat. Rev. Mol. Cell Biol.* **8**, 195–208.

Stoffler, D., *et al.* (2003). Cryo-electron tomography provides novel insights into nuclear pore architecture: Implications for nucleocytoplasmic transport. *J. Mol. Biol.* **328**, 119–130.

Studer, D., *et al.* (2008). Electron microscopy of high pressure frozen samples: Bridging the gap between cellular ultrastructure and atomic resolution. *Histochem. Cell Biol.* **130**, 877–889.

Suntharalingam, M., and Wente, S. R. (2003). Peering through the pore: Nuclear pore complex structure, assembly, and function. *Dev. Cell* **4**, 775–789.

Svitkina, T. M., *et al.* (1997). Analysis of the actin-myosin II system in fish epidermal keratocytes: Mechanism of cell body translocation. *J. Cell Biol.* **139**, 397–415.

Terry, L. J., *et al.* (2007). Crossing the nuclear envelope: Hierarchical regulation of nucleocytoplasmic transport. *Science* **318**, 1412–1416.

Tippit, D. H., *et al.* (1980). Organization of spindle microtubules in *Ochromonas danica*. *J. Cell Biol.* **87**, 531–545.

Urban, E., *et al.* (2010). Electron tomography reveals unbranched networks of actin filaments in lamellipodia. *Nat. Cell Biol.* **12**, 429–435.

van Driel, L. F., *et al.* (2009). Tools for correlative cryo-fluorescence microscopy and cryo-electron tomography applied to whole mitochondria in human endothelial cells. *Eur. J. Cell Biol.* **88**, 669–684.

Vicidomini, G., *et al.* (2010). A novel approach for correlative light electron microscopy analysis. *Microsc. Res. Tech.* **73**, 215–224.

Weston, A. E., *et al.* (2010). Towards native-state imaging in biological context in the electron microscope. *J. Chem. Biol.* **3**, 101–112.

Xue, F. L., *et al.* (2007). Enhancement of intracellular delivery of CdTe quantum dots (QDs) to living cells by Tat conjugation. *J. Fluoresc.* **17**, 149–154.

Yang, Q., *et al.* (1998). Three-dimensional architecture of the isolated yeast nuclear pore complex: Functional and evolutionary implications. *Mol. Cell* **1**, 223–234.

Yuste, R. (2005). Fluorescence microscopy today. *Nat. Methods* **2**, 902–904.

Zhang, P., *et al.* (2007). Direct visualization of *Escherichia coli* chemotaxis receptor arrays using cryo-electron microscopy. *Proc. Natl. Acad. Sci. USA* **104**, 3777–3781.

CHAPTER THIRTEEN

3D Visualization of HIV Virions by Cryoelectron Tomography

Jun Liu,* Elizabeth R. Wright,† *and* Hanspeter Winkler‡

Contents

Abstract

The structure of the human immunodeficiency virus (HIV) and some of its components have been difficult to study in three-dimensions (3D) primarily because of their intrinsic structural variability. Recent advances in cryoelectron tomography (cryo-ET) have provided a new approach for determining the 3D structures of the intact virus, the HIV capsid, and the envelope glycoproteins located on the viral surface. A number of cryo-ET procedures related to specimen preservation, data collection, and image processing are presented in this

* Department of Pathology and Laboratory Medicine, University of Texas Medical School at Houston, Houston, Texas, USA
† Department of Pediatrics, Emory University School of Medicine, Atlanta, Georgia, USA
‡ Institute of Molecular Biophysics, Florida State University, Tallahassee, Florida, USA

Methods in Enzymology, Volume 483
ISSN 0076-6879, DOI: 10.1016/S0076-6879(10)83014-9

chapter. The techniques described herein are well suited for determining the ultrastructure of bacterial and viral pathogens and their associated molecular machines *in situ* at nanometer resolution.

1. INTRODUCTION

Human immunodeficiency virus (HIV) is a spherical retrovirus with a diameter of ∼100–150 nm. The lipid-enveloped virus first buds from the host cell in an immature form (Freed, 1998). The major structural proteins are unprocessed at this point and are referred to as the Gag polyprotein. The proteins are arranged radially in the following order from the lipid bilayer: matrix (MA), capsid (CA), SP1, nucleocapsid (NC), SP2, and p6 (Fuller *et al.*, 1997; Wilk *et al.*, 2001). During the budding process, the viral protease (PR) becomes active and initiates cleavage of the major protein domains of the Gag polyprotein, which results in a major structural rearrangement known as "maturation." It is during this process that the CA domains reorganize to form the bullet-shaped core or "capsid" that encompasses the viral genome of the mature virus. Throughout the entire process of budding and maturation, the virus is enclosed by a viral envelope consisting of a lipid bilayer in which the envelope glycoprotein (Env) is embedded. HIV Env binds CD4 receptors and coreceptors on target cells, thereby initiating viral entry and infection. Native HIV Env is a trimeric complex organized as three heterodimers. The heterodimers consist of a noncovalently associated extracellular subunit gp120 and a transmembrane subunit gp41. Although several crystal structures of the gp41 and the gp120 monomer are known (Chan *et al.*, 1997; Chen *et al.*, 2005, 2009; Huang *et al.*, 2005; Kwong *et al.*, 1998; Zhou *et al.*, 2007), the crystal structure of the HIV Env trimer has not been determined.

Electron microscopy (EM) of sectioned materials and negatively stained virions has contributed significantly to our understanding of the morphology and fine structure of HIV (Gelderblom *et al.*, 1987). However, conventional EM methods have been unable to provide information about the 3D architecture of the virus. The advent of cryoelectron microscopy (cryo-EM) made it possible for the structure of both the immature and mature forms of the virus to be readdressed. In several landmark papers, cryo-EM studies of the immature virus revealed the radial arrangement of the Gag polyprotein domains; determined that the CA domain forms a hexagonal lattice with a distinctive spacing of ∼8.0 nm; and defined the number of individual CA proteins required to compose the core of the mature virus (Briggs *et al.*, 2004, 2006b; Fuller *et al.*, 1997; Nermut *et al.*, 1998). Concurrently, the mature virus was examined by cryo-EM and it was determined that virion size remains constant through the maturation process;

virions can have conical or altered shaped cores; virions can have multiple cores; subunits of the CA protein of the cores form a hexagonal lattice with a ~9.6-nm spacing; and the assembly of the core is most-likely template driven (Briggs *et al.*, 2003). The 2D cryo-EM structures of intact immature and mature HIV virions dramatically changed how researchers thought about HIV, its assembly, and maturation. Simultaneous to studies of the intact virus, research groups have sought to address questions related to the individual proteins as they rearrange during viral maturation. A significant effort has been placed on studies of the CA protein and the dramatic conformation rearrangement it undergoes during maturation. Studies of the individual proteins have been driven forward through the combined use of computational modeling, 2D cryo-EM, electron crystallography, NMR, and X-ray crystallography methods (Byeon *et al.*, 2009; Douglas *et al.*, 2004; Ganser *et al.*, 1999, 2003; Ganser-Pornillos *et al.*, 2007; Lanman *et al.*, 2002, 2003; Li *et al.*, 2000; Massiah *et al.*, 1994, 1996; Pornillos *et al.*, 2009).

Researchers still endeavored to study the 3D structure of intact HIV virions. It was with this in mind that electron tomography was used to visualize the trilobed structure of Env spikes on negatively stained HIV virions (Zhu *et al.*, 2003) and the architecture of the virus–cell contact region from chemically fixed and stained specimens (Sougrat *et al.*, 2007). The leap of using electron tomography to examine an external structure of the intact HIV virions was remarkable; however, studies of the internal architecture of the virus was not possible because of conventional sample preservation artifacts. Soon after the electron tomography results of HIV Env spikes were published, microscope manufactures developed streamlined, computer controlled, automated microscopes. It was due to these significant technological advances that cryoelectron tomography (cryo-ET) became a powerful technology for studying the structure of intact virions and whole cells (Grunewald *et al.*, 2003; Medalia *et al.*, 2002).

Through a number of recent investigations, cryo-ET has now been established as the ideal approach for studying the distinct architecture of HIV. Research groups have recently determined the 3D structure of both immature and mature virions (Benjamin *et al.*, 2005; Briggs *et al.*, 2006a, 2009; Wright *et al.*, 2007). Cryo-ET examinations of mature HIV-1 virions revealed that the conical cores were unique in structure and position, but they also demonstrated certain similarities with respect to size and shape, the distance of the cone's base from the envelope/MA layer, the range of the cone angle. It was also observed that the conical CA shape was preferred *in vivo*, which provided additional evidence to support the template-directed model of CA formation (Benjamin *et al.*, 2005; Briggs *et al.*, 2006a). Structural studies of the immature virion soon followed and some of the methods and results are described in detail in the section below (Briggs *et al.*, 2009; Wright *et al.*, 2007). Concurrent to the cryo-ET studies of HIV maturation, cryo-ET, combined with advanced computational methods, was used to determine the molecular architecture of native Env spikes *in situ* at nanometer resolution (Liu *et al.*, 2008; Zanetti *et al.*, 2006; Zhu *et al.*, 2006). The studies of HIV Env spike structure, arrangement, and Env in-complex with neutralizing antibodies has provided a new driving force for the investigation and development of Env-specific therapeutics.

In this chapter, we present a number of cryo-ET procedures related to specimen preservation, data collection, and image processing that have proved to be useful for establishing a better understanding of HIV viral assembly, viral entry, and antibody neutralization at the nanometer level.

2. CRYOELECTRON TOMOGRAPHY

Cryo-ET is established as an important 3D imaging technique that serves to bridge the information gap between atomic structures and EM reconstructions that reveal the cellular or viral architectures (Baumeister *et al.*, 1999; Lucic *et al.*, 2005; Subramaniam, 2005). In combination with advanced computational methods, cryo-ET is the most promising approach to determine the molecular architecture of nanomachines *in situ*. The potential of cryo-ET lies in its ability to investigate cellular and viral components in their native state without fixation, dehydration, embedding, or sectioning artifacts. However, the poor signal-to-noise ratio (SNR) of the collected image data and the lack of effective protein labeling techniques are major impediments in cryo-ET, which confound researchers' efforts to identify specific molecules reliably within a cell or pleomorphic virus.

Our interest has gravitated around the improvement of cryo-ET methods for determining 3D structure of large macromolecular complexes *in situ*. The goal is to render molecular detail at higher resolution and to automate and expedite the reconstruction process and the subsequent data analysis. The high-throughput cryo-ET processing pipeline includes data acquisition, fast tilt series alignment, 3D reconstruction, subvolume extraction, and analysis. The combination of these techniques has enabled us to determine 3D structures of macromolecular assemblies at resolutions in the range of 2–4 nm (Liu *et al.*, 2008, 2009), which permits an interpretation of structure/function relationships at the nanometer level.

2.1. Sample preparation: Frozen-hydrated specimen

The preparation of frozen-hydrated viral specimens is the key step for the direct visualization of virus particles by cryo-ET. The general procedure for preparing frozen-hydrated biological specimens is well described in detail (Dubochet *et al.*, 1988; Chapter 3, Volume 481). For cryo-EM and cryo-ET, the basic steps for making frozen-hydrated specimens include: EM grid preparation, sample preparation, and plunge freezing.

2.1.1. EM grid preparation

The type of EM holey carbon grid used is an important component for the consistent, high-quality frozen-hydrated viral sample preparation. There are several grid options: lacey or holey grids, Quantifoil grids, and C-flat grids. Quantifoil and C-flat 200 mesh grids are preferred, because of the well-ordered pattern of the holes in the carbon film and the 200 mesh size allows for maximal tilting of the grid in the microscope. It is essential for the formation of a thin layer of ice-suspended virions that the grids are cleaned and processed by glow discharging prior to plunge freezing in order to remove manufacturing by-products and increase the hydrophilic nature of the grid surface.

2.1.2. Sample preparation

If the tilt series are to be aligned by the use of gold markers, there are two basic options for applying the 5 or 10 nm gold particles to the EM grid. (1) A solution of the colloidal gold particles is applied to the EM grid and allowed to air-dry. (2) A concentrated stock of the gold particles is mixed with the viral sample and the mixture is applied to the EM grid during the freezing process. The second procedure is commonly used for tomography applications.

2.1.3. Plunge freezing

A virus suspension of 4 μl is placed on a freshly glow discharged holey carbon grid. The excess solution is removed manually with a piece of filter paper and the grid is then plunged into liquid ethane. Alternatively, the sample can also be frozen using a semiautomatic or automatic plunge freezing apparatus. There are a number of apparatuses available, including the FEI Vitrobot (FEI, Hillsboro, OR) and the Gatan CP3 (Gatan, Pleasanton, CA), and the newly introduced Leica EM GP (Leica Microsystems GmbH, Wetzlar, Germany). Both FEI and Gatan systems as well as a manual plunge freezing apparatus have been used to prepare grids of frozen-hydrated viral samples.

2.2. Dose series: Determining the optimal electron dose

It has been shown that a statistically well-defined 3D reconstruction can be obtained from low-dose and noisy projection images, as long as the total dose is sufficient (McEwen et al., 1995). However, radiation damage is a fundamental problem for frozen-hydrated biological specimens (Chapter 15, Volume 481). Higher total dose and higher defocus increases contrast, but with the introduction of significant radiation damage and a reduction in resolution. Therefore, it is important for each application to find the optimal tradeoff between dose, defocus, contrast, and resolution before collecting tilt series. For this purpose, twenty high-dose (10 e/$\mathring{A}^2$) images at $0°$ tilt were recorded (Fig. 13.1). Comparative analysis or visual inspection of a movie of aligned images provides a good estimation of beam damage and image distortion that occurred during dose series. The tilt series selected for processing were usually exposed to a total dose of 80 e/$\mathring{A}^2$ for HIV Env studies and 120 e/$\mathring{A}^2$ for HIV maturation analysis.

2.3. Data acquisition: Low-dose tilt series

Because the time required for acquiring a tilt series and computing a tomographic reconstruction is significant, it is important to locate as many well-preserved virions as possible within a sample. The regions of interest that exhibit a sufficiently high virion concentration and appropriate vitreous ice thickness are typically selected at low magnification ($4700\times$) and high defocus (300 μm) under low-dose conditions. The low-dose "search" images have reasonable contrast to readily locate the area of interest.

The use of the latest advances in automated image acquisition (Koster *et al.*, 1992, 1997) is essential for data acquisition of frozen-hydrated biological specimens. Currently, there are several software packages available for automatic data acquisition: SerialEM (Mastronarde, 2005), UCSF Tomo (Zheng *et al.*, 2004), TOM (Nickell *et al.*, 2005), Xplore3D (FEI), EMMenu (Tietz), and Leginon (Suloway *et al.*, 2009; chapter 14, Volume 483). In particular, Leginon and batch tomography from FEI Xplore3D were used successfully to

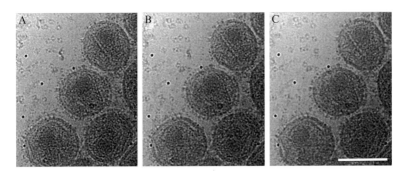

Figure 13.1 Dose series of virus: determining the optimal electron dosage. A series of high-dose (10 e/$\mathring{A}^2$) images of viruses at $0°$ tilt are recorded. Three of them (10e, 80e, and 160e) are shown in (A)–(C), respectively. The difference among these images is subtle. The contrast of image (C) is very good, but the fine detail of viruses (e.g., membrane bilayer) disappears. In order to conserve the fine details of virus, a total electron dose of between 80e and 120e was selected to collected tilt series. The scale bar is 100 nm.

collect large amounts of data (Liu *et al.*, 2008, 2009; Suloway *et al.*, 2009). The major benefit of batch tomography is the automatic collection of data without human intervention. In the FEI Polara electron microscope, the specimen can be maintained at liquid nitrogen temperature for $\sim$24 h, and as many as 3000 2K × 2K CCD images can be taken during this period.

For single-axis tilt series acquisition, the angular range is usually from $-70°$ to $+70°$ with fixed angular increments of $1°$ or $2°$. Batch tomography in FEI Xplore3D and Leginon starts at $0°$, then tilts to $-70°$ with the chosen increment. After reaching the highest tilt angle, the stage will return to $0°$ and then tilt to $+70°$ with the same increment. Two images at $0°$ tilt are collected, one in the beginning and the other one on the return of the stage to $0°$. A comparison of these two images provides information about radiation-induced changes to the sample or to the added gold markers, as well as any changes of other microscope parameters during the image acquisition.

Frozen-hydrated virus specimens were imaged at liquid nitrogen temperatures using a Polara field emission gun electron microscope (FEI) equipped with a 2K × 2K CCD placed at the end of either a GIF 2002 or Tridium energy filter (Gatan). The electron microscopes were operated in two slightly different conditions, but for similar purposes of enhancing image contrast (Liu *et al.*, 2008; Wright *et al.*, 2007). In order to determine $\sim$2.0 nm resolution structures without taking CTF correction into consideration, about 2 μm defocus was chosen to collect low-dose single-axis tilt series for HIV Env studies. At this defocus, energy filtering with a narrow slit (20 eV) becomes critical for enhancing the image contrast.

2.4. 3D tomographic reconstruction

The generation of a 3D reconstruction from a tilt series can be subdivided into three tasks: (1) preprocessing of the raw micrographs, (2) alignment of the tilt series, and (3) the computation of the tomogram. The preprocessing removes image imperfections, such as density gradients or density outliers ("hot" or "cold" pixels of the CCD camera). The images in the tilt series must be aligned prior to the computation of a 3D map, because a common reference frame and accurate geometric parameters must be determined before a 3D map can be computed. The collected images are not in register for various reasons: noneucentricity or mechanical instability of the stage, specimen drift, or poor tracking and recentering. The commonly used alignment methods fall into two categories, marker-based and marker-free methods (Chapter 13, Volume 482). In earlier studies (Zhu *et al.*, 2006, 2008), a purely marker-free approach (Winkler and Taylor, 2006) was used. More recently, a hybrid approach which is based on an initial fiducial alignment with IMOD (Kremer *et al.*, 1996) or Inspect3D (FEI) followed by a marker-free projection-matching refinement was used to analyze large numbers of tilt series (Liu *et al.*, 2008, 2009).

The marker-free method as implemented in the *protomo* software package (Winkler and Taylor, 2006) usually works well if specimens are less than 200 nm in thickness (Dai *et al.*, 2008; Liu *et al.*, 2006; Ye *et al.*, 2008; Zhu *et al.*, 2006, 2008). The advantage of this method is that by using cross-correlation techniques, the alignment is based directly on the signal produced by structural features of the specimen and not on artificially introduced gold markers. The alignment principle is similar to the projection-matching approach in single particle reconstruction and thus permits iterative refinement. A preliminary map computed from already aligned images in the tilt series is reprojected and used to refine the alignment. An additional step is the reevaluation of more accurate geometric parameters. Alignment and geometry refinement are alternated and it generally takes multiple cycles of alignment and geometry refinement to process a tilt series, so that this method is computationally more expensive than the marker-based method.

For fast and reliable marker-based alignment, IMOD (Kremer *et al.*, 1996) and Inspect3D (FEI) are useful packages presently available. The disadvantage of this approach is the relatively large size of gold markers, which are about 5 nm or larger in diameter so that their positions can only be determined with limited accuracy. Furthermore, the markers may not be immobile during the course of data acquisition, especially in frozen-hydrated specimens. Thus, the alignment based on gold markers may not guarantee the optimal alignment of specimen features.

Taking into account these considerations, a hybrid method combining the marker-based and marker-free approaches was developed for the Env project. As a first step, the tilt series were initially aligned with gold markers, and as a second step, projection matching was used to refine the alignment. The cross-correlation in the second step is carried out with a band-pass filter of which the high-pass filter component suppresses very low spatial frequencies, and the low-pass filter component excludes any signal beyond the first zero of the contrast transfer function. Computation of the final tomograms is carried out with *protomo* (Winkler and Taylor, 2006) which uses a weighted backprojection algorithm implementing general weighting functions (Harauz and van Heel, 1986).

2.5. Subvolume analysis

The goal of subvolume analysis is the extraction of smaller volumes from tomograms that contain structural motifs of interest and the rendering of the motifs with improved SNR and resolution for further study. This process may include a classification step if the motifs are structurally heterogeneous, in order to average only similar motifs. The subvolume analysis also needs to take into account the problem of missing information in reciprocal space, which is the cause of image artifacts in the tomograms. The missing information in a single-axis tilt series is usually referred to as the "missing wedge," which arises due to the fact that the data is collected from a limited angular tilt range in the microscope (e.g., from $-70°$ to $+70°$). The effects of the missing wedge can be eliminated in the final averaged image only if the motifs occur in various orientations in the tomograms. In other words, the orientation of the missing wedge relative to the motif structure varies correspondingly, and the gaps in reciprocal space of one motif overlap with sampled regions in other motifs, so that in the final merged image missing regions are effectively eliminated.

The processing pipeline of subvolume analysis starts with locating structural motifs in the tomograms. In general, these motifs are oriented variably within the tomogram; so that the initial search must take into account not only the position but also the orientation. The orientation search considerably increases the computational effort of automatically identifying the motifs. Algorithms are available for 3D template matching which are primarily based on cross-correlation techniques (Frangakis, 2006). Three-dimensional template matching is much more demanding than its 2D counterpart mainly because of effects of the missing wedge, the low contrast and the poor SNR of typical cryotomograms.

In the subvolume analysis of HIV Env, the motifs (the Env spikes) were manually picked using the graphical display program "tomopick" in *protomo* that lets the user scroll through the volume and identify spike locations at various depths in the tomogram (Winkler, 2007). Alternatively, IMOD (Kremer *et al.*, 1996) and Chimera (Pettersen *et al.*, 2004) can be used for this purpose. The orientation in space was derived from the measured spike positions under the assumption that the virions can be approximated by a spherical or ellipsoidal surface. The unknown coefficients of the equation for an ellipsoidal surface are calculated with the measured positions as input by a least squares fit. At each measured point, a surface normal is computed, based on the equation of the fitted ellipsoid. An approximate direction of the spike axis can then be derived from the surface normal. Only the rotation about the spike axis remains unknown, and must be determined later at the alignment stage.

In order to form an average of the raw motifs, subvolumes are extracted from the tomograms and aligned with respect to each other. The extraction or windowing is carried out either explicitly, by copying the subvolume to a new image, or implicitly, by simply recording the motif position and orientation. In the former case, the size of the data set can be reduced if the motifs are widely separated in the original tomograms. The latter case is more flexible, though, since window sizes can be changed easily at any point during processing. Subvolumes are aligned by cross-correlation and effects of the missing wedge are compensated by the use of constrained correlation

(Förster *et al.*, 2005; chapter 11, Volume 483). The rotational alignment is performed with an orientation search by maximizing the cross-correlation coefficient; the translational alignment is simultaneously derived from the position of the correlation peak maximum. If the approximate orientation of the motif is known, as is the case for the HIV Env, the grid search can be limited in accordance with the expected deviation of the orientation estimates; otherwise the full range of the three Euler angles must be scanned, which comes at a higher computational cost and is less reliable. Averaging is carried out with a merging procedure in reciprocal space that takes into account the regions with missing information in the raw motifs that are averaged.

2.6. Alignment strategies

If the examined motifs are heterogeneous, the alignment procedure outlined above cannot be directly applied. In the studies of HIV Env a contributing factor to the structural heterogeneity are orientational differences. The motif orientation relative to the tomogram determines which regions of reciprocal space fall in the sector covered by the missing wedge. The missing information affects the rendering of structural details. For instance, the membranes of spikes picked from the top or bottom of a virion almost disappear, whereas they are clearly delineated in the side views. Unless spikes are selected according to orientation, say from the top or bottom of virions, classification methods must be employed to ensure that only similarly oriented motifs are averaged.

One strategy that is commonly used in single particle analysis is multi-reference alignment and classification to separate various projections of a macromolecule (Penczek *et al.*, 1994). In the case of heterogeneous 3D motifs the procedure of aligning and classifying the motifs is similar: (1) create multiple references for alignment that are representative of the heterogeneous population, (2) align each raw motif to all references and select the best according to cross-correlation peak height, (3) classify the aligned motifs, and (4) average each class separately. Class averages can then be used for further alignment and classification cycles. The number of classes that are computed are chosen according to the number of expected conformations in the heterogeneous data set. This is impossible if there is a continuum of conformations rather than discrete states, for instance, if flexible parts are present in a molecule. Since some clustering algorithms, such as hierarchical ascendant classification, require a specific number of classes as input, several sets of class averages were computed, usually 4, 10, and 20 classes for the Env spike data, which were subsequently compared visually to identify significant structural differences.

The second strategy called "alignment by classification" was developed recently (Winkler *et al.*, 2009) to minimize possible reference bias problem, that is, the choice of the references may influence the outcome of the alignment: (1) classify all motifs, (2) compute class averages, (3) align all class averages with respect to each other, and (4) apply the alignment transformations of each class to the respective class members. This procedure is carried out repeatedly. No bias is introduced since no arbitrary reference selection takes place. Also, the alignment is carried out with class averages that have a higher SNR so that it is more robust than a multireference alignment of raw motifs. The number of classes computed in this procedure should be chosen as high as possible, because the goal is to capture as many different spatial orientations as possible in addition to the conformational differences. There is a tradeoff to consider, however, between a large number of classes versus the number of members per class. The number of averaged class members directly determines the improvement of the SNR, and the optimal choice depends on the total number of motifs in the entire data set. After completion of the alignment procedure, the same criteria as for multireference alignment apply for the final classification.

The initial alignment of the Env spikes warrants a further discussion. As mentioned above, only two of the three Euler angles can be determined with the described fitting method in the beginning, namely those that indicate the direction of the spike axis. In order to determine the third one, the rotation about the spike axis, the directional angles are first refined by an alignment to a global average, that was rotationally averaged about the spike axis, so that only two angles need to be scanned. A classification is then carried out and the class averages are aligned with respect to each other rotationally about the spike axis. This will identify motifs with structural differences and unify similar motifs that differ only in orientation, rotationally. The procedure resembles the second strategy, with the exception that one or two angles are restricted in the alignment.

A further problem is the effect of the missing wedge at the classification stage that tends to group subvolumes according to the orientation of the missing wedge rather than structural differences (Walz et al., 1997). One way to overcome this problem is the use of constrained cross-correlation as a similarity measure for the classification (Förster et al., 2008). A similar approach based on the signal overlap of a pair of subvolumes in reciprocal space is described by Bartesaghi et al. (2008). This method uses a variant of the multireference alignment method described above which includes refinement steps in the formation of class averages. There is evidence that classification without missing wedge compensation does not necessarily lead to an orientation based grouping of the motifs. It has been shown that a judicious choice of the classification mask can essentially eliminate the effects of the missing wedge (Winkler et al., 2009). This was verified by plotting the directions of the tilt axis for each class member, and inspecting the resulting distribution pattern visually. If the missing wedge were the driving factor in the classification, it would result in a particular direction within a class being preferred.

3. 3D Visualization of Intact HIV Virion

3.1. Cryoelectron tomography of HIV virions

A number of approaches have been used for the production and purification of concentrated samples of noninfectious HIV for cryo-ET analysis. For studies of viral maturation, all concentrated HIV preparations were kindly provided by Dr Wesley I. Sundquist's laboratory located at the University of Utah, Salt Lake City, UT. For studies of HIV Env, concentrated virus samples (HIVBaL; from AIDS Vaccine Program, SAIC-Frederick, National Cancer Institute) were purified by sucrose gradient centrifugation and inactivated by treatment with Aldrithiol-2 (AT-2). The use of AT-2 inactivates viral infectivity through covalent modification of internal viral proteins while retaining the functional and structural integrity of HIV Env (Arthur *et al.*, 1998).

3.2. Molecular structure of the gag shells

Tilt series images of immature HIV-1 virions were binned twofold and tomographic reconstructions were generated by weighted backprojection using the IMOD package (Kremer *et al.*, 1996). Individual immature virions were selected and cropped out of the completed tomograms using IMOD tools and were subsequently denoised by nonlinear anisotropic diffusion as implemented in BSOFT (Heymann, 2001).

In order to examine and analyze the structure of the Gag shells within the immature HIV virions, the Amira software package (Visage Imaging, Inc.) was used to generate radial density profiles. The profiles were produced by sampling spherical shells of the virions and plotting the density per unit area as a function of radius. Surface projections were created to better represent the surface area covered by each of the Gag proteins over each spherical shell. The surface projections were created by first generating a sphere to represent each Gag shell. Second, the electron densities observed in each shell were sampled along vectors normal to a triangle mesh surface imposed onto each sphere. Then, the summed densities of the vectors were each assigned vertex points on the surface of each shell. To determine the percentage of the surface area within the CA NTD shell covered by ordered and disordered lattice, the surface area of the sphere was calculated with the Amira surface area module. Boundaries around disordered regions (or patches) were drawn manually and the surface area calculated.

To locate the unit cells, the points on the projected surfaces with the least density, that is, regions without protein, within a specified radius were selected. Those points were then labeled with the standard deviation of the distance to their six nearest neighbors, as a measure of the local degree of order in the hexagonal lattice. Amira extension modules created within the Jensen lab executed the preceding functions. For unit cell averaging, the set of extrema on the projected surface layers having the least variation in the distance to their neighbors was selected. For each of the selected points, a local coordinate system was established comprising the normal vector to the surface (the Z axis), the vector from the point to its nearest neighbor (the X axis), and their cross-product (the Y axis). The entire volume containing the virion was then rotated into alignment with the global axes and translated so as to superimpose all of the selected surface extrema. Density renderings of the CA and SP1 domains and atomic model fitting were performed with the UCSF Chimera (Pettersen *et al.*, 2004).

While in 2D projection, we observed the radial spoke arrangement of Gag, in 3D only the CA and SP1 domains of Gag contained patches of hexagonal order, which were resolved in the tomographic reconstructions (Fig. 13.2). The most significant finding was the arrangement of the double-layered hexagonal lattice formed from the CA and SP1 domains. Upon further analysis, we proposed a model in which individual CA hexamers are stabilized by a bundle of six SP1 helices (Fig. 13.3). At the resolution of our data, we could not determine which portions of the associated CA and NC domains were involved in the formation of the helices. However, this structural discovery suggests why the SP1 spacer is essential for assembly of the Gag lattice and how cleavage between SP1 and CA acts as one of the structural switches controlling maturation. The structural interpretation was supported by evidence demonstrating that the Gag SP1 region is critical for viral assembly (Accola *et al.*, 1998; Gay *et al.*, 1998; Li *et al.*, 2003; Liang *et al.*, 2002) and functions as a maturation "switch" (Accola *et al.*, 1998; Gross *et al.*, 2000). The junction between SP1 and CA of the immature Gag lattice has also become a target for the development of retroviral maturation inhibitors. One group of compounds, the betulinic acid derivatives, has been demonstrated to be successful in blocking the cleavage of the CA-SP1 connection (Li *et al.*, 2003; Zhou *et al.*, 2005). One derivative, PA-457 or Bevirimat, has recently entered Phase 2 clinical trials in order to assess its efficacy for suppression of the virus.

3.3. Molecular architecture of HIV Env

Typically, for each tilt series, a $1900 \times 1900 \times 500$ reconstruction (~7 GB) was generated by weighted backprojection as implemented in *protomo* (Winkler and Taylor, 2006). After $4 \times 4 \times 4$ binning, contrast inversion and low-pass filtering, the tomograms were examined carefully

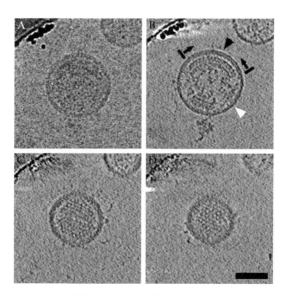

Figure 13.2 Raw image and slices through the 3D reconstruction of an HIV-1 virion. (A) The central image from a tilt series of an immature HIV-1 virion. (B–D) 5.6 nm slices through the 3D reconstruction, displaying the order of the Gag lattice. In (B), large gaps are present in the Gag lattice below the lipid bilayer (black arrows and bars mark one ordered region of the Gag lattice). In regions above visibly ordered Gag, the membrane-MA layer appears bilaminar (black arrowhead). In regions where there is no visibly ordered Gag, it appears unilaminar (white arrowhead). Scale bar 50 nm (from Wright *et al.*, 2007, with permission).

by IMOD and the virions with visible spikes were identified and extracted from the original tomograms, based on the coordinates of their centers. Each extracted 320 × 320 × 320 volume contains one individual virion with an original sampling size of 4.1 Å. The same virion was also stored in an 80 × 80 × 80 volume after 4 × 4 × 4 binning as described above.

The significantly enhanced contrast of the binned maps is the key for visualizing the intact virions and picking Env spikes (Fig. 13.4). Surface spikes on each virion were identified by visual inspection, using "tomo-pick" in *protomo* (Winkler, 2007) or UCSF Chimera (Pettersen *et al.*, 2004). In total, 4741 spikes were selected from 382 HIV virions, 4323 spikes from 306 HIV virions complexed with b12-fab, 4849 spikes from 292 HIV virions complexed with full-length b12 and 4900 spikes from 503 HIV virions complexed with CD4 and 17b-Fab (Liu *et al.*, 2008).

To speed up the image analysis, 2 × 2 × 2 binned subvolumes of the spikes were generated for the initial rounds of alignment and classification. Binning also increases the SNR, and thus results in a more reliable initial alignment and classification. This has proven to be critical for tomographic subvolume analysis, especially if large amounts of low contrast data need to

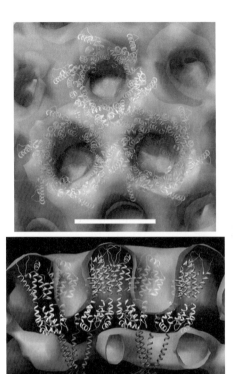

Figure 13.3 A model of the arrangement of Gag domains within the immature lattice. Top and side views (top and bottom, respectively) of the averaged Gag lattice (gray surface) with atomic models of CA NTD (cyan), CA CTD (yellow), and SP1 (magenta) fit into the density. Note that this is one interpretation of the density in terms of gross molecular architecture. Scale bar 8 nm (from Wright *et al.*, 2007, with permission). (For interpretation of the references to color in this figure legend, the reader is referred to the Web version of this chapter.)

be processed. As a first step, a global average of all the extracted volumes (~ 4000) is formed and if the positions and orientations were determined correctly, the average should reveal the membrane and the mushroom shaped spike (Fig. 13.5A). Since only two of the three Euler angles were determined, the spikes are not yet in register axially, thus the mushroom shaped appearance. A translational alignment was carried out based on the center of mass of the spike density, and an ellipsoidal mask (left panels in Fig. 13.5A) that included the spike volume was applied. Initial classification within a mask (defined in right panels of Fig. 13.5A) clearly showed the inherent threefold symmetry in the spike structure (Fig. 13.5B–E). The further alignment, multivariate statistical analysis and classification were performed as described by (Winkler, 2007) and later confirmed with an

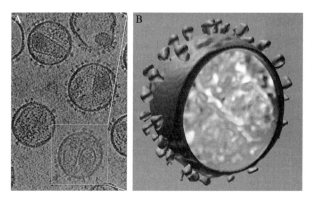

Figure 13.4   The 4 × 4 × 4 binned map of a tomographic reconstruction reveals significant contrast for 3D visualization of intact virion (B) and identification of Env spikes. The square size in (A) is 120 × 120 nm.

alternative approach (Bartesaghi *et al.*, 2008). Once the structures were convincingly ascertained, threefold symmetry was imposed for subsequent rounds of refinement. Fourier shell correlation (FSC) was used to estimate the resolution of the final structures.

The molecular architecture of the HIV Env trimer is shown in Fig. 13.6D. There are several atomic models of monomeric gp120 available in the PDB, and three of them are displayed within their 2.0 nm resolution envelopes which were calculated by "pdb2mrc" in EMAN (Ludtke *et al.*, 1999). At 2.0 nm resolution, CD4 binding and b12 binding conformations of gp120 are very different from the ligand free conformation of gp120 from SIV. Although visual comparison of these maps suggests that two models (CD4 binding and b12 binding conformations) match the corresponding densities in the HIV Env map better, the resolution is insufficient for a quantitative comparison.

3.4. Molecular details of neutralizing antibody b12-Env complex

b12 is a well known broadly cross-reactive, neutralizing antibody (Burton *et al.*, 1994). A recent crystal structure of monomeric gp120 core complexed with Fab fragments from b12 not only reveals the conformationally invariant surface for initial CD4 attachment but also provides atomic-level detail of the b12 epitope (Zhou *et al.*, 2007). Therefore, antibody b12 can also be utilized to map the CD4 binding site on HIV Env. Both b12 and its fab were incubated with intact virions separately. Due to the limited resolution, the resulting tomographic reconstructions show little evidence of antibody binding. However, the class averages of b12 (or b12 fab) and Env complex

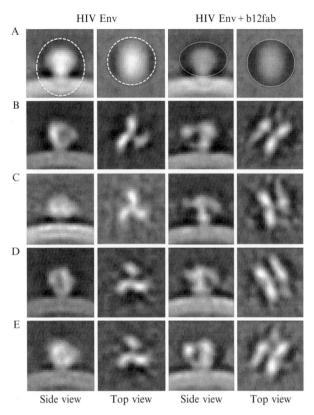

Figure 13.5 Initial classification of HIV Env spikes. Global average shows the membrane and mushroom shaped spike (A). Two different masks were used for 3D alignment (left) and 3D classification (right), respectively. Four class averages from initial classification were generated from HIV Env spikes without (left) and with b12 fab (right). The square size is 30 × 30 nm.

clearly reveal extra density, compared to the native Env without any ligand (Fig. 13.5). The extra density from a refined structure (Fig. 13.7) corresponds very well with one antibody fab. A pseudo-atomic model of the HIV gp120 trimer (Liu *et al.*, 2008) was produced by fitting the crystal structure of b12 in complex with an HIV gp120 core (Zhou *et al.*, 2007) into EM density maps using Chimera (Pettersen *et al.*, 2004). The unassigned densities (Fig. 13.7C) at the apex of HIV Env likely correspond to the variable loops (e.g., V1/V2), which are supported by recent V1/V2 deletion mutant studies (Hu, Liu, Taylor, and Roux, unpublished work). The unassigned densities at the bottom of gp120 represent N/C-termini, which are confirmed by a recent crystal structure (Pancera *et al.*, 2010). This result indicates that cryo-ET is becoming a powerful and reliable technique for

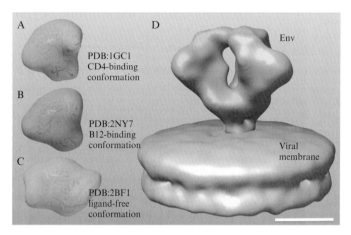

Figure 13.6   3D density map of HIV Env at 2.0 nm resolution (Liu *et al.*, 2008). Comparison of EM map (D) with simulated maps calculated from three gp120 monomer crystal structures: CD4-binding conformation (A), b12-binding conformation (B), and ligand free conformation (C) indicates that the molecular docking remains challenging at 2.0 nm resolution. The scale bar is 5 nm.

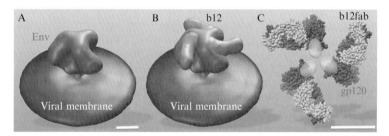

Figure 13.7   3D structure of HIV Env and b12 complex clearly reveal extra density (B), compared to that of the native Env without any ligand (A). The extra density corresponds very well with one antibody fab. A pseudo-atomic model (Liu *et al.*, 2008) of HIV gp120 trimer as shown in (C) is derived based on the rigid body fit of crystal structure (Zhou *et al.*, 2007) from the complex of gp120 and b12 fab. The scale bar is 5 nm.

determining the native structures of HIV Env and its interactions with antibodies at molecular resolution. Currently, there are considerable interests in understanding the interaction between HIV Env and other well known broadly neutralizing antibodies (2G12 against gp120, 2F5, and 4E10 against gp41). It is critical to identify functionally conserved regions of the highly variable HIV Env for better understanding of their structural basis for neutralization. It will certainly enhance the development of novel antibodies capable of neutralizing diverse HIV isolates.

3.5. CD4 induces major conformational change of HIV Env

It was believed that HIV Env undergoes a series of conformational changes when it interacts with receptor (CD4) and coreceptor on the host cell surface, leading to fusion of viral and cellular membranes. In the light of high resolution crystal structures, especially of the ternary complex (gp120 with both CD4 and CD4 induced antibody 17b; Kwong *et al.*, 1998), the 3D structure of native HIV Env in a ternary complex with CD4 and 17b was determined (Liu *et al.*, 2008) within a month from data acquisition to modeling. This could be achieved mainly because of the application of a high-throughput cryo-ET procedure. Most remarkably, the intact crystal structure of the ternary complex can be fitted into EM map as one rigid body (Fig. 13.8A). This structure suggests that the binding of Env to CD4 results in a major reorganization (Fig. 13.8B) of the Env trimer and a close contact between the virus and target cell coreceptor, thus facilitating viral entry (Fig. 13.8C).

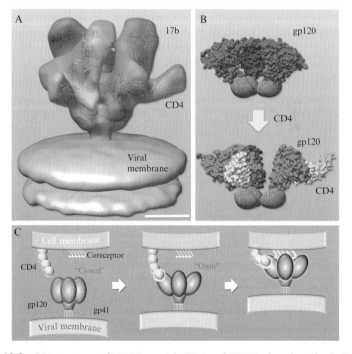

Figure 13.8   3D structure of HIV Env with CD4 and CD4 induced antibody 17b (Liu *et al.*, 2008). The crystal structure from a ternary complex of gp120 with CD4 and 17b (Kwong *et al.*, 1998) was fitted in EM density map as rigid body (A). gp120 was colored in magenta, CD4 in yellow, 17b fab in light green. CD4 induces huge conformational change in the HIV Env spike (B). A model of viral entry was proposed in (C). The scale bar is 5 nm. (See Color Insert.)

4. CONCLUSIONS AND PERSPECTIVE

Cryo-ET is the method of choice to study intact heterogeneous viruses at molecular resolution (Grunewald *et al.*, 2003; Subramaniam *et al.*, 2007). The recent technological innovations have established cryo-ET as the most advanced technique for determining 3D structures of macromolecular complexes *in situ* (Beck *et al.*, 2007; Liu *et al.*, 2008, 2009; Murphy *et al.*, 2006). However, poor SNR and the lack of effective protein labeling techniques remain two significant limitations of cryo-ET methodologies. We believe that further advances in cryo-ET through the development of innovative techniques for increasing throughput and resolution, adapting traditional antibody labeling, and combining genetic approaches with cryo-ET for the specific characterization of macromolecular complexes *in situ* will provide greater insights into the fascinating cellular processes of living organisms.

ACKNOWLEDGMENTS

J. L. thanks Dr James Stoops for comments on and discussion of the manuscript. The authors are thankful to Dr Grant Jensen for helpful suggestions. J. L. thanks Dr Ken Roux for sharing unpublished results. The work reported here was done in the laboratories of Drs Grant Jensen, Sriram Subramanian and Ken Taylor. J. L. is supported by a Welch Foundation Grant AU-1714, NIH grant 1R01AI087946.

REFERENCES

Accola, M. A., Hoglund, S., and Gottlinger, H. G. (1998). A putative alpha-helical structure which overlaps the capsid-p2 boundary in the human immunodeficiency virus type 1 Gag precursor is crucial for viral particle assembly. *J. Virol.* **72,** 2072–2078.

Arthur, L. O., Bess, J. W., Jr., Chertova, E. N., Rossio, J. L., Esser, M. T., Benveniste, R. E., Henderson, L. E., and Lifson, J. D. (1998). Chemical inactivation of retroviral infectivity by targeting nucleocapsid protein zinc fingers: A candidate SIV vaccine. *AIDS Res. Hum. Retroviruses* **14**(Suppl 3), S311–S319.

Bartesaghi, A., Sprechmann, P., Liu, J., Randall, G., Sapiro, G., and Subramaniam, S. (2008). Classification and 3D averaging with missing wedge correction in biological electron tomography. *J. Struct. Biol.* **162,** 436–450.

Baumeister, W., Grimm, R., and Walz, J. (1999). Electron tomography of molecules and cells. *Trends Cell Biol.* **9,** 81–85.

Beck, M., Lucic, V., Förster, F., Baumeister, W., and Medalia, O. (2007). Snapshots of nuclear pore complexes in action captured by cryo-electron tomography. *Nature* **449,** 611–615.

Benjamin, J., Ganser-Pornillos, B. K., Tivol, W. F., Sundquist, W. I., and Jensen, G. J. (2005). Three-dimensional structure of HIV-1 virus-like particles by electron cryotomography. *J. Mol. Biol.* **346,** 577–588.

Briggs, J. A., Wilk, T., Welker, R., Krausslich, H. G., and Fuller, S. D. (2003). Structural organization of authentic, mature HIV-1 virions and cores. *EMBO J.* **22,** 1707–1715.

Briggs, J. A., Simon, M. N., Gross, I., Krausslich, H. G., Fuller, S. D., Vogt, V. M., and Johnson, M. C. (2004). The stoichiometry of Gag protein in HIV-1. *Nat. Struct. Mol. Biol.* **11,** 672–675.

Briggs, J. A., Grunewald, K., Glass, B., Förster, F., Krausslich, H. G., and Fuller, S. D. (2006a). The mechanism of HIV-1 core assembly: Insights from three-dimensional reconstructions of authentic virions. *Structure* **14,** 15–20.

Briggs, J. A., Johnson, M. C., Simon, M. N., Fuller, S. D., and Vogt, V. M. (2006b). Cryo-electron microscopy reveals conserved and divergent features of gag packing in immature particles of Rous sarcoma virus and human immunodeficiency virus. *J. Mol. Biol.* **355,** 157–168.

Briggs, J. A., Riches, J. D., Glass, B., Bartonova, V., Zanetti, G., and Krausslich, H. G. (2009). Structure and assembly of immature HIV. *Proc. Natl. Acad. Sci. USA* **106,** 11090–11095.

Burton, D. R., Pyati, J., Koduri, R., Sharp, S. J., Thornton, G. B., Parren, P. W., Sawyer, L. S., Hendry, R. M., Dunlop, N., Nara, P. L., *et al.* (1994). Efficient neutralization of primary isolates of HIV-1 by a recombinant human monoclonal antibody. *Science* **266,** 1024–1027.

Byeon, I. J., Meng, X., Jung, J., Zhao, G., Yang, R., Ahn, J., Shi, J., Concel, J., Aiken, C., Zhang, P., and Gronenborn, A. M. (2009). Structural convergence between Cryo-EM and NMR reveals intersubunit interactions critical for HIV-1 capsid function. *Cell* **139,** 780–790.

Chan, D. C., Fass, D., Berger, J. M., and Kim, P. S. (1997). Core structure of gp41 from the HIV envelope glycoprotein. *Cell* **89,** 263–273.

Chen, B., Vogan, E. M., Gong, H., Skehel, J. J., Wiley, D. C., and Harrison, S. C. (2005). Structure of an unliganded simian immunodeficiency virus gp120 core. *Nature* **433,** 834–841.

Chen, L., Kwon, Y. D., Zhou, T., Wu, X., O'Dell, S., Cavacini, L., Hessell, A. J., Pancera, M., Tang, M., Xu, L., Yang, Z. Y., Zhang, M. Y., *et al.* (2009). Structural basis of immune evasion at the site of CD4 attachment on HIV-1 gp120. *Science* **326,** 1123–1127.

Dai, W., Jia, Q., Bortz, E., Shah, S., Liu, J., Atanasov, I., Li, X., Taylor, K. A., Sun, R., and Zhou, Z. H. (2008). Unique structures in a tumor herpesvirus revealed by cryo-electron tomography and microscopy. *J. Struct. Biol.* **161,** 428–438.

Douglas, C. C., Thomas, D., Lanman, J., and Prevelige, P. E., Jr. (2004). Investigation of N-terminal domain charged residues on the assembly and stability of HIV-1 CA. *Biochemistry* **43,** 10435–10441.

Dubochet, J., Adrian, M., Chang, J. J., Homo, J. C., Lepault, J., McDowall, A. W., and Schultz, P. (1988). Cryo-electron microscopy of vitrified specimens. *Q. Rev. Biophys.* **21,** 129–228.

Förster, F., Medalia, O., Zauberman, N., Baumeister, W., and Fass, D. (2005). Retrovirus envelope protein complex structure in situ studied by cryo-electron tomography. *Proc. Natl. Acad. Sci. USA* **102,** 4729–4734.

Förster, F., Pruggnaller, S., Seybert, A., and Frangakis, A. S. (2008). Classification of cryo-electron sub-tomograms using constrained correlation. *J. Struct. Biol.* **161,** 276–286.

Frangakis, A. S., and Rath, B. K. (2006). Motif search in electron tomography. *In* "Electron Tomography: Methods for Three-Dimensional Visualization of Structures in the Cell," (J. Frank, ed.). Springer, New York, London.

Freed, E. O. (1998). HIV-1 gag proteins: Diverse functions in the virus life cycle. *Virology* **251,** 1–15.

Fuller, S. D., Wilk, T., Gowen, B. E., Krausslich, H. G., and Vogt, V. M. (1997). Cryo-electron microscopy reveals ordered domains in the immature HIV-1 particle. *Curr. Biol.* **7,** 729–738.

Ganser, B. K., Li, S., Klishko, V. Y., Finch, J. T., and Sundquist, W. I. (1999). Assembly and analysis of conical models for the HIV-1 core. *Science* **283,** 80–83.

Ganser, B. K., Cheng, A., Sundquist, W. I., and Yeager, M. (2003). Three-dimensional structure of the M-MuLV CA protein on a lipid monolayer: A general model for retroviral capsid assembly. *EMBO J.* **22,** 2886–2892.

Ganser-Pornillos, B. K., Cheng, A., and Yeager, M. (2007). Structure of full-length HIV-1 CA: A model for the mature capsid lattice. *Cell* **131,** 70–79.

Gay, B., Tournier, J., Chazal, N., Carriere, C., and Boulanger, P. (1998). Morphopoietic determinants of HIV-1 Gag particles assembled in baculovirus-infected cells. *Virology* **247,** 160–169.

Gelderblom, H. R., Hausmann, E. H., Ozel, M., Pauli, G., and Koch, M. A. (1987). Fine structure of human immunodeficiency virus (HIV) and immunolocalization of structural proteins. *Virology* **156,** 171–176.

Gross, I., Hohenberg, H., Wilk, T., Wiegers, K., Grattinger, M., Muller, B., Fuller, S., and Krausslich, H. G. (2000). A conformational switch controlling HIV-1 morphogenesis. *EMBO J.* **19,** 103–113.

Grunewald, K., Desai, P., Winkler, D. C., Heymann, J. B., Belnap, D. M., Baumeister, W., and Steven, A. C. (2003). Three-dimensional structure of herpes simplex virus from cryo-electron tomography. *Science* **302,** 1396–1398.

Harauz, G., and van Heel, M. (1986). Exact filters for general geometry three dimensional reconstruction. *Optik* **73,** 11.

Heymann, J. B. (2001). Bsoft: Image and molecular processing in electron microscopy. *J. Struct. Biol.* **133,** 156–169.

Huang, C. C., Tang, M., Zhang, M. Y., Majeed, S., Montabana, E., Stanfield, R. L., Dimitrov, D. S., Korber, B., Sodroski, J., Wilson, I. A., Wyatt, R., and Kwong, P. D. (2005). Structure of a V3-containing HIV-1 gp120 core. *Science* **310,** 1025–1028.

Koster, A. J., Chen, H., Sedat, J. W., and Agard, D. A. (1992). Automated microscopy for electron tomography. *Ultramicroscopy* **46,** 207–227.

Koster, A. J., Grimm, R., Typke, D., Hegerl, R., Stoschek, A., Walz, J., and Baumeister, W. (1997). Perspectives of molecular and cellular electron tomography. *J. Struct. Biol.* **120,** 276–308.

Kremer, J. R., Mastronarde, D. N., and McIntosh, J. R. (1996). Computer visualization of three-dimensional image data using IMOD. *J. Struct. Biol.* **116,** 71–76.

Kwong, P. D., Wyatt, R., Robinson, J., Sweet, R. W., Sodroski, J., and Hendrickson, W. A. (1998). Structure of an HIV gp120 envelope glycoprotein in complex with the CD4 receptor and a neutralizing human antibody. *Nature* **393,** 648–659.

Lanman, J., Sexton, J., Sakalian, M., and Prevelige, P. E., Jr. (2002). Kinetic analysis of the role of intersubunit interactions in human immunodeficiency virus type 1 capsid protein assembly in vitro. *J. Virol.* **76,** 6900–6908.

Lanman, J., Lam, T. T., Barnes, S., Sakalian, M., Emmett, M. R., Marshall, A. G., and Prevelige, P. E., Jr. (2003). Identification of novel interactions in HIV-1 capsid protein assembly by high-resolution mass spectrometry. *J. Mol. Biol.* **325,** 759–772.

Li, S., Hill, C. P., Sundquist, W. I., and Finch, J. T. (2000). Image reconstructions of helical assemblies of the HIV-1 CA protein. *Nature* **407,** 409–413.

Li, F., Goila-Gaur, R., Salzwedel, K., Kilgore, N. R., Reddick, M., Matallana, C., Castillo, A., Zoumplis, D., Martin, D. E., Orenstein, J. M., Allaway, G. P., Freed, E. O., *et al.* (2003). PA-457: A potent HIV inhibitor that disrupts core condensation by targeting a late step in Gag processing. *Proc. Natl. Acad. Sci. USA* **100,** 13555–13560.

Liang, C., Hu, J., Russell, R. S., Roldan, A., Kleiman, L., and Wainberg, M. A. (2002). Characterization of a putative alpha-helix across the capsid-SP1 boundary that is critical for the multimerization of human immunodeficiency virus type 1 gag. *J. Virol.* **76,** 11729–11737.

Liu, J., Taylor, D. W., Krementsova, E. B., Trybus, K. M., and Taylor, K. A. (2006). Three-dimensional structure of the myosin V inhibited state by cryoelectron tomography. *Nature* **442,** 208–211.

Liu, J., Bartesaghi, A., Borgnia, M. J., Sapiro, G., and Subramaniam, S. (2008). Molecular architecture of native HIV-1 gp120 trimers. *Nature* **455,** 109–113.

Liu, J., Lin, T., Botkin, D. J., McCrum, E., Winkler, H., and Norris, S. J. (2009). Intact flagellar motor of *Borrelia burgdorferi* revealed by cryo-electron tomography: Evidence for stator ring curvature and rotor/C-ring assembly flexion. *J. Bacteriol.* **191,** 5026–5036.

Lucic, V., Förster, F., and Baumeister, W. (2005). Structural studies by electron tomography: From cells to molecules. *Annu. Rev. Biochem.* **74,** 833–865.

Ludtke, S. J., Baldwin, P. R., and Chiu, W. (1999). EMAN: Semiautomated software for high-resolution single-particle reconstructions. *J. Struct. Biol.* **128,** 82–97.

Massiah, M. A., Starich, M. R., Paschall, C., Summers, M. F., Christensen, A. M., and Sundquist, W. I. (1994). Three-dimensional structure of the human immunodeficiency virus type 1 matrix protein. *J. Mol. Biol.* **244,** 198–223.

Massiah, M. A., Worthylake, D., Christensen, A. M., Sundquist, W. I., Hill, C. P., and Summers, M. F. (1996). Comparison of the NMR and X-ray structures of the HIV-1 matrix protein: Evidence for conformational changes during viral assembly. *Protein Sci.* **5,** 2391–2398.

Mastronarde, D. N. (2005). Automated electron microscope tomography using robust prediction of specimen movements. *J. Struct. Biol.* **152,** 36–51.

McEwen, B. F., Downing, K. H., and Glaeser, R. M. (1995). The relevance of dose-fractionation in tomography of radiation-sensitive specimens. *Ultramicroscopy* **60,** 357–373.

Medalia, O., Weber, I., Frangakis, A. S., Nicastro, D., Gerisch, G., and Baumeister, W. (2002). Macromolecular architecture in eukaryotic cells visualized by cryoelectron tomography. *Science* **298,** 1209–1213.

Murphy, G. E., Leadbetter, J. R., and Jensen, G. J. (2006). *In situ* structure of the complete *Treponema primitia* flagellar motor. *Nature* **442,** 1062–1064.

Nermut, M. V., Hockley, D. J., Bron, P., Thomas, D., Zhang, W. H., and Jones, I. M. (1998). Further evidence for hexagonal organization of HIV gag protein in prebudding assemblies and immature virus-like particles. *J. Struct. Biol.* **123,** 143–149.

Nickell, S., Förster, F., Linaroudis, A., Net, W. D., Beck, F., Hegerl, R., Baumeister, W., and Plitzko, J. M. (2005). TOM software toolbox: Acquisition and analysis for electron tomography. *J. Struct. Biol.* **149,** 227–234.

Pancera, M., Majeed, S., Ban, Y. E., Chen, L., Huang, C. C., Kong, L., Kwon, Y. D., Stuckey, J., Zhou, T., Robinson, J. E., Schief, W. R., Sodroski, J., *et al.* (2010). Structure of HIV-1 gp120 with gp41-interactive region reveals layered envelope architecture and basis of conformational mobility. *Proc. Natl. Acad. Sci. USA* **107,** 1166–1171.

Penczek, P. A., Grassucci, R. A., and Frank, J. (1994). The ribosome at improved resolution: New techniques for merging and orientation refinement in 3D cryo-electron microscopy of biological particles. *Ultramicroscopy* **53,** 251–270.

Pettersen, E. F., Goddard, T. D., Huang, C. C., Couch, G. S., Greenblatt, D. M., Meng, E. C., and Ferrin, T. E. (2004). UCSF Chimera—A visualization system for exploratory research and analysis. *J. Comput. Chem.* **25,** 1605–1612.

Pornillos, O., Ganser-Pornillos, B. K., Kelly, B. N., Hua, Y., Whitby, F. G., Stout, C. D., Sundquist, W. I., Hill, C. P., and Yeager, M. (2009). X-ray structures of the hexameric building block of the HIV capsid. *Cell* **137,** 1282–1292.

Sougrat, R., Bartesaghi, A., Lifson, J. D., Bennett, A. E., Bess, J. W., Zabransky, D. J., and Subramaniam, S. (2007). Electron tomography of the contact between T cells and SIV/HIV-1: Implications for viral entry. *PLoS Pathog.* **3,** e63.

Subramaniam, S. (2005). Bridging the imaging gap: Visualizing subcellular architecture with electron tomography. *Curr. Opin. Microbiol.* **8,** 316–322.

Subramaniam, S., Bartesaghi, A., Liu, J., Bennett, A. E., and Sougrat, R. (2007). Electron tomography of viruses. *Curr. Opin. Struct. Biol.* **17,** 596–602.

Suloway, C., Shi, J., Cheng, A., Pulokas, J., Carragher, B., Potter, C. S., Zheng, S. Q., Agard, D. A., and Jensen, G. J. (2009). Fully automated, sequential tilt-series acquisition with Leginon. *J. Struct. Biol.* **167,** 11–18.

Walz, J., Typke, D., Nitsch, M., Koster, A. J., Hegerl, R., and Baumeister, W. (1997). Electron tomography of single ice-embedded macromolecules: Three-dimensional alignment and classification. *J. Struct. Biol.* **120,** 387–395.

Wilk, T., Gross, I., Gowen, B. E., Rutten, T., de Haas, F., Welker, R., Krausslich, H. G., Boulanger, P., and Fuller, S. D. (2001). Organization of immature human immunodeficiency virus type 1. *J. Virol.* **75,** 759–771.

Winkler, H. (2007). 3D reconstruction and processing of volumetric data in cryo-electron tomography. *J. Struct. Biol.* **157,** 126–137.

Winkler, H., and Taylor, K. A. (2006). Accurate marker-free alignment with simultaneous geometry determination and reconstruction of tilt series in electron tomography. *Ultramicroscopy* **106,** 240–254.

Winkler, H., Zhu, P., Liu, J., Ye, F., Roux, K. H., and Taylor, K. A. (2009). Tomographic subvolume alignment and subvolume classification applied to myosin V and SIV envelope spikes. *J. Struct. Biol.* **165,** 64–77.

Wright, E. R., Schooler, J. B., Ding, H. J., Kieffer, C., Fillmore, C., Sundquist, W. I., and Jensen, G. J. (2007). Electron cryotomography of immature HIV-1 virions reveals the structure of the CA and SP1 Gag shells. *EMBO J.* **26,** 2218–2226.

Ye, F., Liu, J., Winkler, H., and Taylor, K. A. (2008). Integrin alpha $_{IIb}$ beta $_3$ in a membrane environment remains the same height after Mn^{2+} activation when observed by cryoelectron tomography. *J. Mol. Biol.* **378,** 976–986.

Zanetti, G., Briggs, J. A., Grunewald, K., Sattentau, Q. J., and Fuller, S. D. (2006). Cryoelectron tomographic structure of an immunodeficiency virus envelope complex in situ. *PLoS Pathog.* **2,** e83.

Zheng, Q. S., Braunfeld, M. B., Sedat, J. W., and Agard, D. A. (2004). An improved strategy for automated electron microscopic tomography. *J. Struct. Biol.* **147,** 91–101.

Zhou, J., Huang, L., Hachey, D. L., Chen, C. H., and Aiken, C. (2005). Inhibition of HIV-1 maturation via drug association with the viral Gag protein in immature HIV-1 particles. *J. Biol. Chem.* **280,** 42149–42155.

Zhou, T., Xu, L., Dey, B., Hessell, A. J., Van Ryk, D., Xiang, S. H., Yang, X., Zhang, M. Y., Zwick, M. B., Arthos, J., Burton, D. R., Dimitrov, D. S., *et al.* (2007). Structural definition of a conserved neutralization epitope on HIV-1 gp120. *Nature* **445,** 732–737.

Zhu, P., Chertova, E., Bess, J., Jr., Lifson, J. D., Arthur, L. O., Liu, J., Taylor, K. A., and Roux, K. H. (2003). Electron tomography analysis of envelope glycoprotein trimers on HIV and simian immunodeficiency virus virions. *Proc. Natl. Acad. Sci. USA* **100,** 15812–15817.

Zhu, P., Liu, J., Bess, J., Jr., Chertova, E., Lifson, J. D., Grise, H., Ofek, G. A., Taylor, K. A., and Roux, K. H. (2006). Distribution and three-dimensional structure of AIDS virus envelope spikes. *Nature* **441,** 847–852.

Zhu, P., Winkler, H., Chertova, E., Taylor, K. A., and Roux, K. H. (2008). Cryoelectron tomography of HIV-1 envelope spikes: Further evidence for tripod-like legs. *PLoS Pathog.* **4,** e1000203.

CHAPTER FOURTEEN

AUTOMATION IN SINGLE-PARTICLE ELECTRON MICROSCOPY: CONNECTING THE PIECES

Dmitry Lyumkis, Arne Moeller, Anchi Cheng, Amber Herold, Eric Hou, Christopher Irving, Erica L. Jacovetty, Pick-Wei Lau, Anke M. Mulder, James Pulokas, Joel D. Quispe, Neil R. Voss, Clinton S. Potter, *and* Bridget Carragher

Contents

Abstract

Throughout the history of single-particle electron microscopy (EM), automated technologies have seen varying degrees of emphasis and development, usually depending upon the contemporary demands of the field. We are currently faced with increasingly sophisticated devices for specimen preparation, vast increases in the size of collected data sets, comprehensive algorithms for image processing, sophisticated tools for quality assessment, and an influx of interested scientists from outside the field who might lack the skills of experienced microscopists. This situation places automated techniques in high

National Resource for Automated Molecular Microscopy, Department of Cell Biology, The Scripps Research Institute, La Jolla, California, USA

Methods in Enzymology, Volume 483
ISSN 0076-6879, DOI: 10.1016/S0076-6879(10)83015-0

demand. In this chapter, we provide a generic definition of and discuss some of the most important advances in automated approaches to specimen preparation, grid handling, robotic screening, microscope calibrations, data acquisition, image processing, and computational infrastructure. Each section describes the general problem and then provides examples of how that problem has been addressed through automation, highlighting available processing packages, and sometimes describing the particular approach at the National Resource for Automated Molecular Microscopy (NRAMM). We contrast the more familiar manual procedures with automated approaches, emphasizing breakthroughs as well as current limitations. Finally, we speculate on future directions and improvements in automated technologies. Our overall goal is to present automation as more than simply a tool to save time. Rather, we aim to illustrate that automation is a comprehensive and versatile strategy that can deliver biological information on an unprecedented scale beyond the scope available with classical manual approaches.

1. INTRODUCTION

The field of cryo-electron microscopy (cryo-EM) has witnessed an explosion of activity during the recent past with the emergence of increasingly sophisticated, automated technologies for performing the steps required to obtain a refined electron density map. As a result, cryo-EM is rapidly moving from a method practiced by a small group of experts to one that is applicable to a variety of scientific disciplines. Our increased ability to collect, store, and process previously unimaginable amounts of data has not only pushed the limits of questions addressable by the electron microscopy (EM) community, but has also attracted significant interest from experts outside our field, requiring structural insights but lacking the time and resources available to experienced EM practitioners. The success and general scientific credibility of our field will largely depend on how we handle the impending transition from EM as an esoteric methodology to a readily accessible technique in the scientific toolbox.

As a research group, our goal has been to develop automated data collection and processing software that aid both expert and novice users by combining automation and streamlined user-friendly interfaces with an underlying transparent framework that is both accessible to neophytes and extendable by seasoned EM practitioners. Our field is in a phase of expansion, and new software packages are being released at a remarkable rate, some contributed by outsiders to the field. It is our belief that the future of cryo-EM lies in bridging the gap between expert users developing sophisticated technologies and novices who benefit from user-friendly automation. We have always acknowledged, however, that "black box" automation of any aspects of cryo-EM is undesirable. The user must maintain control over automated routines, and a careful balance must be maintained between manual control and automated efficiency so as to pose minimal risks to the versatility, expandability, and quality of existing and future methods. Automation should strive to be more than a mere elimination of repetitive tasks. It should also serve as an aid to expert users, encouraging development of next generation algorithms and technical advances.

Our overall focus has been on the construction of a "pipeline" for single-particle EM to provide a streamlined but transparent route from specimen preparation to the reconstruction of a refined electron density map. In an earlier chapter describing our automated data collection software, we quoted the following definition of automation: "Automation is the use of computers to control machinery and processes, replacing human operators. It is a step beyond mechanization, where human operators are provided with machinery to help them do their jobs. Some advantages are repeatability, tighter quality control, waste reduction, integration with systems, increased productivity, and reduction of labor. Some disadvantages are high initial costs and increased dependence on maintenance" (Suloway *et al.*, 2005). Given our recent focus on automated image processing, we expand on the definition to add that: automation in EM enables the elimination of repetitive, predictable user interactions required to manipulate, process, translate or move data between image-processing modules, using an intuitive mode of user learning. In addressing the problems posed by inter-package incompatibility of data conventions and definitions, we are not simply constructing a *de novo* pipeline of novel image-processing algorithms. Rather, we are assembling existing modules, interpreting their functionality, and identifying inter-connecting paths, as if putting together pieces of a large and complicated puzzle. As new components are added, the pipeline grows, but not necessarily in a linear fashion. There are many paths to solving structures using EM, and our goal is to supply the user with the means to connect any selected set of pieces together, while also ensuring that interchangeable parts fit together in a seamless way and providing the ability to add new components as needed (Fig. 14.1).

Given the breadth of the topic of automation, we have adopted a "problem-centric" rather than "package-centric" approach to our description of the various areas of cryo-EM where automation has a distinct contribution. The manuscript is divided into sections on: (1) sample preparation, (2) robotic loading and screening, (3) electron microscope operation, alignment, and data collection, (4) single-particle image processing, and (5) assessment and integration. We begin each section with a general description of the problem and follow with a summary of the automated, as contrasted to manual, approaches that have been undertaken to address the particular issue, providing references and highlights to specific packages where appropriate. We discuss some of the major advantages and limitations of the existing techniques and hypothesize how future questions or

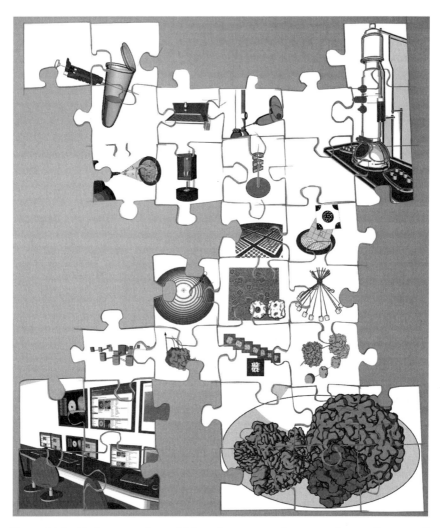

Figure 14.1 Each aspect of cryo-EM can be automated to some degree. Our overall goal is to show that automation can be more than simply a tool to save time. Rather, we aim to present automation as a comprehensive and versatile strategy, one where the overall picture becomes more valuable than the sum of its individual parts. When the pieces are connected, electron microscopy becomes a mature and sophisticated technology that is increasingly capable of addressing complex biological questions.

problems might be addressed via automation. Finally, in the method boxes, we provide specific step-by-step protocols for how these procedures are implemented at the National Resource for Automated Molecular Microscopy (NRAMM). We regret any omissions, errors, or oversights that are almost inevitable in a review of this wide scope.

2. Specimen Preparation

Reproducibility is the primary concern of experimentalists, and yet, in the field of single-particle EM, reproducing a good data set is far from trivial.

Specimen preparation, still considered an art form perfected by few, presents the major bottleneck to extracting identical structural information from independently prepared biological samples. Poor specimen preparation will adversely complicate data collection, image processing, and interpretation of resulting structures. The grid, of course, is a very large piece of real estate compared to the size of an average macromolecule, and when only a small number of fields needs to be acquired, a well embedded and suitably distributed specimen can likely be found somewhere on the grid, even if at the expense of time. However, the push toward higher resolution and the need to examine conformationally variable macromolecules poses higher demands for the extent of the grid that must be suitable for data collection and encourages the development of more consistent and reproducible specimen preparation methods.

Traditionally, heavy metal stains have been frequently used to embed the sample for viewing inside the electron microscope, providing a "footprint" of

the specimen on the adsorption substrate. Following optional modification of the substrate's hydrophobic properties, several microliters of sample are applied to the grid, which is then followed by application of the stain of choice. Specific protocols for negative staining have varied among the EM community (e.g. reviewed in Ohi *et al.*, 2004). Deviations from conventional practices involve differences in the adsorption substrate on the grid (e.g., Quantifoil, lacey carbon, or C-flat holey-carbon grids), the method of substrate preparation (e.g., glow-discharging or plasma-cleaning), and the specifics of the technique for applying the specimen (e.g. single-blot or deep-staining). Much is still a matter of controversy and possibly also depends on the specimen itself. A special protocol for negative staining has been developed that avoids the need for blotting with filter paper by making use of a liquid-handling system to control the application of solutions to the grid (Kühlbrandt *et al.*, 2003) and has enabled the automation of screens for 2D crystallization conditions of membrane proteins (Cheng *et al.*, 2007; Iacovache *et al.*, 2010; Vink *et al.*, 2007). For more general purposes, such systems could be used to investigate the effect of buffer conditions, test different sample dilutions (ideal for a liquid-handling robot with a small well), or collect time points in a time-resolved study.

Vitrification has presented more opportunities for automated technology development (Chapter 3, Vol. 481). Since the early days of manual cryo-blotting and freeze-plunging (Dubochet *et al.*, 1982; Taylor and Glaeser, 1976) the entire process has

evolved to one that is reasonably controllable and reproducible with automated systems. In contrast to manual methods where frequent practice is important to maintain the skill set, an automated process can be learned quickly and returned to with reasonable facility even after a long hiatus. The first and most widely recognized of the automated grid preparation devices is the FEI Vitrobot, a commercially available robot based on earlier work by several groups (Frederik and Hubert, 2005; White *et al.*, 1998, 2003). More recently, two newer models have come onto the market—the Gatan Cp3 (Melanson, 2009) and the Leica EM GP (Microsystems, 2009). Several homemade automated devices have also been developed, but will not be discussed here due to their lack of public availability. They all perform the same basic operation. The grid is held in place inside an isolated chamber that can be adjusted to a desired temperature and humidity by the user. After sample application, a robotic mechanism applies filter paper to the grid and blots away excess liquid using parameters specified by the user (e.g., blotting time, number of blots, time between blots). Finally, the grid is plunged into a liquid ethane medium at a temperature of about − 185 °C. This combination of conditions creates vitreous, rather than crystalline, ice and effectively preserves the sample in its natural state by avoiding deformation caused by intruding ice crystals (a comprehensive review on sample vitrification can be found in Dubochet *et al.*, 1988). Each preparation device possesses distinct advantages. For example, Gatan Cp3 allows for one-sided or two-sided blotting. Newer models of the FEI Vitrobot allow for automated transfer of the grid from the vitrification medium into liquid nitrogen. The Leica EM GP is capable of sensor controlled blot timing and promises to provide the most accurate and reproducible sample preparation parameters. When accurately calibrated, these devices all increase the reproducibility of cryo-freezing and can provide grids with large areas of well-embedded particles.

One interesting area of recent developments in automated vitrification technologies is time-resolved cryo-EM, wherein reactants essential to a structural event are sprayed onto the sample within a few milliseconds prior to plunging. This allows for a range of reactants to be delivered as aerosols to

the system of interest, whose conformational and/or compositional changes can then be studied. Time-resolved cryo-EM has been successfully used on a number of occasions (Berriman and Unwin, 1994; Walker *et al.*, 1995, 1999; White *et al.*, 1998, 2003), but the intricacy of the technique, and perhaps the lack of an easily controlled automated device, has precluded its widespread use in the EM community. Two major challenges for automation of time-resolved vitrification have arisen. From a mechanical standpoint, there is a need for a system that enables reproducible production of a thin film of primary solution immediately before applying activating solution. From a computational standpoint, additional complexity introduced by compositional heterogeneity in reactant mixing regions places high demands on classification and sorting algorithms in later image-processing steps. To address these needs, a first-generation monolithic device has recently been described that enables mixing of two reactant solutions in a microfluidic channel immediately prior to being sprayed onto a conventional EM grid as it is being plunged into cryogen (Lu *et al.*, 2009). This setup allows for adjustable combinations of reactants and requires no blotting, removing variability in contact between the grid and filter paper. While this device has yet to be commercially manufactured, the preliminary results look promising, making it a likely candidate for future automation.

When optimizing specimen preparation, the grid substrate is not only an important factor for embedding the sample but also affects the distribution of the particles and the ease with which we can automate the subsequent stage of data collection. For cryo-EM, samples are typically spread over either contin-

uous or fenestrated carbon substrates. Automated data acquisition requires identifying areas of ice of suitable thickness, either across the continuous carbon or suspended across the holes. Maximizing the efficiency of data acquisition requires that the distribution of particles is neither too crowded nor too sparse, a factor that is influenced by the size of the holes and the thickness of the carbon at the edge of the hole. Methods for producing fenestrated carbon have existed for half a century (Drahos and Delong, 1960; Harris, 1962), but they produce holes in a wide array of sizes and shapes. This poses challenges (although not insurmountable ones) for automatic identification of a suitable area for focusing, drift monitoring, and overcoming charging effects. If holes are arranged in a regular geometry, these issues are more easily addressed. Quantifoil grids (Ermantraut *et al.*, 1998) were the first commercially offered grids with predictably sized and spaced holes. They were rapidly adopted by the community and continue to be widely used. About a decade later, C-flat grids were developed at NRAMM (Quispe *et al.*, 2007) and have been commercialized by Protochips, Inc. The C-flat grids are composed purely of carbon, avoiding the need to remove underlying substrate, and providing a smaller lip around the edge of the hole. These grids are particularly well suited to ensuring an even distribution of small particles embedded in thin ice, which might otherwise be swept to the edge of the hole. More recently, Yoshioka *et al.* tested a new prototype fenestrated grid from Protochips, Inc., called Cryomesh, which is composed of doped silicon carbide using processes adapted from the semiconductor industry. The improved low temperature conductivity and extreme rigidity of these grids appear to have a significant impact on reducing beam-induced movement in TEM images of tilted cryo-preserved samples, and thus has the potential for improving the resolution of three-dimensional (3D) cryo-reconstructions in general (Yoshioka *et al.*, 2010). Similarly, Pantelic *et al.* (2010) showed that graphene oxide may provide a substrate superior to amorphous carbon. Another interesting class of grids that has emerged, termed affinity grids (Kelly *et al.*, 2008), contains a monolayer of lipids spread over the holes (Chapter 4, Vol. 481). The lipids carry a Ni-NTA head group, thereby providing a rapid and convenient technique for concentrating His-tagged particles out of solution onto the surface of the grid without biochemical intervention. Using a similar strategy, Dune Sciences has now started to commercialize SMART Grids™ with specific types of affinities for various biomolecules. In the near future, we can expect plenty of interesting biological uses arising directly from these developing technologies.

We envision that the future of EM sample and grid preparation will follow the path of X-ray crystallography, where robotics and microfluidic devices are increasingly used for the preliminary stages of optimizing conditions for crystal formation. Although a standard protocol can be followed for the majority of analyzed samples (Box 1), improvements can be conceived for various aspects of specimen preparation. For negative stain, robotic machines that precisely monitor (1) the volume placed on the grid, (2) the number of "washes," (3) the amount of time allotted to its absorbance, and (4) the amount of remaining stain after wicking, can be automated to provide increased reproducibility and efficiency. For vitreous ice, we might optimize the blotting procedure, or develop a new method entirely, to decrease variability in ice thickness across the sample grid. The automated transfer of grids after staining or vitrification would limit the handling (and mangling) of grids and reduce exposure of cryo-grids to contamination. Automated or robotic screening procedures (a few of which will be discussed below) would also facilitate comparisons between different preparation conditions.

Box 1   NRAMM Specimen Preparation Protocol

Purpose: Prepare a sample for imaging in the electron microscope.

1. *Select the adsorption substrate*: While carbon films are the most commonly available substrate, others are gaining popularity (see text). For cryo-frozen specimens, the most commonly used substrates are holey-carbon films (typically C-flat or Quantifoils), thin continuous carbon over holey carbon, or continuous carbon. We typically choose to use C-flats, as these often provide the most well distributed particles embedded in large uniform areas of vitreous ice. Thin continuous carbon over holes can be used for small particles that prefer to adhere to carbon, and continuous carbon can be suitable for larger particles and potentially provides for better estimation of the CTF and reduction in beam-induced specimen movement at high tilt.

2. *Prepare the adsorption substrate*: Prior to sample application the hydrophobicity properties of the adsorption substrate for the specimen must be modified. Plasma cleaning or glow discharging the grid will modify the hydrophilic properties of the substrate, so as to improve buffer dispersion and sample adherence to the grid. We use the Gatan Solarus Plasma cleaner, typically set for 5–10 s, 25% O_2, 75% argon.

3. *Apply the sample to the grid*: The standard protocol is to apply 3–5 µL of sample and allow for 30–60 s adherence time, but these parameters are specimen dependent. Prior to embedding the sample in either heavy metal stain or vitrified ice, filter paper is used to wick away any excess sample. In the case of automated freeze-plunging, this step is performed by the FEI Vitrobot.

4. *Embed the sample*: The sample is fixed using negative staining or freeze-plunging protocols. Automated negative staining protocols are available but are not widely used, and in order to maintain the fidelity of the true structure, cryo-freezing is often preferred. Samples are vitrified using the FEI Vitrobot (typically set to 4C, 95% relative humidity). Our usual protocol is to use a range of blotting times, typically from 3 to 9 seconds, to make approximately four grids. This provides a bracket for ice thickness that frequently produces at least one grid that is suitable for data collection (Fig. 14.2).

5. *Store the sample:* Negatively stained grids are stored in standard grid boxes in a dessicator. Vitrified grids are stored in smaller four-grid boxes, which are in

---

(continued)

Box 1   (*continued*)

turn placed into storage tubes suspended under liquid nitrogen inside long-term storage dewars.

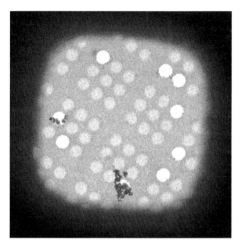

Figure 14.2   Vitrified sample prepared over C-flat grid, freshly cleaned using Gatan Solarus Plasma cleaner and vitrified using the FEI Vitrobot.

3. AUTOLOADING AND ROBOTIC SCREENING

Since the introduction of the electron microscope, physical exchange of specimen grids has been the job of a human operator. For small-scale operations where only a few grids are exchanged in a daily session, automation may not prove more efficient than a human being. However, for large-scale projects requiring routine screening procedures, automation pays off profoundly. This section will focus on technologies that relieve the burden of routine and highly repetitive grid loading and screening procedures.

Software-controlled grid-exchange appliances have been developed in

both academic and commercial environments. The first, developed at NRAMM (Cheng *et al.*, 2007; Potter *et al.*, 2004), uses a six-axis articulate robot arm mounted separately from the microscope to mimic the human operation of inserting the loaded grid holder through the air lock. In the current design, suction created at the tip of a fine nozzle attached to the same robot arm is used to pick up grids from a preloaded 96 grid tray and load them into the specimen holder aided by an actuator to raise and lower the specimen holder lever arm. A second implementation, developed at the New York Structural Biology Center (NYSBC) shares the same principle but uses a Cartesian robot (which moves back and forth along a single, fixed axis) for the task of inserting the loaded grid holder and a Selective Compliant Assembly Robot Arm (SCARA, limited to motion within a defined xy-plane) for loading grids from the grid tray to the holder (Hu et al., 2010). Some advantages of subdividing the grid loading and holder insertion tasks include a decrease in the requirements for robotic alignments, an increase in fine-tuned motion specific to EM grid loading, and an overall decrease in cost. The grid pickup mechanism on this robot has additionally evolved to consist of three spaced-out vacuum nozzles instead of one. A third implementation (Lefman et al., 2007), developed by Gatan Inc., mounts its grid tray and insertion mechanism directly on the microscope column and under vacuum. The grid "tray" is a transfer cylinder in a Gatling gun design where grids are preloaded in individual cartridge-type grid holders, effectively bypassing the air lock entrance and exit for individual grids. The fourth implementation uses a modified FEI "autoloader" that can accept, from a cylindrical grid reservoir, one of the 8 Titan autoloader cassettes with pre-loaded grids (Coudray, et al., 2010). It can then use the mechanism of the autoloader to put the grid into the imaging path. All four systems are used exclusively at room temperature. For cryo-specimens, JEOL microscopes currently provide side-entry grid holders and cartridge-based cryo-grid loaders that allow for their exchange in and out of the imaging path without operating the air lock. The user loads a cryo-grid cartridge, thereby eliminating possible atmosphere-based grid contamination during transfer. However, the task of physically exchanging grids still requires manual mechanical operations. Two instruments developed by FEI (the Polara and the Titan Krios) allow for the exchange of multiple grids, preloaded into cartridges, using a specially designed cryo-grid loader controlled by a software interface. This is a highly beneficial development for improving the throughput when acquiring large cryo-data sets, where multiple grids might be used during the collection process. The current alternative—individual grids mounted in manually inserted side-entry stages—is slow, inefficient, and leads to contamination of the grid and the microscope vacuum. While the cost for the current generation of automated cryo-grid exchange instruments is extremely high, the obvious advantages should lead to further developments in this area.

Grid-exchange robots, when effectively coupled to automated techniques for target selection and image acquisition, can drastically increase the throughput of screening projects. To date, the only processes that have been set up are for performing initial screens of 2D crystallization conditions for membrane proteins (Cheng *et al.*, 2007; Vink *et al.*, 2007). The Leginon package (Suloway *et al.*, 2005) provides an application for communicating with a robotic specimen loader and randomly sampling selected areas of the grid without requiring user intervention (Box 2). Apart from its usefulness in large screen-based studies, the large number of high magnification images output by the application can subsequently focus data collection to a specified region. Simple image-processing calculations can be employed to determine, in a more quantitative manner, optimal imaging conditions. For example, regions exhibiting variations in ice thickness, tendency to specimen drift, or impurities are quantifiable data that have already been extensively analyzed with simple algorithms (Stagg *et al.*, 2006). When coupled to "smart" screening protocols that learn to distinguish different conditions, we could limit image acquisition to optimal areas of the grid, thereby systematizing data collection and reducing the amount of micrograph post-collection processing.

Provided that we can improve automated loading and screening procedures, such that these operations could be performed quickly, efficiently, and routinely, we would open up opportunities for conducting experiments that would otherwise remain undone. For example, development of robotic devices and screening protocols will greatly facilitate the examination of more heterogeneous samples in a high-throughput fashion, perhaps with the intent of examining the effects of different conditions on specimen heterogeneity. Such screening capabilities, coupled with finely tuned time-resolved methods could be used for increased resolution and efficiency in large-scale analyses (Mulder *et al.*, 2010). Whole cell and cell section studies might benefit from the decrease in operator time required for finding the optimal area of the grid for imaging structures of interest in electron tomographic experiments. Given the increasing role that EM is playing in addressing various biological questions, we can envision that the number of laboratories requiring automated screening procedures is likely to increase in the future, which should drive the development of commercial systems to facilitate these applications.

Box 2   NRAMM Autoloading and Robotic Screening

Purpose: Utilize robotic grid insertion with an automated screening application to systematically survey multiple areas on different grids. At NRAMM, this procedure is largely limited to negatively stained grids but the protocol and software can be easily extended to enable microscopes offering cryo-grid exchange to perform iterative screening protocols (Fig. 14.3).

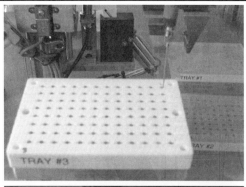

Figure 14.3 *Top*: Grid tray loaded with 96 grids. *Bottom*: The robotic arm is mounted from the ceiling next to the Tecnai F20 electron microscope.

1. *Prepare the sample tray*: at NRAMM, up to 96 individual grids are placed onto a tray. Without a mode for preserving the controlled environment necessary for cryo-frozen specimens, this procedure is largely limited to negatively stained grids. The MSI-Robot application available in Leginon will enable microscopes offering cryo-grid exchange to perform iterative screening protocols.

2. *Calibrate the insertion robot*: when the grids are not preloaded into a specially designed cartridge, the robotic mechanism for insertion requires proper calibration. Allowable errors in translational movement vary with the specifics of the robot design.

(continued)

Box 2 (*continued*)

3. *Test grid insertion*: it is crucial to make sure that the first grid insertion by the robot proceeds smoothly, ensuring that subsequent grids do not damage the microscope when the human operator is absent.
4. *Establish Leginon/robot communication*: Start Leginon and start robot server.
5. *Start the MSI-Robot 1st Pass application*: this Leginon application allows the user to systematically screen multiple areas on different grids.
6. *Select the grids*: this is carried out through the "Robot" node within Leginon.
7. The robot inserts the first selected grid before returning the control back to Leginon.
8. *Obtain an atlas*: a comprehensive illustration of the specimen grid can be obtained within the "Grid_Survey_Targeting" node. Modifying the settings will determine the number of acquired fields for constructing an atlas, and thus the area to be screened for the given grid.
9. *Obtain higher magnification images*: the procedure for obtaining high-magnification images is analogous to standard Leginon protocols (see Box 3). The "Square_Targeting" node samples grid squares, while the "Mid_Mag_Survey_Targeting" node specifies the areas for surveying the grid through higher magnification images.
10. Robot returns the grid to the tray.
11. Repeat steps 7–10 for remaining grids.

4. MICROSCOPY

Traditional methods for collecting micrographs from a TEM require careful hand-eye coordination—the operator views the image on a phosphor screen or video display and makes adjustments to parameters by turning knobs

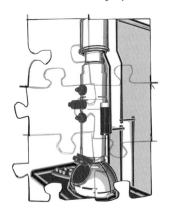

and pushing buttons. In this way, areas of interest on the grid are selected and the imaging conditions calibrated using adjustments to stage position, magnification, defocus, and other microscope parameters. Once the conditions are acceptable, a final micrograph is recorded to film or a digital camera. These manual procedures have several drawbacks. For one, visual surveys of a specimen grid can be tedious, and impractical when acquiring thousands of micrographs. Additionally, as a general rule for radiation-sensitive specimens, searching for suitable target areas using the same microscope parameters as used for the final exposure will subject the sample to excessive radiation. This was realized early on with tomographic data collection, where automation is critical to reducing the overall specimen radiation. Computational approaches have since greatly increase the efficiency of the process (Chapter 12, Vol. 481). Single-particle data collection, though much simpler than tomography, can benefit from similar automation.

Prior to the development of automated image acquisition packages, the first protocols for automation focused on electron beam calibrations, with

 particular emphases placed on the ability to set the defocus, correct astigmatism, and achieve coma-free alignment of beam tilt. Several practical procedures have been implemented for determining these three image aberrations (Kimoto *et al.*, 2003; Koster *et al.*, 1987; Saxton *et al.*, 1983; Typke and Dierksen, 1995; Typke and Köstler, 1977; Zemlin, 1979; Zemlin *et al.*, 1978). The crucial findings impacting automation protocols involved the realization

that all three aberrations can be efficiently determined by examining either diffractograms of the images (Typke and Köstler, 1977; Zemlin, 1979) or the images themselves (Koster *et al.*, 1987) using induced beam-tilt image pairs. The latter idea was further clarified to describe the calibration matrices for measuring defocus, astigmatism, and beam-tilt misalignment using beam-tilt-induced image displacement (Koster and de Ruijter, 1992) and quickly adopted in a number of early automation packages for defocus measurements (Dierksen *et al.*, 1992, 1993; Fung *et al.*, 1996; Kisseberth *et al.*, 1997; Rath *et al.*, 1997). Today, as the intended resolution during data collection increases, these established methods for microscope alignments remain central to defocus measurements and astigmatism correction, but the push for near-atomic resolution structures has demanded further improvements. Coma-free beam-tilt alignments, which primarily impact high-resolution information captured on the micrograph, have become the next area of development. To our knowledge, only the current release of the Leginon package (Suloway *et al.*, 2005) has incorporated a fully automated procedure for correcting axial coma, based on the diffractogram tableau analysis outlined by Zemlin (1979).

Automation packages have been established for acquiring tilt-series images in both academic (Marsh *et al.*, 2007; Mastronarde, 2005; Nickell *et al.*, 2005; Suloway *et al.*, 2009; Zheng *et al.*, 2007a,b, 2009) and commercial releases (software from FEI, JEOL, Gatan, TVIPS, and others), for acquiring random-conical tilt (RCT) (Zheng *et al.*, 2007b) or orthogonal tilt (OT) image pairs (Yoshioka *et al.*, 2007), for screening 2D crystallization conditions (Cheng *et al.*, 2007; Coudray *et al.*, 2008; Hu *et al.*, 2010), for acquiring montages of serial sections of biological tissues (Cardona *et al.*, 2010; Mastronarde, 2005), and for routine single-particle, helical, and 2D-crystal analyses employing either single or multiple exposures at a fixed tilt angle (Carragher *et al.*, 2000; Lei and Frank, 2005; Marsh *et al.*, 2007; Potter *et al.*, 1999; Shi *et al.*, 2008; Suloway *et al.*, 2005; Zhang *et al.*, 2009, 2001, 2003). For tomography and RCT/OT applications, the need for acquiring multiple images of the same final targets under defined, but different microscope parameters requires, by itself, a high level of automation for tracking features and defocus. Untilted data collection packages have expanded upon the ideas initially developed for tomography with improvements to methods for selecting and locating specified targets. By examining the approaches enabling low-dose targeting, we can divide the software developed in the last ten years into three distinct levels.

At the lowest level, the software provides multiple user-defined microscope states for the different tasks that need to be performed, such as search, focus, and exposure. Target selection and centering are performed by direct

user feedback while observing the image on the microscope viewing screen. Microscope manufacturers routinely provide this level of automation in the form of a low-dose kit. With CCD cameras gaining popularity, several packages also provide the option of saving the acquired digital images (e.g., EMmenu from TVIPS, Digital Micrograph from Gatan, or TEM Imaging and Analysis

(TIA) from FEI). The JAMES package (Marsh *et al.*, 2007), developed specifically to unify software from JEOL microscopes and Gatan CCD cameras into a single Python integrator, provides a few additional capabilities for directly communicating with the microscope. At the next level of automation, targets are selected on a CCD image acquired at a low magnification and the stage is driven to the target location using a transformation calibration, as in SerialEM (Mastronarde, 2005), TOM (Nickell *et al.*, 2005), and SAM (Shi *et al.*, 2008). However specimen drift, common with cryo-samples, is not monitored, thus presenting the possibility that selected targets will have moved by the time the final high magnification image is acquired. Algorithms for drift monitoring or compensation improve the likelihood that targets selected at low magnification will be centered in the field of view during the final high magnification image acquisition. AutoEM (Lei and Frank, 2005; Zhang *et al.*, 2001, 2003) monitors drift, while JADAS (Zhang *et al.*, 2009) compensates for it using a piezo drive that ensures greater precision in stage movement. Even with these additional features, targeting algorithms based purely on a transformation matrix cannot predict long-term changes and, thus, are only reliable for immediate selections on a lower magnification parent image. At the highest level of automation, individualized adjustment for each target allows the user to queue up a large set of targets and extend data collection over many hours or even days. For example, the user might queue up several low magnification squares on the grid that contain multiple high magnification acquisition targets. If, during the processing of the first square, the software detects specimen drift, it measures the exact distance by which the specimen has shifted, and adjusts all low and high magnification targets by that value. This advantage enables batch queuing and processing, as is done in Leginon (Suloway *et al.*, 2005), UCSFTomo (Zheng *et al.*, 2009), and FEI tomography. Multiple-scale adjustment also allows UCSFTomo to return to the same area after an in-plane rotation of the stage and for Leginon to return to the same target after reinsertion of a grid by a grid-loading robot (Cheng *et al.*, 2007).

At the current state, automation in image acquisition has relieved users from many tedious tasks such as estimating the defocus, tracking micrograph features, and monitoring specimen drift. In many cases, it has also improved the accuracy of targeting and reduced the accumulated dose on the specimen. However, one cannot argue that the power of automation has offset the importance of specimen and grid preparation, nor that automation can replace a good operator in controlling the quality of the final images. For one, most current packages lack controlled feedback loops for target selection, which would aim to emulate the sophistication and reliability of a human brain in adjusting the targets selected at low magnification based on an analysis of the quality of the images acquired at high magnification. Second, the high level of automation in packages such as Leginon, comes with the price of demanding both accurate and precise microscope alignments in order to maintain consistent calibrations. Effectively, the sensitivity of the performance of automated procedures to microscope alignments and calibrations often requires the user to possess significant knowledge about these issues. With the development of algorithms that automatically align and calibrate the microscope using digital analysis, one can envision surpassing these obstacles. Thus far, only beam shift, rotation center, and coma-free beam tilt can be adjusted by the click of a button in Leginon. Other alignment routines are limited by the availability of adjustment functions supplied by the microscope manufacturers, and, crucially, quality control feedback loops. For example, the task of moving the stage to a target of interest currently requires the operator to interrupt the process for recalibration if the target is not properly centered in the field of view. With feedback loops, errors in the acquisition process could be detected automatically and invoke the required calibrations. In its optimal form, automation software should perform a thorough alignment and calibration inspection before, and periodically during, data acquisition, and then optimize alignments as needed. Such automated monitoring would also provide an objective log of the exact parameter settings at the time, which can be used to reproduce the behavior of the microscope in a future session.

We implemented much of our vision for automation when we began redesigning our original Leginon package in 2002 (Suloway *et al.*, 2005) (Box 3). We believe that a generalized image acquisition automation package should be: (1) extendable to new functionality and applications, (2) flexible for different imaging and specimen conditions, (3) portable for use with different microscopes, cameras, grid-exchange robots, and novel devices, and (4) tightly integrated with a database to allow for data management and tracking. For extendibility and flexibility, Leginon is written using object-oriented modules called nodes that are combined into applications. Individual users can customize their specific sequence of operations (e.g., different methods of focusing) through a graphical user interface, and are always provided with the option of stepping down from full automation, when desired. A similar concept is implemented in JADAS (Zhang *et al.* 2009), where a recipe gathers modular operations and conditions to complete a series of image acquisitions and target selections, and TOM (Nickell *et al.*, 2005), which is created under the toolbox concept in Matlab. For portability, Leginon employs the pyScope python extension—a thin wrapper around an application-based programming interface distributed by the device manufacturers. Leginon can adapt to new microscopes and cameras by modification of a small set of functions contained within the pyScope library. The software is in use on FEI Tecnai, Polara, Titan Krios, and CM series platforms as well as selected JEOL instruments (Hu *et al.*, 2010), and supports a variety of digital cameras (manufactured by TVIPS, Gatan, and FEI). The extensive use of a relational database, together with a web-based data viewer has made it possible for us to promptly locate any of the more than 1 million images acquired by Leginon since we began data collection in 2003. Each is associated with metadata that provides a comprehensive summary of the imaging conditions and related parameters at the time. The database and web-based infrastructure developed for Leginon has subsequently been extended to the construction of our image-processing pipeline, Appion (Lander *et al.*, 2009), which provides full processing capabilities to any session at the microscope.

Box 3   NRAMM Automated Data Acquisition

Purpose: Collect an untilted data set using Leginon (Fig. 14.4).

1. *Align the microscope*: Leginon employs beam–tilt-induced image shift for several operations, and thus relies on a well-aligned microscope. Prior to initializing Leginon, it is necessary to (a) calibrate the electron gun, (b) align and stigmate condenser and objective lenses, (c) calibrate the electron beam, and (d) identify the focal plane.
2. *Initialize Leginon*: Make sure that Leginon Client is connected to the microscope.

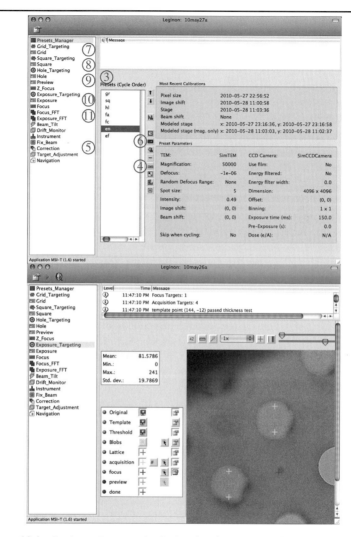

Figure 14.4 Leginon Layout. (top) Graphical User Interface for the Leginon software package. Each number refers to the description in Box 3. (bottom) Magnified view of the "Exposure_Targeting" node, displaying the log file, image statistics, the hole-finder routine, and the hole image with chosen targets.

3. *Import or manually define magnification presets*: For standard data collection, we use the "gr" preset for ∼200× magnification, the "sq" preset for ∼500× magnification, the "hl" preset for ∼5000× magnification, the "fa" preset for autofocusing at 50,000× magnification, and the "en" preset for acquiring final images at a specimen-dependent magnification.

(continued)

Box 3 (*continued*)

4. *Align image shift between magnifications*: Adjustments to beam-tilt-induced image shift between different magnification settings can be done through Leginon with the "align presets to each other" function within the "Presets_Manager" node. This ensures that, regardless of the magnification, stage-movement coordinates correspond to user-defined targets from a parent image.

5. *Acquire dark and bright image references*: Baseline standards for electron counts with "en" images can be set within the "Correction" node.

6. *Determine the electron dose*: This is done within the "Presets_Manager" node using an empty area of the grid.

7. *Acquire an atlas*: The atlas represents a composite view of the entire grid and is acquired using the "Grid_Targeting" node. It is critical to make sure that the objective aperture is removed for this step, so that it does not obstruct the view for lower magnification images.

8. *Select a square*: Within the "Square_Targeting" node, the user can specify square targets. While an automatic square detection algorithm is available, it is usually faster to perform this manually, as only a limited set of squares will likely be used for data acquisition.

9. *Select a hole*: The image of the selected square will appear within the "Hole_Targeting" node. In an analogous procedure to square targeting, the user can queue targets at this magnification and select a target for automatic high magnification focusing, along with a focus point for setting the Z-height of the stage. Alternatively, the hole-finder algorithm guides the user through setting up an automated hole-detection procedure.

10. *Select high magnification images*: The image of the selected hole will appear within the "Exposure_Targeting" node, where final acquisition images are selected. A single autofocus image is usually defined for all acquisition images within this node. As in the "Hole_Targeting" procedure, an automated hole-detection protocol can be used here.

11. *Monitor initial data collection for accuracy*: In the initial stages of data collection, it is important to make sure that every procedure executes in the intended manner. For example, monitoring the "Z_Focus" node, the "Focus" node, and the "Drift_Monitor" node is critical to make sure that all automated algorithms are proceeding smoothly. It might also be necessary to monitor the accuracy of the hole-finder, and adjust if necessary. When this has been ascertained, Leginon can be left unaltered until all targets have been acquired.

12. *Refill liquid nitrogen*: For cryo-specimens, liquid nitrogen must be added to the stage every 3 h. The cold trap can be refilled at the same time or on a 6 hourly schedule. For negative stain specimens a specially adapted cold trap is available that can last for 12 h or an autofill system can be setup.

5. IMAGE PROCESSING

A question often faced by single-particle electron microscopists is: what is the best manner in which to process large amount of data? In our lab, rather than relying on a single package, we have developed Appion, a wrapper software for high-throughput image processing in single-particle EM (Lander et al., 2009). Appion incorporates individual routines from a number of packages and integrates them into a standardized Python-scripted wrapper. It is intimately connected to a graphical web-based user interface, as well as a centralized SQL database, which stores all input and output parameters and enables complex queries of the metadata. Our goal in developing Appion is to provide a streamlined, intuitive, integrated, and transparent process, to thoroughly monitor and track relevant parameters and results associated with processing runs, and to enable users with various levels of expertise to run procedures and analyze the outcome without the need to decipher myriads of cryptic data files (see Box 4 for a conventional protocol for obtaining a refined electron density map using Appion). This transparency is fundamentally important for the advancement of automated EM techniques, but it remains subordinate to the actual content underneath, which consist of image-processing and computational algorithms that were developed by hundreds of scientists in the field (Baxter et al., 2007; Crowther et al., 1996; Frank et al., 1996; Heymann, 2001; Heymann and Belnap, 2007; Heymann et al., 2008; Hohn et al., 2007; Lander et al., 2009; Ludtke et al., 1999; Marabini et al., 1996; Sorzano et al., 2004; Tang et al., 2007; van Heel and Keegstra, 1981; van Heel et al., 1996; also see Chapter 15, Vol. 482). In this section, we describe the automated advancements dealing with the some of the basic areas of single-particle image processing. Additionally, we highlight recent advances in making each particular task more transparent, integrated, and streamlined, thereby contributing to the overall usability of the separate operations.

Box 4 NRAMM Image Processing

Purpose: Reconstruct a 3D electron density map in Appion using a streamlined protocol (Fig. 14.5).

Appion presents the user with a menu of options for image processing that is dynamically updated as each step is completed. When the user clicks on one of the menu options, Appion generates a new web page specific to the selected operation that requests inputs and allows the user to launch jobs on one of several processing machines or clusters. The job progress is monitored by updates to the menu. Once a completed job shows up in the menu, the user

(continued)

Box 4   (*continued*)

may click on its entry to generate a web page that reports on the results. Most input options are provided with defaults. Help options for each input are provided as pop-ups on the Appion web pages. Detailed step-by-step instructions for most of the procedures are available within the Appion documentation.

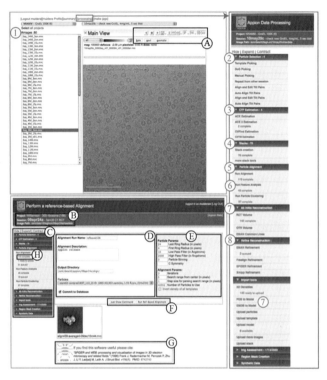

Figure 14.5   Appion Webpage Layout (A) Immediately above images in the imageviewer are tools for viewing particle picks, FFT and ACE results, downloading images in jpeg, mrc, or tiff format, making on-image measurements, and sorting images using the "hide" and "exemplar" buttons. All processing pages contain (B) a header with project name, session name, and file path information and (C) the Appion sidebar with available processing functionalities. Clicking on a processing functionality opens a web-based form for launching jobs. (D) General information and parameters that must be changed by the user are displayed on the left hand side, (E) whereas more specialized input parameters are displayed on the right hand side with default values specified. (F) Buttons at the bottom of the form will show the command or launch the job from the web page. (G) Citations are provided for the software packages and/or algorithms contained within a particular functionality. (H) Running, complete, and queued jobs can be tracked within in the Appion sidebar, and job log-files are accessible by clicking on the "# running " link.

1. *View the raw micrographs*: Images may be acquired using Leginon or uploaded to the database using the "Upload images" functionality in Appion. Once images are uploaded they are automatically tracked by the database and can be viewed using a web-based Imageviewer. Clicking the "processing" button in the Imageviewer takes the user to the main processing page of Appion.

2. *Select particles*: Appion provides several methods: "DoG Picker" is a reference-free approach, "Template Correlator" uses the reference-based approach in FindEM, and "Manual Picker" provides for interactive particle picking by the user. An example workflow for particle selection includes: (a) Selecting a subset of particles via DoG Picker or Manual Picker, (b) Aligning and classifying the subset to produce class averages, (c) Using selected class averages for Template Correlator. Users always have the option of cleaning up picks from any of the automated picking runs using Manual Picker. Overall results are provided on Appion summary pages or can be viewed overlaid on the individual images in the Image viewer by selecting the "P" button. As with CTF estimation, particle picking can be started concurrently with image acquisition.

3. *Estimate the CTF*: ACE, and ACE 2 provide fairly robust algorithms for CTF estimation on untilted images that generally require no adjustments to the provided default settings. CTFTilt can be used for CTF estimates on tilted micrographs. A summary of results can be viewed by clicking on the "complete" CTF items in the Appion menu. Results for individual images can also be viewed on the Imageviewer pages, by selecting the "ACE" button. Individual results include graphical overlays, estimated parameters and associated fitness values. We generally find that fitness values of > 0.8 are acceptable. CTF correction can be started concurrently with image acquisition; the estimation program, once started, will keep querying for new images as they come in.

4. *Create a particle stack*: The "Stack Creation" page is used to extract particles from the images based on the picks from a particle-selection run, or the picks associated with a previously created and modified stack. Inputs include options for filtering, binning, CTF correction, etc. Particles can be rejected based on CTF fitness parameters, particle correlation values, or defocus range. Results pages provide a summary of the stack, a link to view the stack as individual particles, and further options to clean up the stack using a variety of filters.

5. *Align the raw particles*: Reference-free procedures include Xmipp maximum-likelihood and SPIDER reference-free alignment, which can be used to create references for subsequent reference-based alignment. Reference-based procedures include Xmipp reference-based maximum likelihood, SPIDER multi-reference, IMAGIC multi-

(continued)

Box 4 (*continued*)

> reference, and EMAN multi-reference alignments. Most procedures can be run initially using default input parameters. The aligned particles can be examined and manipulated further from the summary pages.
>
> 6. *Classify the aligned particles*: As with the alignment routines, Appion provides several different options of classification that can be applied to any alignment run. These include: SPIDER Correspondence Analysis, IMAGIC Multivariate Statistics Analysis, or the Xmipp Kerden Self-Organizing Map routine. The specified feature analysis routine locks the user into a clustering procedure, which generates summed class averages. Class averages can be viewed and manipulated further from the summary pages after requesting "View montage as a stack for further processing." Options include viewing the raw particles associated with each class, creating templates or substacks from selected classes or running common-lines procedures to create an initial model form selected classes.
>
> 7. *Generate an initial model*: Many options are available. Models can be uploaded from the PDB or EMDB, read in from a file, or imported from previously reconstructed datasets. If tilted data has been acquired, it is possible to perform "one-click" RCT or OTR, and tomographic reconstructions from selected 2D class averages or Z-projected subtomogram averages. These options are presented when viewing the stacks or class averages of appropriate datasets. Other options include common-lines approaches either utilizing EMAN's cross-common-lines protocol or an automated version of IMAGIC's angular reconstitution. These options are available when viewing class averages or from the *ab initio* reconstruction menu option.
>
> 8. *Refine*: Options include procedures using EMAN1, Frealign, or Spider application packages. Results can be viewed on summary pages, which includes more detailed information for each iteration, such as data and graphical output for Resolution curves, Euler angle distributions, snapshots of 3D maps, class averages, and particles contributing to the map.

5.1. CTF estimation and correction

Image formation in EM is distorted by the modulation of a contrast transfer function (CTF), which affects the amplitude of the Fourier coefficients at all spatial frequencies of the electrons contributing to the projection of the object. The distortion depends upon the physical parameters of the electron microscope, such as accelerating voltage, lens aberrations, and, crucially, the displacement of the focal plane at the time of imaging. Correcting for this aberration is critical for high-resolution reconstructions. It has long been

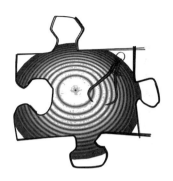 known that power spectra representing the spatial frequencies present in an electron micrograph can be used as robust estimators for the CTF (Welch, 1967), allowing the user to manually generate a theoretical CTF with the intent of finding the highest correlation to the experimentally observed power spectral density (PSD). This task, however, becomes burdensome, error-prone, and simply impractical when very large numbers of micrographs are collected on a daily basis, when Thon rings in the PSD are not clearly delineated, or when the presence of astigmatism or stage tilt introduces additional unknowns into the equation.

The requirement of accurate CTF correction for high-resolution reconstructions has encouraged a number of automated approaches to CTF estimation. In its most fundamental form, algorithms for defining a theoretical CTF model require optimizing some correlation function between a predicted and an experimentally observed PSD for each electron micrograph (Fernandez *et al.*, 1997; Huang *et al.*, 2003; Mallick *et al.*, 2005; Mindell and Grigorieff, 2003; Sander *et al.*, 2003; Sorzano *et al.*, 2007; Velasquez-Muriel *et al.*, 2003; Welch, 1967; Zhou *et al.*, 1996). Any optimization of an automated CTF estimation algorithm must focus on minimizing estimation errors inherent in the algorithm itself or the input data. Such errors can result from the low signal-to-noise ratio of cryo-micrographs (Baxter *et al.*, 2009), inconveniently located local minima in CTF estimation functions, and cross-talk between different CTF parameters that lead to identical results. Recently, Sorzano *et al.* published a study in which a comprehensive mathematical description was devised to analyze the effects of different CTF parameters and corresponding estimation errors (Sorzano *et al.*, 2009b). In doing so, the authors provide a quantitative model which describes the way in which errors in the input parameter values translate into errors in the output values (i.e., parameter sensitivity). Studies like this pave the road for future improvements in automated CTF estimation algorithms, demonstrating the need for reliable quality control criteria that take into account parameter sensitivities, lens aberrations and astigmatism, and low SNR inherent to cryo-micrographs. Comprehensive confidence scores, rather than simple correlation values between an experimental and a theoretical PSD, provide a more robust metric for assessing the accuracy of CTF estimation. They can be used to not only systematize a challenging manual task, but to accelerate the pace at which beginning stages of data processing can be completed.

5.2. Particle selection and stack creation

Ideally, the specimen of interest resides on the grid in a uniform and homogeneous array. However, we can think of many examples that deviate  from this scenario. These might include structural specimen heterogeneity arising from compositional or conformational variations, foreign substances resulting from inadequate specimen purification, specimen aggregation, or artifacts and contaminants arising from the preparation procedures. The outcome is a conglomeration of "stuff," which must be sorted to identify the particles of interest while keeping the false positive and false negative detection rates as low as possible. Once factors such as variation in ice thickness (and thus particle contrast), particle orientation, specimen heterogeneity, and the sheer number of potential picks are considered, the task of distinguishing the "good" from the "bad" becomes far from trivial.

Depending on the stage of image processing, different approaches can be taken to automate particle selection. Particle selection algorithms can be loosely classified into reference-free or reference-based categories, with an occasional hybrid of the two (for a review see Zhu *et al.*, 2004). In working with a novel sample, it may be wise to choose a reference-free algorithm wherein particles are segmented based on a measure of "saliency," such as variance, Markovian fields, or segmented edges. Such approaches (Ogura and Sato, 2004a; Plaisier *et al.*, 2004; Singh *et al.*, 2004; Umesh Adiga *et al.*, 2004; Voss *et al.*, 2009; Zhu *et al.*, 2003) need not necessarily be template-free, and several of the algorithms effectively apply a template-matching strategy, in which the template is generated without *a priori* knowledge of the sample. An alternative strategy is to supply a reference based on some prior structure (Huang and Penczek, 2004; Rath and Frank, 2004; Roseman, 2004; Wong *et al.*, 2004). In this case, rather than detecting salient features, the primary task is to compute an optimum similarity measurement between the supplied reference and all possible particle orientations, a procedure that works well for later stages of processing where the primary structures and directional views have been established and the goal is to optimize the raw particle stack for refinement. A third approach combines reference-free and reference-based methods, whereby the algorithm "learns" which particles to select in a supervised manner (Hall and Patwardhan, 2004; Mallick *et al.*, 2004; Ogura and Sato, 2004b; Plaisier *et al.*, 2004; Short, 2004; Sorzano *et al.*, 2009c; Volkmann, 2004). Naturally, combinations of these algorithms are often employed. For example, one can begin with a reference-free approach, then align and classify the particles to obtain templates for subsequent rounds of reference-based picking. The crucial step lies in careful visual examination of the actual picks in order to assess the algorithm's applicability to the particularities of the data. Classification schemes (Chen and Grigorieff, 2007; Shaikh *et al.*, 2008), filtering tactics employing "trap templates" for false-negative detection (Chen and Grigorieff, 2007), multistage picking classifiers (Sorzano *et al.*, 2009c), and other strategies often achieve equivalent results to rigorous manual optimization of the selected particles. On the other hand, one might want to completely eliminate bias by utilizing a reference-free approach to pick any specimen-resembling species, as was successfully performed by Mulder *et al.* to collect over a million ribosomal particles at various stages of assembly (Mulder *et al.*, 2010)—an inconceivable task without automated particle selection. Sometimes, therefore, the best strategy might be to simply pick everything.

5.3. 2D alignment and classification

In order to extract quantitative information out of the inherently low SNR data obtained by cryo-EM, 2D averaging must be applied to homogenous subsets of single particles. This requires the particles to be brought into alignment with one another, so that the signal of common motifs is amplified. Alignment protocols typically operate by shifting, rotating, and mirroring each particle in the data set in order to find the orientation of particle A that maximizes some similarity function (of which many variations exist; Chen and Grigorieff, 2007; Roseman, 2004; Saxton and Frank, 1977; Stewart and Grigorieff, 2004; van Heel *et al.*, 1992) with particle B (reviewed in Frank, 2006b; Joyeux and Penczek, 2002). As with particle-selection algorithms, alignment procedures are separated into reference-free and reference-based approaches, depending upon the existence of a template obtained from *a priori* information about the specimen. Within automated alignment procedures the "do loop" has long freed the user from manual calculation of similarity functions and tracking of parameters for thousands of images, thereby setting the scene for iterative alignments, wherein vectorial addition of parameters is tracked and applied to each particle after all iterations. Despite their utility for fine comparisons, the iterative strategy is sometimes hindered by over-fitting of noise, which presents a critical drawback and

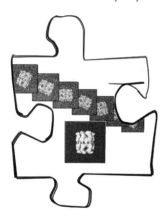

requires particular scrutiny when dealing with small or conformationally heterogeneous specimens (Stewart and Grigorieff, 2004). Maximum-likelihood algorithms (Scheres *et al.*, 2005b; Sigworth, 1998; Sjors and Sigworth *et al.*, this volume) have gained popularity with numerous incorporations into the Xmipp processing package (Marabini *et al.*, 1996; Scheres *et al.*, 2008; Sorzano *et al.*, 2004). These techniques integrate the two disparate branches of alignment by iteratively extracting references directly from the data without any *a priori* knowledge. They create a formalism for the task of aligning images to this set of references that includes estimates for the noise and angular distribution of particles and have shown the ability to overcome false minima and reference bias (Sigworth, 1998). These methods also require significantly more computing power that can be partly addressed by reducing the search space over the alignment parameters (Scheres *et al.*, 2005a), but in the end relies upon advances in computing and processing power.

Following alignment, the particles require grouping into classes in accordance with compositional, conformational, and orientational similarity. Classification separates a heterogeneous data set into homogenous subsets of classes with minimal intraclass variation based on a pixel-by-pixel comparison of its members. If *a priori* knowledge about the specimen includes a reference, supervised classification allows grouping of data in accordance with similarity to the references (van Heel and Stoffler-Meilicke, 1985). However, in cases where a reference is not available, unsupervised classification is required. Given the massive amount of information contained in large data sets, classification programs typically make use of data reduction techniques before proceeding with a mathematical description of inter-image variance. There are many variations on the initial approach proposed by Marin van Heel and Joachim Frank (Frank and van Heel, 1982; van Heel, 1984; van Heel and Frank, 1980, 1981, reviewed in Bonnet, 1998; Bonnet, 2000; Frank, 2006b), but all are conceptually similar to principal components analysis (PCA) (Pearson, 1901), a technique that has become the ideal basis for algorithms dealing with complicated image analysis such as facial recognition. Available classification modules have, for the most part, automated data reduction and the description of inter-image variance, requiring little to no manual input from the user and are available in most processing packages. After reducing the dimensionality, particles are ordered and summed according to their relative proximity within the reduced multidimensional space, traditionally using k-means or hierarchical clustering methods (Frank *et al.*, 1988). Other classification techniques have included "fuzzy" definitions of class memberships (Carazo *et al.*, 1990; Scheres *et al.*, 2005b) and neural networks, which find patterns in a data set through a learning process and heuristically sort images onto a 2D self-organizing map (SOM) (Marabini and Carazo, 1994; Pascual-Montano *et al.*, 2001). All of these methods still require manual user input to specify the number of desired classes. One promising recent development uses an algorithm entitled affinity propagation (Frey and Dueck, 2007) that no longer requires this specification and has been implemented for single-particle analysis within the latest version of Appion (Lander *et al.*, 2009).

Existing alignment and classification modules have done much to mathematically describe variance within the data, but have not always succeeded in placing identically oriented single particles into homogeneous classes using automation. In theory, this means eliminating rotational and translational errors during alignment, and minimizing intraclass variance while maximizing interclass variance during classification. Currently, this combinatorial procedure remains the manual task of the user, who might iteratively reclassify the aligned stack using modified parameters, carefully inspect and subclassify distinct classes to account for additional variations, or remove and realign raw particles based on classification results. Efforts toward aiding the user in the decision-making process have been made in the form of computational methods for estimating alignment errors (Baldwin and Penczek, 2005) and graphical user interfaces that display information relevant to quality assessment (Lander *et al.*, 2009), such as spectral signal-to-noise ratio or Fourier ring correlation values, correlation distributions, variance analyses, and comprehensive particle tracking, which permit "one-click" particle manipulations based on class affiliation. Ideally, these decisions can be encoded into algorithms that would intelligently seek the most likely set of outcomes.

5.4. *Ab initio* reconstructions

With an absence of prior knowledge about specimen structure, the precise orientation of the object with respect to a stationary camera remains 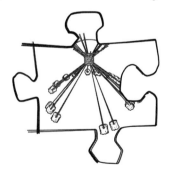 unknown. This is one of the central problems in EM and poses the question— how do we determine the correct orientational Euler angles to each recorded 2D projection and combine these into a 3D reconstructed map? It is interesting to note that the human mind is fully capable of extrapolating 3D structures when only 2D information is present. By examining different views (2D projections) of an unknown object, one might superimpose the projection images in various orientations with one another, as if putting together a 3D puzzle, a task that was performed with remarkable accuracy for early structures of the ribosome (Lake, 1976). This type of manual interpretation, even when accurate, depends on the predilection of the user. Automated methods for performing *ab initio* reconstructions largely reduce the bias associated with manual analyses and fall into one of two general categories—those that do and those that do not require physical specimen tilting inside the microscope. Specific advantages and disadvantages to each will be mentioned below, for which comprehensive discussions can be found in (Cheng *et al.*, 2006; Voss *et al.*, 2010).

Tomographic reconstructions (Frank, 2006a), random-conical tilt (RCT) reconstructions (Radermacher *et al.*, 1986), and orthogonal tilt (OT) reconstructions (Leschziner and Nogales, 2006; Leschziner *et al.*, this volume) rely upon physical specimen tilt. With the tomographic technique, the specimen stage is sequentially tilted from about $-60°$ to $+60°$ using a small angular increment, providing nearly all directional views with the exception of the missing wedge of information. A limitation of this straightforward technique is that the biological specimen acquires radiation damage with each electron dose causing damage to the specimen. The combination of serial electron dosing and the resulting missing wedge of information currently limits the resolution of tomography to ~ 30 Å, although algorithmic improvements for tilted CTF estimation (Fernandez *et al.*, 2006; Philippsen *et al.*, 2007; Sorzano *et al.*, 2009a), tilt-series alignment (Sorzano *et al.*, 2009a; Winkler and Taylor, 2006), and subtomogram averaging (Winkler *et al.*, 2009) are pushing this limit. RCT and OT methods also utilize specimen tilt, but require only the collection of image pairs. In RCT, images are taken at $0°$ and $45–60°$. This allows the former to be used for 2D alignment and classification, and the latter for 3D reconstruction. OT differs from RCT in that images are collected at $+45°$ and $-45°$, providing the equivalent of a $90°$ rotation and eliminating any missing Fourier information. In practice, all of these tilted acquisition methods are challenging to perform manually, primarily due to the significant amount of bookkeeping required to track the precise alignment shifts and rotations, classification dependencies, and tilt angles for each particle. In Appion, a complete record of all particle parameters is retained within the database, and therefore RCT, OT, and tomographic reconstructions can be achieved in a single step following automated alignment and classification (Voss *et al.*, 2010). Thus, with appropriate checkpoints to provide for user intervention as needed, all of the *ab initio* methods that rely on titled image acquisition are amenable to full automation.

When physical specimen tilt is not an option, it is possible to obtain an initial model exclusively from the transmitted information composing each image. Common-lines based methods (Crowther *et al.*, 1970; Farrow and Ottensmeyer, 1992; Goncharov and Vainshtein, 1986; Penczek *et al.*, 1996; Singer *et al.*, 2010; van Heel, 1987) rely on the central section theorem, which permits the identification of identical intersecting 1D lines for all combinatorial pairs of projections and the resulting assignment of Euler angles, although this strategy is only viable when the particles do not exhibit preferred orientation. The problem with any common-lines based approach is that, given a specific set of input values, a single globally or locally "optimal" structure will inevitably emerge and there is no reliable way to distinguish between correct and incorrect reconstructions. Some groups have relied on random methods of generating initial models, initializing either from random noise or a symmetric Gaussian sphere (Ludtke *et al.*, 1999; Yan *et al.*, 2007), on the principle that iterative convergence to a common motif indicates an appropriate starting model and avoids the introduction of bias into the reconstruction procedure. This impartial approach is sometimes successful when applied to large and highly symmetric particles (Liu *et al.*, 2007; Ochoa *et al.*, 2008; Pan *et al.*, 2009; Parent *et al.*, 2010) but is limited by the unfavorable, low signal-to-noise nature of cryo-EM images and can lead to incorrect structures within local minima for smaller, less symmetric objects. Automation and increased computing power has the potential to address the issue of structural reliability by applying statistical validation and the ability to repeat reconstructions using different starting models. Nevertheless, the benefit of simply gathering auxiliary information to validate 3D reconstructions as in (Singer *et al.*, 2010), is crucial.

5.5. 3D refinement

Most initial models establish nothing more than a preliminary sense of the overall shape of the biological specimen. In order to reveal structural information that can answer specific biological questions, the model requires refining. In single-particle analysis, a refinement is an iterative procedure, which sequentially aligns the raw parti-cles, assign to them appropriate spatial orientations (Euler angles) by comparing them against the model, and then back-projects them into 3D space to form a new model. Effectively, a full refinement takes as input a raw particle stack and an initial model and is usually carried out until no further improvement of the structure can be observed, often measured by convergence to some resolution criterion (Frank *et al.*, 1981; Harauz and van Heel, 1986; Penczek *et al.*, 1994; Sousa and Grigorieff, 2007; Unser *et al.*, 1987, 1989). Iterative refinement of a 3D model relies on the incremental adjustment of various parameters for full exploitation of the information content of the data. For example, decreas-ing the angular projection increment of the input model allows for a finer

sampling of the data during classification and back-projection. Resolution-dependent band-pass filtering procedures are routinely applied in order to eliminate irrelevant information that exceeds the resolution limit of the map. Density sharpening using b-factor weighting or low-pass filtering (Fernandez *et al.*, 2008; Rosenthal and Henderson, 2003) boosts the amplitudes of the high-frequency information within the map, lost during image acquisition. Many packages also offer automasking procedures in which a structure-specific mask is applied based on densities in the digital image and the particle size (Grigorieff, 2007; Ludtke *et al.*, 1999; Tang *et al.*, 2007; van Heel *et al.*, 1996). Analysis of the Euler angles and correlation values assigned to raw particles can ensure that only the most reliable particles contribute to the final reconstruction (Stagg *et al.*, 2006). The ability to identify and appropriately apply these parameters is what distinguishes a "manual" from an "automated" iterative refinement. In the manual approach, user intervention partitions the large combination of steps pertaining to each iteration into distinct protocols involving frequent decision-making (e.g., how many particles to exclude or how to filter the stack for the next iteration). Automated iterative refinements take such decisions into account, and have already been variously implemented into the major EM packages, as for example the "semiauto-mated" projection-matching reconstruction scheme within EMAN (Ludtke *et al.*, 1999). Similar intelligent algorithms that take into consideration information limits or inconsistencies in the data before proceeding with the next step of a refinement can be expected in the future. The key is to impose enough transparency on the process such that tweaking parameters at any step remains feasible without sacrificing streamlined efficiency.

6. Assessment and Integration

Perhaps the most important aspect of any automated, or for that matter, manual, operation is the means to assess its outcome (a list of automated quality assessment steps for both Leginon and Appion is described in Box 5). Automation, certainly at this stage of its development, should not be thought of as a replacement for intelligent and critical evaluation by an operator. Thus, a major goal of any automated procedure should be to provide feedback to the end-user in a transparent and easily accessible form. In areas such as specimen preparation,

this kind of feedback is not readily available until the grid is in the electron microscope. This has led to proposals for devices that examine vitrified grids using light microscopy, but so far their use is limited to fluorescently labeled specimens (Sartori *et al.*, 2007). However, if specimen transfer itself becomes more automated, the need for pre-screening devices will be reduced by an improvement in the transfer operation and resulting reduction in grid and vacuum contamination. Once the grid is in the microscope and automated image acquisition begins, quality assessment becomes a critical component at multiple points in the process. For example, without a user constantly monitoring the instrument, as during manual acquisition, the automated software is responsible for monitoring changes in parameters such as beam intensity, image and beam shifts, and microscope alignments. If these parameters depart from expected ranges, they must either be automatically adjusted or user intervention must be requested. The user should also be able to readily examine the images being acquired, preferably from any remote computer. For example, in Leginon, this form of monitoring is available with web-based viewing tools that provide immediate access to images as they are acquired. Using any standard web browser, the user can parse through images, examine the relationship between the quality of high-magnification images and targets selected on lower magnification images (a hierarchical viewing strategy), compute power spectra to evaluate the defocus and astigmatism settings, view particle picking and CTF estimation results overlaid on the raw micrographs, and perform simple measurements on the data.

Box 5 NRAMM Quality Assessment

Purpose: Describe tools used for routine quality assessment in Leginon and Appion.

1. *Assess the quality of automated data collection*: In Leginon, the results of any operation can be tested within the interactive GUI prior to target queuing, as, for example, when assessing the hole-finder algorithm. This ensures that optimal criteria are set prior to data collection. Leginon displays output to the user when it encounters errors during data collection, as in the case when an autofocus procedure fails or when the specimen drifts for an extended period of time. At all times, the functional state of the software and all relevant data collection parameters are monitored and tracked using the database to ensure reproducibility in future sessions.

2. *Assess the quality of the collected data*: Each Leginon session keeps track of statistics during data collection, which are used to assess the quality of the images. For each experiment and for each collected image, Leginon

(continued)

Box 5 (*continued*)

monitors: (a) all instrument parameters associated with the microscope, (b) the duration of collection, (c) camera readout statistics and electron dose, (d) the magnitude of specimen drift over time, (e) the ice thickness for each hole, (f) image and beam shift values, (g) the accuracy of the autofocus procedure, and other additional parameters. This type of record keeping allows us to track the state of the instruments and analyze the quality of the collected data.

3. *Eliminate undesired micrographs or regions*: Images can be manually selected to be "hidden" from view using buttons on the Imageviewer. Note that images can also be marked as "Exemplars." In Appion, "Junk Finding" and "Manual Masking" options exclude undesired areas of the micrograph for particle selection. The "Multi-Image Assessment" tool allows the user to reject entire micrographs, usually after a particle-selection run, while the "Image Rejector" function automatically rejects images that either do not have CTF parameters, particle picks, or associated tilt pairs, or are outside a specified fitness factor of defocus range after CTF estimation. Images that have been "rejected" or "hidden" can be optionally excluded from processing runs. Processing runs can also be set to use only "exemplar" images, often useful for testing. Rejected or hidden images, or exemplars, can still be viewed in the Imageviewer by selecting them form the pulldown menu options. Hidden images can be restored while viewing them by selecting the hidden button again. Images are never permanently removed from the database.

4. *Assess the quality of processing algorithms*: Appion's imageviewer allows the user to assess some of the initial steps of image processing. The "ACE" button provides graphical displays of CTF estimation as well as fitting parameters and fitness values. The user can visualize images of the edges detected by the algorithm and compare them to the Thon rings present in the PSD of the image. The "P" button displays all particle picks from the current micrograph for the particle-selection run specified by the user. Histograms of confidence scores and correlation values are displayed on the Appion summary pages for CTF estimation and particle picking.

5. *Assess the quality of particle stacks*: Stacks can be directly examined as a montage of particles. Summary pages provide graphical summaries of the intensity and standard deviation of the particle images and these outputs can be useful for cleaning up the stack by rejecting particles on the edge of a hole or over the carbon. The "Xmipp_sort_by_statistics" algorithm can also be used to identify junk particles. During subsequent image-processing steps, particles may be rejected based on poor scores within 2D alignment and classification steps, high Euler jumper statistics, or poor correlations for class assignment during 3D refinement.

6. *Assess the accuracy of 2D alignment and classification*: Appion report pages display Eigenimages, variance averages, and/or correlation histograms for each alignment and classification run; some outputs are algorithm dependent. All translations, rotations, and mirror reflections applied to the images, as well as the coordinates within the original micrograph are stored in the database.

7. *Assess veracity of an electron density map*: Appion displays a variety of output related to the reconstruction refinement as an aid in determining the consistency and veracity of the reconstructed map. Each method for generating initial models provides combinatorial assessment criteria consistent with the routine (exemplified by the random-conical reconstruction method in Fig. 14.6). Later refinement runs automatically generate summary pages providing output at each iteration that includes resolution derived from the Fourier Shell and Rmeasure criteria, distribution of Euler angles, Euler angle differences between the current and previous iteration as well as the average median Euler difference for all iterations, side-by-side comparison of input projections with associated reprojections of the map, and snapshots of the reconstructed model.

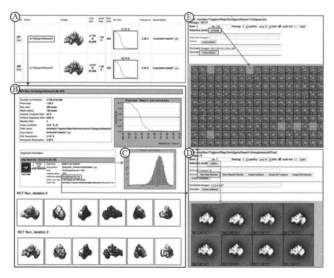

Figure 14.6   Assessment Pages for a random-conical tilt reconstruction (A) A table summarizes the number of particles, FSC, R-measure resolution, and description for all available 3D reconstructions. (B) Clicking on a reconstruction name opens a detailed summary page that includes iterative snapshots and links to (C) 2D alignment plots and (D) class averages used for 3D reconstruction. Within the viewer, individual 2D class averages can be selected and their raw particles viewed. (E) Individual raw particles can be marked for exclusion during sub-stack creation.

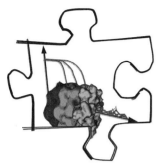

Similarly, during processing and analysis of the images the requirement for assessment and critical evaluation along all steps of the procedure is obvious. Sometimes assessment might be fairly straightforward. For example, when picking particles on a micrograph or estimating the position of Thon rings in the micrograph PSD, the user has ready access to graphical overlays on the images and can visually inspect the result of the automated procedures. However, many of the most exciting structural challenges of the future will likely require the processing of hundreds of thousands, or millions, of particles, and in this case visual assessment of every image or every particle becomes unreasonable. Instead, the automated algorithms must also provide figures of merit that provide the means to accept or reject the results, ideally in conjunction with visual inspection of a limited number of images. We note here that automated algorithms are unlikely to ever be in 100% agreement with user decisions and thus must be reevaluated at several points along the pipeline. Ultimately, the goal of the automated algorithms is not to precisely emulate a human operator, but rather to provide the highest quality results using the given data. At the later processing stages of alignment, classification, and 3D reconstruction a significant level of automation exists, and the current goal is to provide the user with as much information as possible to facilitate decisions about the validity of the outcome. For example, it would be dangerous to believe a 3D map based only on an inspection of the map itself without a close, critical examination of all associated data. This last should include Euler angle statistics, resolution curves, comparisons between classes and reprojections, summaries of selected vs. rejected particles, variance maps, and other validations. All of these factors should also be assessed over the entire course of the iterative procedures. One of our goals in developing Appion (Lander *et al.*, 2009) was to make sure that the user has ready access to this information no matter what set of procedures is used for processing, analysis, and refinement. To this end, Appion uses a web browser to present summaries of relevant data in a single format regardless of the individual programs used for processing or refinement. This frees the user from the need for detailed knowledge of how and where each of the different software suites stores and presents data.

While the final evaluation of a calculated map is currently left almost entirely to the judgment of an experienced user, we can envisage future figures of merit based on an integrated evaluation of the entire set of outcomes, or comparisons to previous results for similar structures. The idea of an integrated system that has access to all data and meta data for all processing steps has been a significant driving force in the development of

both Leginon and Appion. Both systems are intimately connected to a relational database that tracks every item of data (Fellmann *et al.*, 2002). For Leginon this includes the calibration and setup parameters, and a complete record of the state of the instrument for every image acquired. In Appion, the database tracks every procedure used and all input and output data and parameters, creating links and calculating statistics where necessary. This provides a complete and permanent record of the processing steps applied to any data set that can be accessed at any time in the future, independent of an individual's record keeping skills. For example, in Appion and Leginon, the exact time, stage position inside the microscope, and ambient air and column temperature are known for each particle in a given 3D reconstruction. We should also note that the term "database," as used in Appion and Leginon, refers to more than just a set of linearly searchable, well-organized files. Rather, it is a collection of logically structured, related data, together with a system to query and update that data in an efficient, non-linear, manner. This implementation combines the database management system and the database itself, providing the tools to create schemas and perform complex information queries using the Structured Query Language (SQL). Mining the database allows us to efficiently identify relevant parameters in a processing run, readily repeat precise processing conditions, output valuable quantitative statistics, and inter-connect paths from different processing stages. An important goal for Appion is to use this infrastructure and the resulting accessibility of meta-data for individual experiments to provide intelligent feedback during processing and analysis.

7. THE FUTURE OF AUTOMATION

The unfortunate limitation of our field is that the theoretical Nyquist frequency of an electron micrograph does not predict the achievable resolu-

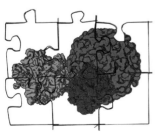

tion for the specimen of interest. Early in the development of the field, impressive illustrations of the potential of EM to attain atomic detailed reconstructions were provided using 2D electron crystallography (e.g., Henderson *et al.*, 1990; Nogales *et al.*, 1998). While these methods obviously benefited from the inherent averaging imposed by specimen crystallization, they provided hope for attaining high-resolution reconstructions for single particles. "Chasing the train," a pun that indicated the potential to trace the polypeptide chain into an EM density map, became somewhat popular at the turn of the century (van Heel *et al.*, 2000). Less than a decade later, backbone traces and atomic resolution maps for single particles have begun to appear, although still primarily limited to specimens with high symmetry and large molecular weight (Chen *et al.*, 2009; Cong *et al.*, 2010; Jiang *et al.*, 2008; Ludtke *et al.*, 2008; Yu *et al.*, 2008; Zhang *et al.*, 2008). Significant progress has also been made in improving the resolution of particles with low symmetry and molecular weight; the last several years have seen 3D reconstructions of the 290 kDa Transferrin–Transferrin Receptor complex at 7.5 Å resolution (Cheng *et al.*, 2004), the 180 kDa tetrameric P53 tumor suppressor at 13.7 Å resolution (Okorokov *et al.*, 2006), and a DNA nanostructure at just 78 kDa solved to 12 Å resolution (Kato *et al.*, 2009). But while the theoretical potential of the technology seems increasingly within reach, the ability to achieve high resolution remains the exception rather than the norm. In practice, most specimens are unlikely to achieve a backbone trace, especially given the conformational and compositional heterogeneity of many biological samples. Recently, in a seminal paper covering the last 30 years of single-particle reconstructions, Joachim Frank proposed that finding recipes to achieve resolutions beyond what has already been achieved, particularly for asymmetric particles, remains a daunting task, which will require pushing the quantity of data being both collected and processed (Frank, 2009). We believe that automation has a crucial role to play in this arena.

We return once again to our original definition of automation and investigate why this term is so appropriate to EM. In the most fundamental case, automation allows for the acquisition of thousands of micrographs and hundreds of thousands of single particles. Beyond this, we can ensure reproducibility when preparing multiple vitrified samples in an identical fashion, tighter quality control by collecting images exclusively within a specific range of ice thickness, waste reduction by focusing solely on the particles that have been grouped into homogeneous subsets through classification, and integration with systems by enabling inter-package processing compatibility. With the incorporation of machine-learning algorithms, comprehensive graphical user interfaces, and integrated centralized databases for image processing, automated technologies begin to resemble human inductive methods of learning. They begin to adapt to the particularities of each novel dataset, allow for user-based intervention, and contain, at their core, the connective glue to interlink the assortment of operations from initial data collection through construction of a final map. The result of this is increased productivity and reduction of labor. Automation, when done right, opens up possibilities for performing experiments that might otherwise remain out of reach.

In an effort to push the limits of EM technologies, many laboratories around the world, both in academia and industry, are addressing current technical challenges with sophisticated microscope and computational accessories. Spherical and chromatic aberration correctors will allow finer sampling by reducing contrast delocalization in high-resolution areas. Energy filters eliminate inelastically scattered electrons contributing to background noise. Phase plates positioned in the back focal plane of the objective lens modify the CTF and amplify contrast of vitrified specimens (Danev and Nagayama, 2001; Nagayama and Danev, 2008; Danev et al., this volume), potentially allowing for reconstruction of very small structures. Cameras equipped with direct electron detectors could soon surplant current fiber optic-coupled CCD cameras and improve the modulation transfer function (MTF) of acquired images (Faruqi and Henderson, 2007). General-purpose computing on graphics processing units (GPUs) (reviewed in Schmeisser et al., 2009) have demonstrated the capability to provide more than 50-fold speedups of matrix multiplication, fast Fourier transforms, and multireference particle comparisons (Castano Diez et al., 2007, 2008; Govindaraju et al., 2008) and may replace CPU processing for computationally expensive tasks like maximum-likelihood or bootstrap calculations. Collectively, these technical improvements promise to expand the scope of questions addressable by single-particle EM, but will likely increase the complexity of our instruments. Automation will undoubtedly play a role in providing maintenance, alignment, and calibration of these and other increasingly sophisticated devices.

While technical advancements in many aspects of EM may signify that atomic resolution will be achievable for some suitable macromolecules, there are many interesting biological questions that can be answered at considerably lower resolutions. As a resource facility, we aim to provide technology to a wide array of researchers, including biologists, chemists, material scientists, and other practitioners from outside the cryo-EM field and a high level of automation and streamlining is critical to this mission. We are increasingly seeing EM used as one of a wide variety of hybrid methods. For example, cryo-EM has been combined with multiangle laser light scattering (MALLS), small-angle X-ray scattering (SAXS), mass-spectrometry, and biochemical mutational analyses to identify the dimerization domain of ALIX during its involvement in membrane remodeling (Pires et al., 2009). It has been used in combination with nuclear magnetic resonance (NMR) spectroscopy, X-ray crystallography, SAXS, and computational docking, to determine the structure of the actin-binding domain of Talin and provide direct evidence for its interaction with cytoskeletal actin (Gingras et al., 2008). It has been combined with data from ultracentrifugation, quantitative immunoblotting, affinity purification, overlay assays, immuno-labeling, and bioinformatics in a tour-de-force structural analysis of the entire nuclear pore complex (Alber et al.,

2007a,b). EM, as a field, has long since ceased to belong purely to the physicists and is rapidly transforming into a mature technology embraced by many scientific disciplines. Groups around the world are assembling the puzzle pieces for all of the distinct operations, collaboratively putting together a streamlined EM pipeline, polishing the building materials, repairing the leaks, and broadening the bottlenecks in the pipeline. The successful future of EM does not simply depend upon its ability to rival contenders in attaining atomic resolution, but rather the broader integration of the technology as a hybrid approach to addressing increasingly complex questions.

ACKNOWLEDGMENTS

We acknowledge primary support from the National Institutes of Health (NIH) through the National Center for Research Resources (NCRR) P41 program (Grants RR17573 and RR023093). We are also grateful for the support provided to Arne Moeller by the Joint Center for Innovation in Membrane Protein Production for Structure Determination (Grant RFA-RM-08-019) and to Pick-Wei Lau by the American Heart Association. We ask for forgiveness for any omissions, errors, or oversights that are almost inevitable in a review of this wide scope.

REFERENCES

Alber, F., *et al.* (2007a). Determining the architectures of macromolecular assemblies. *Nature* **450,** 683–694.
Alber, F., *et al.* (2007b). The molecular architecture of the nuclear pore complex. *Nature* **450,** 695–701.
Baldwin, P. R., and Penczek, P. A. (2005). Estimating alignment errors in sets of 2-D images. *J. Struct. Biol.* **150,** 211–225.
Baxter, W. T., *et al.* (2007). SPIRE: The SPIDER reconstruction engine. *J. Struct. Biol.* **157,** 56–63.
Baxter, W. T., *et al.* (2009). Determination of signal-to-noise ratios and spectral SNRs in cryo-EM low-dose imaging of molecules. *J. Struct. Biol.* **166,** 126–132.
Berriman, J., and Unwin, N. (1994). Analysis of transient structures by cryo-microscopy combined with rapid mixing of spray droplets. *Ultramicroscopy* **56,** 241–252.
Bonnet, N. (1998). Multivariate statistical methods for the analysis of microscope image series: Applications in materials science. *J. Microsc.* **190,** 2–18.
Bonnet, N. (2000). Artificial intelligence and pattern recognition techniques in microscope image processing and analysis. *Adv. Imaging Electron Phys.* **114,** 1–77.
Carazo, J. M., *et al.* (1990). Fuzzy sets-based classification of electron microscopy images of biological macromolecules with an application to ribosomal particles. *J. Microsc.* **157,** 187–203.
Cardona, A., *et al.* (2010). An integrated micro- and macroarchitectural analysis of the Drosophila brain by computer-assisted serial section electron microscopy (submitted).
Carragher, B., *et al.* (2000). Leginon: An automated system for acquisition of images from vitreous ice specimens. *J. Struct. Biol.* **132,** 33–45.
Castano Diez, D., *et al.* (2007). Implementation and performance evaluation of reconstruction algorithms on graphics processors. *J. Struct. Biol.* **157,** 288–295.

Castano-Diez, D., *et al.* (2008). Performance evaluation of image processing algorithms on the GPU. *J. Struct. Biol.* **164,** 153–160.

Chen, J. Z., and Grigorieff, N. (2007). SIGNATURE: A single-particle selection system for molecular electron microscopy. *J. Struct. Biol.* **157,** 168–173.

Chen, J. Z., *et al.* (2009). Molecular interactions in rotavirus assembly and uncoating seen by high-resolution cryo-EM. *Proc. Natl. Acad. Sci. USA* **106,** 10644–10648.

Cheng, Y., *et al.* (2004). Structure of the human transferrin receptor-transferrin complex. *Cell* **116,** 565–576.

Cheng, Y., *et al.* (2006). Single particle reconstructions of the transferrin-transferrin receptor complex obtained with different specimen preparation techniques. *J. Mol. Biol.* **355,** 1048–1065.

Cheng, A., *et al.* (2007). Towards automated screening of two-dimensional crystals. *J. Struct. Biol.* **160,** 324–331.

Cong, Y., *et al.* (2010). 4.0-A resolution cryo-EM structure of the mammalian chaperonin TRiC/CCT reveals its unique subunit arrangement. *Proc. Natl. Acad. Sci. USA* **107,** 4967–4972.

Coudray, N., *et al.* (2008). Automatic acquisition and image analysis of 2D crystals. *Micros. Today* **16,** 48–49.

Coudray, N., *et al.* (2010). Automated Screening of 2D crystallization trials using transmission electron microscopy: A high-throughput tool-chain for sample preparation and microscopic analysis. Submitted.

Crowther, R. A., *et al.* (1996). Three dimensional reconstructions of spherical viruses by Fourier synthesis from electron micrographs. *Nature* **226,** 421–425.

Crowther, R. A., *et al.* (1970). MRC image processing programs. *J. Struct. Biol.* **116,** 9–16.

Danev, R., and Nagayama, K. (2001). Transmission electron microscopy with Zernike phase plate. *Ultramicroscopy* **88,** 243–252.

Dierksen, K., *et al.* (1992). Towards automatic electron tomography. *Ultramicroscopy* **40,** 71–87.

Dierksen, K., *et al.* (1993). Towards automatic electron tomography 2. Implementation of autofocus and low-dose procedures. *Ultramicroscopy* **49,** 109–120.

Drahos, V., and Delong, A. (1960). A simple method for obtaining perforated supporting membranes for electron microscopy. *Nature* **186,** 104.

Dubochet, J., *et al.* (1982). Electron microscopy of frozen water and aqueous solutions. *J. Microsc.* **128,** 219–237.

Dubochet, J., *et al.* (1988). Cryo-electron microscopy of vitrified specimens. *Q. Rev. Biophys.* **21,** 129–228.

Ermantraut, E., *et al.* (1998). Perforated support foils with re-defined hole size, shape and arrangement. *Ultramicroscopy* **74,** 75–81.

Farrow, N. A., and Ottensmeyer, F. P. (1992). A posteriori determination of relative projection directions of arbitrarily oriented macromolecules. *J. Opt. Soc. Am. A* **9,** 1749–1760.

Faruqi, A. R., and Henderson, R. (2007). Electronic detectors for electron microscopy. *Curr. Opin. Struct. Biol.* **17,** 549–555.

Fellmann, D., *et al.* (2002). A relational database for cryoEM: Experience at one year and 50000 images. *J. Struct. Biol.* **137,** 273–282.

Fernandez, J. J., *et al.* (1997). A spectral estimation approach to contrast transfer function detection in electron microscopy. *Ultramicroscopy* **68,** 267–295.

Fernandez, J. J., *et al.* (2006). CTF determination and correction in electron cryotomography. *Ultramicroscopy* **106,** 587–596.

Fernandez, J. J., *et al.* (2008). Sharpening high resolution information in single particle electron cryomicroscopy. *J. Struct. Biol.* **164,** 170–175.

Frank, J. (ed.) (2006). Electron Tomography—Methods for Three-Dimensional Visualization of Structures in the Cell, Springer, New York.

Frank, J. (2006b). *Three-dimensional electron microscopy of macromolecular assemblies: Visualization of biological molecules in their native state* Oxford University Press, New York.

Frank, J. (2009). Single-particle reconstruction of biological macromolecules in electron microscopy—30 years. *Q. Rev. Biophys.* **42,** 139–158.

Frank, J., and van Heel, M. (1982). Correspondence analysis of aligned images of biological particles. *J. Mol. Biol.* **161,** 134–137.

Frank, J., *et al.* (1981). Computer averaging of electron micrographs of 40S ribosomal subunits. *Science* **214,** 1353–1355.

Frank, J., *et al.* (1988). Classification of images of biomolecular assemblies: A study of ribosomes and ribosomal subunits of Escherichia coli. *J. Microsc.* **150,** 99–115.

Frank, J., *et al.* (1996). SPIDER and WEB: Processing and visualization of images in 3D electron microscopy and related fields. *J. Struct. Biol.* **116,** 190–199.

Frederik, P. M., and Hubert, D. H. W. (2005). Cryoelectron microscopy of liposomes. *Methods in Enzymol.* **391,** 431–448.

Frey, B. J., and Dueck, D. (2007). Clustering by passing messages between data points. *Science* **315,** 972–976.

Fung, J. C., *et al.* (1996). Toward fully automated high-resolution electron tomography. *J. Struct. Biol.* **116,** 181–189.

Gingras, A. R., *et al.* (2008). The structure of the C-terminal actin-binding domain of talin. *EMBO J.* **27,** 458–469.

Goncharov, A., and Vainshtein, B. K. (1986). Determining the spatial orientation of arbitrarily arranged particles given their projections. *Dokl. Acad. Sci. USSR* **287,** 1131–1134.

Govindaraju, N. K., *et al.* (2008). High performance discrete fourier transforms on graphics processors. Proceedings of the 2008 ACM/IEEE Conference on Supercomputing, pp. 1–12. IEEE Press, Austin, TX.

Grigorieff, N. (2007). FREALIGN: High-resolution refinement of single particle structures. *J. Struct. Biol.* **157,** 117–125.

Hall, R. J., and Patwardhan, A. (2004). A two step approach for semi-automated particle selection from low contrast cryo-electron micrographs. *J. Struct. Biol.* **145,** 19–28.

Harauz, G., and van Heel, M. (1986). Exact filters for general geometry 3-dimensional reconstruction. *Optik* **73,** 146–156.

Harris, W. J. (1962). Holey films for electron microscopy. *Nature* **196,** 499–500.

Henderson, R., *et al.* (1990). Model for the structure of bacteriorhodopsin based on high-resolution electron cryo-microscopy. *J. Mol. Biol.* **213,** 899–929.

Heymann, J. B. (2001). Bsoft: Image and molecular processing in electron microscopy. *J. Struct. Biol.* **133,** 156–169.

Heymann, J. B., and Belnap, D. M. (2007). Bsoft: Image processing and molecular modeling for electron microscopy. *J. Struct. Biol.* **157,** 3–18.

Heymann, J. B., *et al.* (2008). Computational resources for cryo-electron tomography in Bsoft. *J. Struct. Biol.* **161,** 232–242.

Hohn, M., *et al.* (2007). SPARX, a new environment for Cryo-EM image processing. *J. Struct. Biol.* **157,** 47–55.

Hu, M., *et al.* (2010). Automated electron microscopy for evaluating two-dimensional crystallization of membrane proteins. *J. Struct. Biol.* (submitted).

Huang, Z., and Penczek, P. A. (2004). Application of template matching technique to particle detection in electron micrographs. *J. Struct. Biol.* **145,** 29–40.

Huang, Z., *et al.* (2003). Automated determination of parameters describing power spectra of micrograph images in electron microscopy. *J. Struct. Biol.* **144,** 79–94.

Iacovache, I., *et al.* (2010). The 2DX robot: A membrane protein 2D crystallization Swiss Army knife. *J. Struct. Biol.* **169**, 370–378.

Jiang, W., *et al.* (2008). Backbone structure of the infectious epsilon15 virus capsid revealed by electron cryomicroscopy. *Nature* **451**, 1130–1134.

Joyeux, L., and Penczek, P. A. (2002). Efficiency of 2D alignment methods. *Ultramicroscopy* **92**, 33–46.

Kato, T., *et al.* (2009). High-resolution structural analysis of a DNA nanostructure by cryoEM. *Nano Lett.* **9**, 2747–2750.

Kelly, D. F., *et al.* (2008). The Affinity Grid: A pre-fabricated EM grid for monolayer purification. *J. Mol. Biol.* **382**, 423–433.

Kimoto, K., *et al.* (2003). Practical procedure for coma-free alignment using caustic figure. *Ultramicroscopy* **96**, 219–227.

Kisseberth, N., *et al.* (1997). emScope: A tool kit for control and automation of a remote electron microscope. *J. Struct. Biol.* **120**, 309–319.

Koster, A. J., and de Ruijter, W. J. (1992). Practical autoalignment of transmission electron microscopes. *Ultramicroscopy* **40**, 89–107.

Koster, A. J., *et al.* (1987). An autofocus method for a TEM. *Ultramicroscopy* **21**, 209–222.

Kühlbrandt, W. (2003). Two-dimensional crystallization of membrane proteins: A practical guide. *In* "Membrane Protein Purification and Crystallization: A Practical Guide," (C. Hunte, *et al.*, eds.), pp. 253–284. Academic Press, San Diego.

Lake, J. A. (1976). Ribosome structure determined by electron microscopy of Escherichia coli small subunits, large subunits and monomeric ribosomes. *J. Mol. Biol.* **105**, 131–139.

Lander, G. C., *et al.* (2009). Appion: An integrated, database-driven pipeline to facilitate EM image processing. *J. Struct. Biol.* **166**, 95–102.

Lefman, J., *et al.* (2007). Automated 100-position specimen loader and image acquisition system for transmission electron microscopy. *J. Struct. Biol.* **158**, 318–326.

Lei, J., and Frank, J. (2005). Automated acquisition of cryo-electron micrographs for single particle reconstruction on an FEI Tecnai electron microscope. *J. Struct. Biol.* **150**, 69–80.

Leschziner, A. E., and Nogales, E. (2006). The orthogonal tilt reconstruction method: An approach to generating single-class volumes with no missing cone for ab initio reconstruction of asymmetric particles. *J. Struct. Biol.* **153**, 284–299.

Liu, X., *et al.* (2007). Averaging tens to hundreds of icosahedral particle images to resolve protein secondary structure elements using a Multi-Path Simulated Annealing optimization algorithm. *J. Struct. Biol.* **160**, 11–27.

Lu, Z., *et al.* (2009). Monolithic microfluidic mixing-spraying devices for time-resolved cryo-electron microscopy. *J. Struct. Biol.* **168**, 388–395.

Ludtke, S. J., *et al.* (1999). EMAN: Semiautomated Software for High-Resolution Single-Particle Reconstructions. *J. Struct. Biol.* **128**, 82–97.

Ludtke, S. J., *et al.* (2008). De novo backbone trace of GroEL from single particle electron cryomicroscopy. *Structure* **16**, 441–448.

Mallick, S. P., *et al.* (2004). Detecting particles in cryo-EM micrographs using learned features. *J. Struct. Biol.* **145**, 52–62.

Mallick, S. P., *et al.* (2005). ACE: Automated CTF estimation. *Ultramicroscopy* **104**, 8–29.

Marabini, R., and Carazo, J. M. (1994). Pattern recognition and classification of images of biological macromolecules using artificial neural networks. *Biophys. J.* **66**, 1804–1814.

Marabini, R., *et al.* (1996). Xmipp an image processing package for electron microscopy. *J. Struct. Biol.* **116**, 237–240.

Marsh, M. P., *et al.* (2007). Modular software platform for low-dose electron microscopy and tomography. *J. Microsc.* **228**, 384–389.

Mastronarde, D. N. (2005). Automated electron microscope tomography using robust prediction of specimen movements. *J. Struct. Biol.* **152**, 36–51.

Melanson, L. A. (2009). A versatile and affordable plunge freezing instrument for preparing specimens for cryo transmission electron microscopy (cryoEM). *Micros. Today* **2**, 14–17.

Microsystems (2009). L., Leica EM GP Product Brochure.

Mindell, J. A., and Grigorieff, N. (2003). Accurate determination of local defocus and specimen tilt in electron microscopy. *J. Struct. Biol.* **142**, 334–347.

Mulder, A. M., *et al.* (2010). Time resolved single particle EM reveals structure-based mechanism for assembly of the 30S ribosomal subunit (submitted)..

Nagayama, K., and Danev, R. (2008). Phase contrast electron microscopy: Development of thin-film phase plates and biological applications. *Philos. Trans. R. Soc. Lond. B Biol. Sci.* **363**, 2153–2162.

Nickell, S., *et al.* (2005). TOM software toolbox: Acquisition and analysis for electron tomography. *J. Struct. Biol.* **149**, 227–234.

Nogales, E., *et al.* (1998). Structure of the alpha beta tubulin dimer by electron crystallography. *Nature* **391**, 199–203.

Ochoa, W. F., *et al.* (2008). Partitivirus structure reveals a 120-subunit, helix-rich capsid with distinctive surface arches formed by quasisymmetric coat-protein dimers. *Structure* **16**, 776–786.

Ogura, T., and Sato, C. (2004a). Auto-accumulation method using simulated annealing enables fully automatic particle pickup completely free from a matching template or learning data. *J. Struct. Biol.* **146**, 344–358.

Ogura, T., and Sato, C. (2004b). Automatic particle pickup method using a neural network has high accuracy by applying an initial weight derived from eigenimages: A new reference free method for single-particle analysis. *J. Struct. Biol.* **145**, 63–75.

Ohi, M., *et al.* (2004). Negative staining and image classification—Powerful tools in modern electron microscopy. *Biol. Proced. Online* **6**, 23–34.

Okorokov, A. L., *et al.* (2006). The structure of p53 tumour suppressor protein reveals the basis for its functional plasticity. *EMBO J.* **25**, 5191–5200.

Pan, J., *et al.* (2009). Atomic structure reveals the unique capsid organization of a dsRNA virus. *Proc. Natl. Acad. Sci. USA* **106**, 4225–4230.

Pantelic, R. S., *et al.* (2010). Graphene oxide: A substrate for optimizing preparations of frozen-hydrated samples. *J. Struct. Biol.* **170**, 152–156.

Parent, K. N., *et al.* (2010). P22 coat protein structures reveal a novel mechanism for capsid maturation: Stability without auxiliary proteins or chemical crosslinks. *Structure* **18**, 390–401.

Pascual-Montano, A., *et al.* (2001). A novel neural network technique for analysis and classification of EM single-particle images. *J. Struct. Biol.* **133**, 233–245.

Pearson, K. (1901). On lines and planes of closest fit to systems of points in space. *Philos. Mag.* **2**, 559–572.

Penczek, P. A., *et al.* (1994). The ribosome at improved resolution: New techniques for merging and orientation refinement in 3D cryo-electron microscopy of biological particles. *Ultramicroscopy* **53**, 251–270.

Penczek, P. A., Zhu, J., and Frank, J. (1996). A common-lines based method for determining orientations for N > 3 particle projections simultaneously. *Ultramicroscopy* **63**, 205–218.

Philippsen, A., *et al.* (2007). The contrast-imaging function for tilted specimens. *Ultramicroscopy* **107**, 202–212.

Pires, R., *et al.* (2009). A crescent-shaped ALIX dimer targets ESCRT-III CHMP4 filaments. *Structure* **17**, 843–856.

Plaisier, J. R., *et al.* (2004). TYSON: Robust searching, sorting, and selecting of single particles in electron micrographs. *J. Struct. Biol.* **145**, 76–83.

Potter, C. S., *et al.* (1999). Leginon: A system for fully automated acquisition of 1000 electron micrographs a day. *Ultramicroscopy* **77**, 153–161.

Potter, C. S., *et al.* (2004). Robotic grid loading system for a transmission electron microscope. *J. Struct. Biol.* **146**, 431–440.

Quispe, J., *et al.* (2007). An improved holey carbon film for cryo-electron microscopy. *Microsc. Microanal.* **13**, 365–371.

Radermacher, M., *et al.* (1986). A new 3-D reconstruction scheme applied to the 50 S ribosomal subunit of E.coli. *J. Microsc.* **141**, RP1–RP2.

Rath, B. K., and Frank, J. (2004). Fast automatic particle picking from cryo-electron micrographs using a locally normalized cross-correlation function: A case study. *J. Struct. Biol.* **145**, 84–90.

Rath, B. K., *et al.* (1997). Low-dose automated electron tomography: A recent implementation. *J. Struct. Biol.* **120**, 210–218.

Roseman, A. M. (2004). FindEM—A fast, efficient program for automatic selection of particles from electron micrographs. *J. Struct. Biol.* **145**, 91–99.

Rosenthal, P. B., and Henderson, R. (2003). Optimal determination of particle orientation, absolute hand, and contrast loss in single-particle electron cryomicroscopy. *J. Mol. Biol.* **333**, 721–745.

Sander, B., *et al.* (2003). Automatic CTF correction for single particles based upon multivariate statistical analysis of individual power spectra. *J. Struct. Biol.* **142**, 392–401.

Sartori, A., *et al.* (2007). Correlative microscopy: Bridging the gap between fluorescence light microscopy and cryo-electron tomography. *J. Struct. Biol.* **160**, 135–145.

Saxton, W. O., and Frank, J. (1977). Motif detection in quantum noise-limited electron micrographs by cross-correlation. *Ultramicroscopy* **2**, 219–227.

Saxton, W. O., *et al.* (1983). Procedures for focusing, stigmating, and alignment in high-resolution electron microscopy. *J. Microsc.* **130**, 187–201.

Scheres, S. H., *et al.* (2005a). Fast maximum-likelihood refinement of electron microscopy images. *Bioinformatics* **21**(Suppl. 2), ii243–ii244.

Scheres, S. H. W., *et al.* (2005b). Maximum-likelihood multi-reference refinement for electron microscopy images. *J. Mol. Biol.* **348**, 139–149.

Scheres, S. H., *et al.* (2008). Image processing for electron microscopy single-particle analysis using XMIPP. *Nat. Protoc.* **3**, 977–990.

Schmeisser, M., *et al.* (2009). Parallel, distributed and GPU computing technologies in single-particle electron microscopy. *Acta Crystallogr. D Biol. Crystallogr.* **65**, 659–671.

Shaikh, T. R., *et al.* (2008). Particle-verification for single-particle, reference-based reconstruction using multivariate data analysis and classification. *J. Struct. Biol.* **164**, 41–48.

Shi, J., *et al.* (2008). A Script-Assisted Microscopy (SAM) package to improve data acquisition rates on FEI Tecnai electron microscopes equipped with Gatan CCD cameras. *J. Struct. Biol.* **164**, 166–169.

Short, J. M. (2004). SLEUTH—A fast computer program for automatically detecting particles in electron microscope images. *J. Struct. Biol.* **145**, 100–110.

Sigworth, F. J. (1998). A maximum-likelihood approach to single-particle image refinement. *J. Struct. Biol.* **122**, 328–339.

Singer, A., *et al.* (2010). Detecting consistent common lines in cryo-EM by voting. *J. Struct. Biol.* (in press; corrected proof).

Singh, V., *et al.* (2004). Image segmentation for automatic particle identification in electron micrographs based on hidden Markov random field models and expectation maximization. *J. Struct. Biol.* **145**, 123–141.

Sorzano, C. O., *et al.* (2004). Xmipp a new generation of an open-source image processing package for electron microscopy. *J. Struct. Biol.* **148**, 194–204.

Sorzano, C. O. S., *et al.* (2007). Fast, robust, and accurate determination of transmission electron microscopy contrast transfer function. *J. Struct. Biol.* **160**, 249–262.

Sorzano, C. O., *et al.* (2009a). Marker-free image registration of electron tomography tilt-series. *BMC Bioinform.* **10**, 124.

Sorzano, C. O., *et al.* (2009b). Error analysis in the determination of the electron microscopical contrast transfer function parameters from experimental power Spectra. *BMC Struct. Biol.* **9**, 18.

Sorzano, C. O. S., *et al.* (2009c). Automatic particle selection from electron micrographs using machine learning techniques. *J. Struct. Biol.* **167**, 252–260.

Sousa, D., and Grigorieff, N. (2007). Ab initio resolution measurement for single particle structures. *J. Struct. Biol.* **157**, 201–210.

Stagg, S. M., *et al.* (2006). Automated cryoEM data acquisition and analysis of 284, 742 particles of GroEL. *J. Struct. Biol.* **155**, 470–481.

Stewart, A., and Grigorieff, N. (2004). Noise bias in the refinement of structures derived from single particles. *Ultramicroscopy* **102**, 67–84.

Suloway, C., *et al.* (2005). Automated molecular microscopy: The new Leginon system. *J. Struct. Biol.* **151**, 41–60.

Suloway, C., *et al.* (2009). Fully automated, sequential tilt-series acquisition with Leginon. *J. Struct. Biol.* **167**, 11–18.

Tang, G., *et al.* (2007). EMAN2: An extensible image processing suite for electron microscopy. *J. Struct. Biol.* **157**, 38–46.

Taylor, K., and Glaeser, R. M. (1976). Electron microscopy of frozen-hydrated biological specimens. *J. Ultrastruct. Res.* **55**, 448–456.

Typke, D., and Dierksen, K. (1995). Determination of image aberrations in high-resolution electron microscopy using diffractogram and cross-correlation methods. *Optik* **99**, 155–166.

Typke, D., and Köstler, D. (1977). Determination of the wave aberration of electron lenses from superposition diffractograms of images with differently tilted illumination. *Ultramicroscopy* **2**, 285–295.

Umesh Adiga, P. S., *et al.* (2004). A binary segmentation approach for boxing ribosome particles in cryo EM micrographs. *J. Struct. Biol.* **145**, 142–151.

Unser, M., *et al.* (1987). A new resolution criterion based on spectral signal-to-noise ratios. *Ultramicroscopy* **23**, 39–51.

Unser, M., *et al.* (1989). The spectral signal-to-noise ratio resolution criterion: Computational efficiency and statistical precision. *Ultramicroscopy* **30**, 429–433.

van Heel, M. (1984). Multivariate statistical classification of noisy images (randomly oriented biological macromolecules). *Ultramicroscopy* **13**, 165–183.

van Heel, M. (1987). Angular reconstitution: A posteriori assignment of projection directions for 3D reconstruction. *Ultramicroscopy* **21**, 111–124.

van Heel, M., and Frank, J. (1980). Classification of particles in noisy electron micrographs using correspondence analysis. (E. S. Gelsema and L. N. Kanal, eds.), vol. 1, pp. 235–243. Elsevier, Amsterdam.

van Heel, M., and Frank, J. (1981). Use of multivariate statistics in analysing the images of biological macromolecules. *Ultramicroscopy* **6**, 187–194.

van Heel, M., and Keegstra, W. (1981). IMAGIC: A fast flexible and friendly image analysis software system. *Ultramicroscopy* **7**, 113–130.

van Heel, M., and Stoffler-Meilicke, M. (1985). Characteristic views of *E. coli* and *B. stearothermophilus* 30S ribosomal subunits in the electron microscope. *EMBO J.* **4**, 2389–2395.

van Heel, M., *et al.* (1992). Correlation functions revisited. *Ultramicroscopy* **46**, 307–316.

van Heel, M., *et al.* (1996). A new generation of the IMAGIC image processing system. *J. Struct. Biol.* **116**, 17–24.

van Heel, M., *et al.* (2000). Single-particle electron cryo-microscopy: Towards atomic resolution. *Q. Rev. Biophys.* **33**, 307–369.

Velasquez-Muriel, J. A., *et al.* (2003). A method for estimating the CTF in electron microscopy based on ARMA models and parameter adjustment. *Ultramicroscopy* **96**, 17–35.

Vink, M., *et al.* (2007). A high-throughput strategy to screen 2D crystallization trials of membrane proteins. *J. Struct. Biol.* **160**, 295–304.

Volkmann, N. (2004). An approach to automated particle picking from electron micrographs based on reduced representation templates. *J. Struct. Biol.* **145**, 152–156.

Voss, N. R., *et al.* (2009). DoG Picker and TiltPicker: Software tools to facilitate particle selection in single particle electron microscopy. *J. Struct. Biol.* **166**, 205–213.

Voss, N. R., *et al.* (2010). A toolbox for ab initio 3-D reconstructions in single-particle electron microscopy. *J. Struct. Biol.* **169**, 389–398.

Walker, M., *et al.* (1995). Millisecond time resolution electron cryo-microscopy of the M-ATP transient kinetic state of the acto-myosin ATPase. *Biophys. J.* **68**, 87S–91S.

Walker, M., *et al.* (1999). Observation of transient disorder during myosin subfragment-1 binding to actin by stopped-flow fluorescence and millisecond time resolution electron cryomicroscopy: Evidence that the start of the crossbridge power stroke in muscle has variable geometry. *Proc. Natl. Acad. Sci. USA* **96**, 465–470.

Welch, P. D. (1967). Use of Fast Fourier Transform for Estimation of Power Spectra - a Method Based on Time Averaging over Short Modified Periodograms. *IEEE Trans. Audio Electroacoustics* 70.

White, H. D., *et al.* (1998). A computer-controlled spraying-freezing apparatus for millisecond time-resolution electron cryomicroscopy. *J. Struct. Biol.* **121**, 306–313.

White, H. D., *et al.* (2003). A second generation apparatus for time-resolved electron cryomicroscopy using stepper motors and electrospray. *J. Struct. Biol.* **144**, 246–252.

Winkler, H., and Taylor, K. A. (2006). Accurate marker-free alignment with simultaneous geometry determination and reconstruction of tilt series in electron tomography. *Ultramicroscopy* **106**, 240–254.

Winkler, H., *et al.* (2009). Tomographic subvolume alignment and subvolume classification applied to myosin V and SIV envelope spikes. *J. Struct. Biol.* **165**, 64–77.

Wong, H. C., *et al.* (2004). Model-based particle picking for cryo-electron microscopy. *J. Struct. Biol.* **145**, 157–167.

Yan, X., *et al.* (2007). Ab initio random model method facilitates 3D reconstruction of icosahedral particles. *J. Struct. Biol.* **157**, 211–225.

Yoshioka, C., *et al.* (2007). Automation of random conical tilt and orthogonal tilt data collection using feature-based correlation. *J. Struct. Biol.* **159**, 335–346.

Yoshioka, C., *et al.* (2010). Cryomesh: A new substrate for cryo-electron microscopy. *Microsc. Microanal.* **16**, 43–53.

Yu, X., *et al.* (2008). 3.88 A structure of cytoplasmic polyhedrosis virus by cryo-electron microscopy. *Nature* **453**, 415–419.

Zemlin, F. (1979). A practical procedure for alignment of a high resolution electron microscope. *Ultramicroscopy* **4**, 241–245.

Zemlin, F., *et al.* (1978). Coma-free alignment of high resolution electron microscopes with the aid of optical diffractograms. *Ultramicroscopy* **3**, 49–60.

Zhang, P., *et al.* (2001). Automated data collection with a Tecnai 12 electron microscope: Applications for molecular imaging by cryomicroscopy. *J. Struct. Biol.* **135**, 251–261.

Zhang, P., *et al.* (2003). Automated image acquisition and processing using a new generation of 4K × 4K CCD cameras for cryo electron microscopic studies of macromolecular assemblies. *J. Struct. Biol.* **143**, 135–144.

Zhang, X., *et al.* (2008). Near-atomic resolution using electron cryomicroscopy and single-particle reconstruction. *Proc. Natl. Acad. Sci.* **105**, 1867–1872.

Zhang, J., *et al.* (2009). JADAS: A customizable automated data acquisition system and its application to ice-embedded single particles. *J. Struct. Biol.* **165**, 1–9.

Zheng, S. Q., *et al.* (2007a). UCSF tomography: An integrated software suite for real-time electron microscopic tomographic data collection, alignment, and reconstruction. *J. Struct. Biol.* **157**, 138–147.

Zheng, S. Q., *et al.* (2007b). Automated acquisition of electron microscopic random conical tilt sets. *J. Struct. Biol.* **157**, 148–155.

Zheng, S. Q., *et al.* (2009). Dual-axis target mapping and automated sequential acquisition of dual-axis EM tomographic data. *J. Struct. Biol.* **168**, 323–331.

Zhou, Z. H., *et al.* (1996). CTF determination of images of ice-embedded single particles using a graphics interface. *J. Struct. Biol.* **116**, 216–222.

Zhu, Y., *et al.* (2003). Automatic particle detection through efficient Hough transforms. *IEEE Trans. Med. Imaging* **22**, 1053–1062.

Zhu, Y., *et al.* (2004). Automatic particle selection: Results of a comparative study. *J. Struct. Biol.* **145**, 3–14.

Author Index

Subject Index

A

AAA–ATPase
models, building and assessment, 63–64
proteasomal hexamer, 53–55
Actin cytoskeletal, tomography
barbed end, 204
binding protein, 205
eukaryotic, 204
heterogeneity, 210–212
high-resolution light microscopy techniques (HRLM), 205
image analysis
dual-axis tilt, 208
lamella hypothesis, test, 208–209
two-dimensional projection, 209
lamella hypothesis, testing, 205–206
pointed end, 204
rapid transfer apparatus (RTS), 210
sample preparation
correlative light, EM, 207
cryo-EM, 206–207
cryo-ET, 207–208
transience, 209–210
transmission electron cryomicroscopy (cryo-TEM), 211
Animal fatty acid synthase
analytical method
active-site mutants, 195
catalytic domains, 181
classification, 192–193
class merging, 188–189
conical tilt, 193–195
2D class average, 189–190
focused alignment, 190–191
image alignment, preliminary, 185
image preprocessing, particle, 185
microscopy and data collection, 184
particle images, realignment of, 188
particle selection, 184
preliminary image classification, 185–188
specimen preparation, 182–183
catalytic cycle, 180–182
electron microscopy (EM), 180
FAS homodimer, 181
flexible macromolecule
catalytic cycle, 197
single-particle EM, 196–198

Aquaporins (AQPs). *See also* Electron crystallography and aquaporins
adhesive properties, AQP0, 95
AQP4M1 and AQP4M23, 96
AQP1 structure, 95
bR structure, 100
data collection, benefits, 99
dioleyl phosphatidylcholine (DOPC), 94
electron crystallographic studies, 94
AQP2, 98–99
AQP9, 99
AQPZ, 97
GlpF, 97
SoPIP2, 98
α-TIP, 98
image processing library and toolbox (IPLT), 106
mean square deviations (RMSD), 111
methyl-b-cyclodextrin (MBCD), 103
MIP family, 93
signal-to-noise ratio (SNR), 106
Atomic and protein–protein interaction data. *See* Cryoelectron microscopy (cryo-EM) integration
Atomic model protocol
annotation, secondary structure, 11–12
atomic map
template generation, 225–227
TOM toolbox, 225
Cα, 18
atomic model optimization, Rosetta, 19–20
optimization, 17
placement, 16
positions, helices, 16
positions, sheets and loops, 16–17
fitting atomic models, 14
fixing of, 17
flowchart, 9
macromolecular model, 17–18
map rescaling, 18
model optimization, 18–19, 21
model quality monitoring, 19
secondary structure elements identification, 11
segmentation, 10–11
SSEs
correspondence, 14–16
prediction, sequence, 14
structural homologues identification, 13

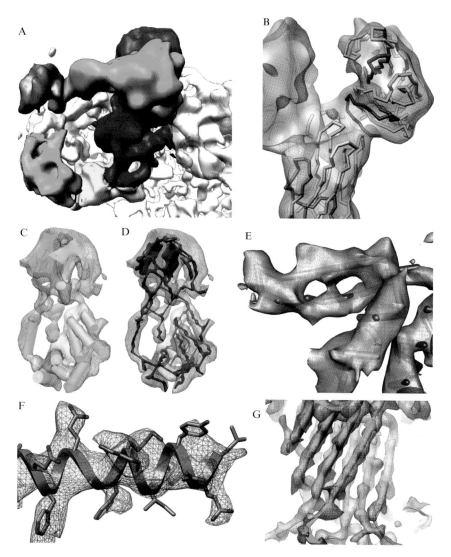

Matthew L. Baker *et al.*, Figure 1.1   Features at subnanometer resolutions. A gallery of structural features from cryo-EM reconstructions is shown. (A) Domains in the clamp region of the 9.5 Å resolution reconstruction of RyR1 can be observed (Serysheva *et al.*, 2008). (B) The atomic models of VP5★ (lower left) and VP8★ (upper right) are fit to the density map corresponding to the VP4 spikes in the 9.5 Å resolution structure of rotavirus (Li *et al.*, 2009). (C) At slightly higher resolutions, secondary structures (α-helices are depicted as cylinders and β-sheets are depicted as planes) can be clearly seen in the capsid proteins of rice dwarf virus at 6.8 Å resolution (Zhou *et al.*, 2001). (D) Around this resolution, possible connections between secondary structure elements can be identified computationally using density skeletonization (red), again as seen in the 6.8 Å resolution rice dwarf virus capsid protein. (E) Increasing resolution reveals the separation of β-strands in GroEL at 4.2 Å resolution (Ludtke *et al.*, 2008). (F) Large, bulky sidechains begin to appear in TriC reconstruction at 4.0 Å resolution (Cong *et al.*, 2010). (G) An unambiguous backbone is apparent in VP6 of rotavirus at ∼3.8 Å resolution (Zhang *et al.*, 2008).

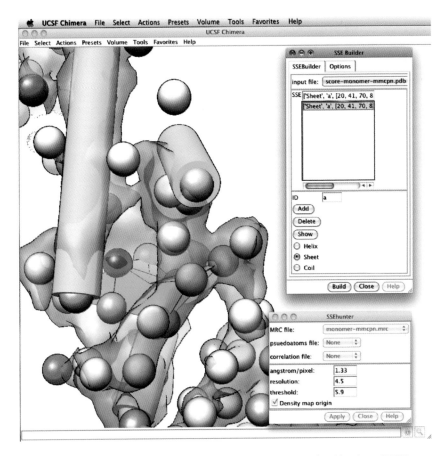

Matthew L. Baker *et al.,* Figure 1.3 Secondary structure identification. SSEHunter and SSEBuilder, both EMAN programs, can be run as plug-ins to UCSF's Chimera. The results for the apical domain of Mm-cpn are shown: red spheres represent helix like regions and the blue spheres represent sheet like regions. Regions of similar scoring pseudoatoms from SSEHunter are grouped and built using SSEBuilder. Helices are depicted as cylinders and sheets are depicted as planes.

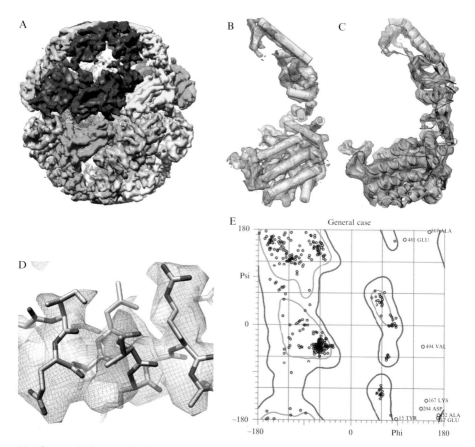

Matthew L. Baker *et al.*, Figure 1.6   Structure of Mm-cpn. (A) The 4.3-Å resolution structure of Mm-cpn is shown (Zhang *et al.*, 2010). (B) Using SSEHunter, the secondary structure elements in the Mm-cpn subunit were identified: α-helices are shown as cylinders and β-strands are shown as planes. (C) Using the *de novo* modeling approach, an atomic model (residues 1–532) was constructed for one subunit of Mm-cpn. (D) Large, bulky sidechains in the model could be seen in the density. (E) The Ramachandran plot of an Mm-cpn monomer shows greater than 98% of all residues with allowable phi–psi angles.

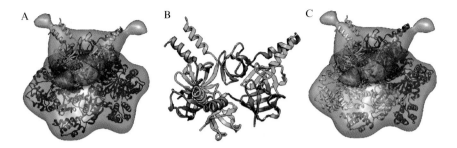

Friedrich Förster and Elizabeth Villa, Figure 3.5 (A) N-PAN (green, cyan) and AAA-PAN (blue). (B) N-PAN hexamer. *cis-* (cyan) and *trans-*positions (green) of the N-PAN monomers in the N-ring. (C) Best-scoring configuration of comparative models (Rpt1/Rpt2/Rpt6/Rpt3/Rpt4/Rpt5) fitted into the EM map.

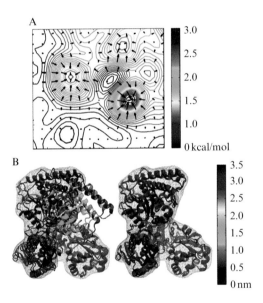

Friedrich Förster and Elizabeth Villa, Figure 3.6 Molecular dynamics flexible fitting. (A) Cross-section of a 2D slice of the potential derived from the EM density of elongation factor Tu (EMD-5036) represented as a contour plot. Arrows represent forces driving the atomic structure toward high-density regions. The circular areas correspond to cross-sections of alpha helices. (B) We use carbon monoxide dehydrogenase to illustrate MDFF: carbon monoxide dehydrogenase adopts two different conformations in the same crystal (PDB code: 1OAO). The EM density of the first structure (gray ribbon) is simulated to $(10 \text{ Å})^{-1}$ resolution (gray mesh). We use the second crystal structure (left; color scheme corresponds to root-mean-squared deviation to the target structure) as a starting model for flexible fitting an atomic model to the EM density. The fitted model (right) exhibits dramatically reduced RMSD deviation.

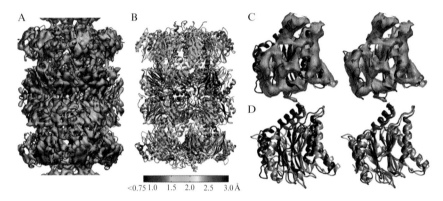

<0.75 1.0 1.5 2.0 2.5 3.0 Å

Friedrich Förster and Elizabeth Villa, Figure 3.7 Flexible fitting of the 20S CP. (A) A comparative model is flexibly fitted to a 7.5 $\mathring{A}^{-1}$ resolution EM map of the *T. acidophilum* 20S proteasome (EMDB-5130). (B) The fitted model is colored by the root-mean-squared deviation (RMSD) of each residue, showing the changes undergone during the fit (initial vs. refined model). (C) One of the α-subunits is shown before (left, red) and after (right, blue) the fit. The segmented EM density for this region is shown in gray. The *CCC* improves from 0.75 to 0.84. (D) The same α-subunit compared to the crystal structure (PDB code: 3IPM). The RMSD of model and crystal structure improves from 5.06 to 2.59 Å for this subunit during the fit.

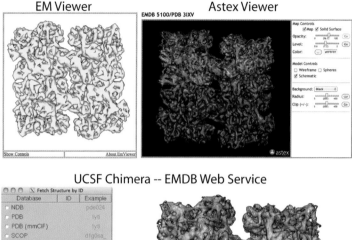

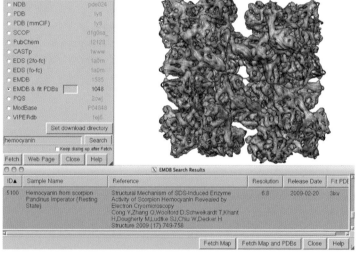

Catherine L. Lawson, Figure 4.5 Tools for 3D visualization of EM structural data. EM Viewer (top left) and Astex Viewer (top right) are Java-based 3D viewers that can be launched from the EMDB atlas "Visualization" pages. EM Viewer employs a compact single contour-level mesh representation for lightweight map viewing; Astex Viewer enables exploration of maps and their fitted models with adjustable map contour level. The EMDB web service implemented with in UCSF Chimera (bottom panel) expedites search of EMDB and subsequent map + model download. Scorpion hemocyanin (Cong *et al.*, 2009, EMD-5100 and PDB id 3ixv) is the example shown in each of the panels.

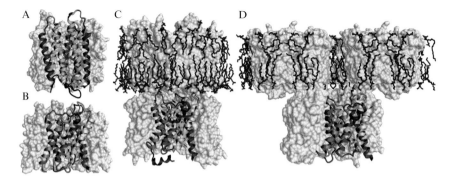

Andreas D. Schenk *et al.*, Figure 5.1 Atomic models of bacteriorhodopsin and aquaporins determined by electron crystallography. (A) Atomic model of the bacteriorhodopsin trimer (PDB entry 2AT9). Two subunits are shown as molecular surfaces and the front most subunit is shown in ribbon representation. The bound retinal is represented by an orange stick model. (B) Atomic model of the AQP1 tetramer (PDB entry 1FQY). Three subunits are shown as molecular surfaces and the front most subunit is shown in ribbon representation. (C, D) Atomic models of the membrane junctions formed by AQP0 (PDB entry 2B60) (C) and AQP4 (PDB entry 2ZZ9) (D). Three subunits of the bottom tetramers are shown as white molecular surfaces and the front most subunits are shown in ribbon representation. The top tetramers are shown as yellow molecular surfaces, together with ordered lipids represented by red stick models. Figures were created using *OpenStructure* (www.openstructure.org). The surfaces were calculated using *MSMS* (Sanner *et al.*, 1996).

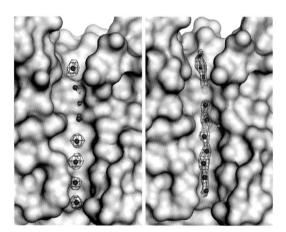

Andreas D. Schenk *et al.*, Figure 5.8 Comparison of the solute densities in the maps of AQP4 determined by electron and X-ray crystallography. The EM structure (PDB entry 2ZZ9) (A) and the X-ray structure (PDB entry 3GD8) (B) of AQP4 are represented as molecular surfaces. The water molecules in the channel are shown as red spheres. The densities for water and glycerol molecules are displayed as blue ($2F_o - F_c$) and green ($F_o - F_c$) wireframes. The water densities appear better defined in the EM map than in the X-ray map.

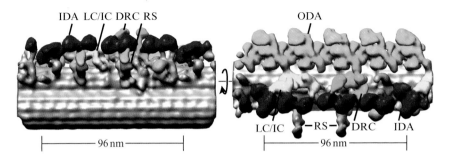

Kenneth H. Downing and Eva Nogales, Figure 6.8 Tomography of *Chlamydomonas* axoneme. A section of a microtubule doublet obtained by averaging densities around and along axonemes is shown, encompassing the basic 96 nm repeat of dyneins and associated proteins. Views are as seen from the center of the axoneme (left) and tangential to the doublets (right). Densities attached to the microtubule are mainly components of the inner dynein arms, aside from the radial spoke components seen pointing down in the view at right. ODA, outer dynein arms (light blue in color figure); IDA, (red) inner dynein arms; DRC (green), dynein regulatory complex; LC (yellow), light chain of dynein 7; RS (blue), radial spokes. The view is similar to that in Bui *et al.* (2009) but represents the majority of dynein arrangements, not the uncommon one between doublets 1 and 9. Figure kindly provided by Drs K. H. Bui and T. Ishikawa.

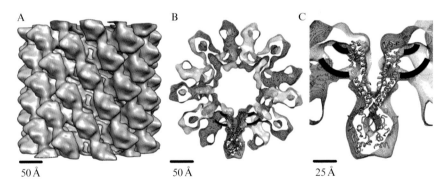

John Paul Glaves *et al.*, Figure 7.3 The density maps of VC-MsbA generated by helical reconstruction to 15 Å resolution. (A) Longitudinal view of the density map (gray surface), which is equivalent to the direction of view of the helical crystal shown in Fig. 7.2B. Each subunit in the density map corresponds to a dimer of MsbA. (B) Cross-sectional view of the helical reconstruction (gray mesh). The X-ray structure of the MsbA dimer (yellow cartoon) from *Salmonella typhimurium* (PDB 3B60) has been docked into the EM density map. (C) Close-up of the fitted MsbA X-ray structure and corresponding EM density map. The putative location of the lipid bilayer is delineated by the black lines.

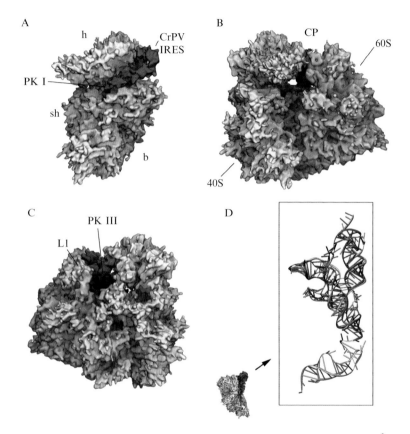

Justus Loerke *et al.*, Figure 8.3   80S ribosome with bound CrPV IRES. 7.3 Å cryo-EM map of the CrPV·IRES·RNA in complex with the yeast ribosome. The map is shown (A) from the back if the 60S subunit with the 60S subunit removed, (B) from the A-site, and (C) from the L1 protuberance (40S subunit, yellow; 60S subunit, blue; CrPV·IRES, magenta). Landmarks for the 40S subunit: b, body; h, head; sh, shoulder. Landmarks for the 60S subunit: CP, central protuberance; L1, L1 protuberance; PKI, pseudoknot I; PKIII, pseudoknot III. In (D) the structure of the CrPV·IRES·RNA molecular model (gray and magenta ribbons) from the cryo-EM map is shown. Aligned to this structure is the model of the PSIV IRES based on a 3.1-Å X-ray structure (Pfingsten *et al.*, 2006; orange ribbons).

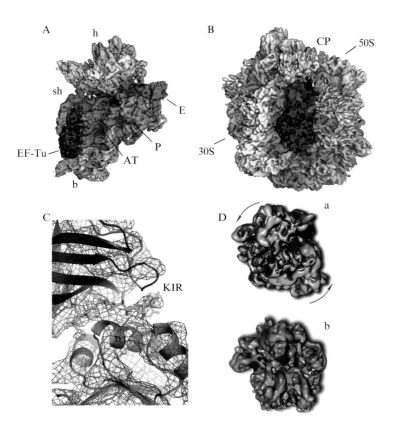

Justus Loerke _et al._, Figure 8.4 Overview of the 70S·EF-Tu·Phe-tRNA·GDP· kirromycin complex. A surface representation of the cryo-EM map is shown (A) from the 50S side, with 50S removed and (B) from the A-site (30S subunit, yellow; 50S subunit, blue; EF-Tu, red; A/T-tRNA, light magenta; P-tRNA, green; E-tRNA, orange; mRNA, dark blue). Landmarks for the 40S subunit: b, body; h, head; sh, shoulder. Landmarks for the 60S subunit: CP, central protuberance, L1, L1 protuberance. (C) Density for the low-molecular weight kirromycin seen between the domains of EF-Tu. (D) (a) Superposition of Population I (gold) compared to Population II (pink, transparent). The reconstructions are shown from the top with the 50S subunit above the 30S subunit. An arrow indicates an intersubunit rotation of the 30S subunit with respect to the 50S subunit. (b) Superposition of Population I (gold) compared to Population III (blue, transparent). The reconstructions are shown from the side. The stalk base of Population III is abbreviated (red circle).

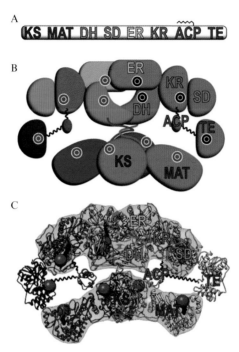

Edward J. Brignole and Francisco Asturias, Figure 9.1 (A) The animal FAS polypeptide contains six catalytic domains: ketoacyl synthase (KS), malonyl/acetyl transferase (MAT), dehydrase (DH), enoyl reductase (ER), ketoacyl reductase (KR), and thioesterase (TE), as well as a nonenzymatic structural domain (SD) and an acyl-carrier domain (ACP). (B) Domain organization of the coiled FAS homodimer (one monomer in color, the other in gray) highlighting the active sites that can be visited by the red carrier domain (red targets) and those that the red carrier domain cannot access (black targets). Flexible linkage between KR-ACP and ACP-TE is indicated as wavy lines. (C) The N-terminal KS domain through the KR domain were observed in the FAS crystal structure (gray outline, PDB: 2vz9). Structures of the carrier domain (red, PDB: 2png) and TE (maroon, PDB: 1xkt) are positioned for illustrative purposes connected by flexible interdomain linkages indicated by wavy lines as in (B). Red balls indicate the docking site on each catalytic domain that can be accessed by the red carrier domain. The black structure extending from each red ball represents the substrate cargo delivered by the carrier domain into the active site. Two catalytic sites that can be contacted by the red ACP are located in the opposite reaction chamber, apparently beyond reach of the red carrier domain due to its short 10-residue tether to the KR domain. Using EM, we uncovered dramatic structural flexibility of FAS that permits contacts between carrier and catalytic domains.

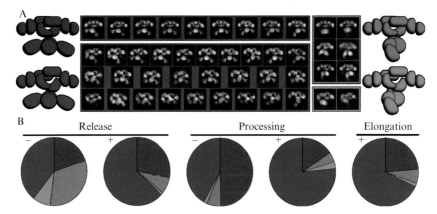

Edward J. Brignole and Francisco Asturias, Figure 9.7 Comparison of conformational distributions between FAS preparations. (A) After particles were aligned and classified, class averages were sorted into categories, according to symmetry (red) or asymmetry (blue) in the upper portion of the structure, and according to lower domains in-plane (bright) or perpendicular (faded) with the upper portion. (B) The conformational redistribution of particles for each mutant FAS in the presence (+) or absence (−) of substrates suggests that catalysis is facilitated by conformations with the lower portion in-plane with an asymmetric upper portion (bright blue). The mutations examined resulted in defective acyl-chain release, processing, or elongation.

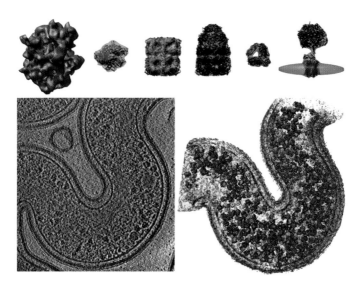

Friedrich Förster *et al.*, Figure 11.1 Visual proteomics of the human pathogen *Leptospira interrogans*. Reference structures are shown in the top panel. The ribosome, RNA polymerase II, GroEL, GroEL-ES, Hsp, and ATP synthase were template matched in the tomogram shown as centered slice through the reconstruction on the left. The templates assigned into this volume are shown surface rendered on the right (membrane in blue, cell wall in brown).

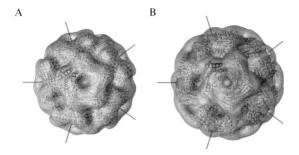

Friedrich Förster *et al.*, Figure 11.2   Comparison of two types of icosahedral structures of lumazine synthase (riboflavin synthase beta subunit) formed by the proteins from *Aquifex aeolicus* (A) and *D. vulgaris* Hildenborough (B). The positions and directions of some of the fivefold axes are indicated with red lines to help the comparison of two structures. A single pentameric ring subunit is shown as a yellow ribbon representation with a single monomer shown in red. Note that the vertices of the pentamers are rotated by different amounts in the two complexes, resulting in two structures with different diameters. (A) Semitransparent isosurface representation of the complex from *A. aeolicus*, computed at the same resolution as that estimated for the structure obtained by electron microscopy for the complex from *Dv*H. A ribbon diagram of the atomic model of the complex (PDB: 1HQK) is embedded in the low-resolution isosurface. (B) Semitransparent isosurface representation of the complex, obtained by electron microscopy, is shown together with the ribbon diagram of a *Dv*H homology model. The pentameric ring is rotated by ≈30° around the icosahedral fivefold axis to produce a good fit within the EM density map. The homology model was obtained by using the MODBASE (Pieper *et al.*, 2006) server located at http://modbase.compbio.ucsf.edu/ modbase-cgi/index.cgi. This figure is adapted from Fig. 2 in Han *et al.* (2009).

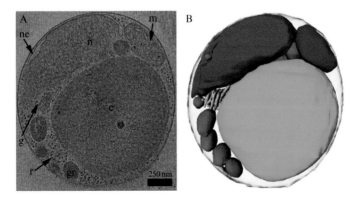

Asaf Mader *et al.*, Figure 12.1   **Cryo-ET of entire eukaryotic cell.** (A) Single *x-y* slice through the middle of a single *O. tauri* cell. In these experiments, the cells were found to be mainly non-dividing, 21.6-nm. Shown are nuclei (n), nuclear envelope (ne), chloroplasts (c), mitochondria (m), Golgi bodies (g), granules (gr), and ribosome-like particles (r). (B) 3-D rendered view of *O. tauri* cell, 36-nm. Adapted from Henderson *et al.* (2007).

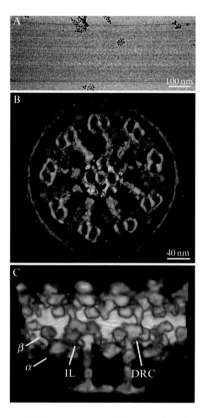

Asaf Mader et al., Figure 12.2   Cryo-ET of sperm flagellum from sea urchin and *Chlamydomonas*. (A) Transmission electron micrograph of an intact quiescent, frozen-hydrated sea urchin sperm flagellum. (B) A cross-section through a surface-rendered view of an axoneme. The structures were translationally averaged using 96-nm axial periodicity. Shown are the plasma membrane (brown), microtubules (gray), outer dynein arms (ODAs) and inner dynein arms (IDAs; red), radial spokes (orange), central pair protrusions (yellow), and bipartite bridges (pink). Scale bar: 40 nm. (C) Volume rendering presents the organization of the wild-type *Chlamydomonas* dynein arm viewed from the B tubule of the adjacent doublet, with the proximal end on the left. The 1-alpha and 1-beta motor domains of the I1complex, the IL which is the DIC/DLC-tail complex (dynein intermediate chains /dynein light chains) and the dynein regulatory complex (DRC) that regulates phosphatase and kinase activities on the doublet MTs are indicated. Adapted from Nicastro *et al.* (2005, 2006).

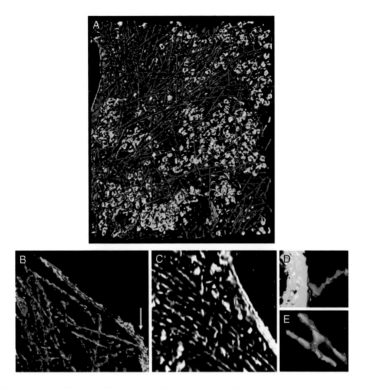

Asaf Mader *et al.*, Figure 12.4   Visualizing the cytoplasm of a *Dictyostelium discoideum* cell. (A) Surface-rendered view of a volume of 815 nm × 870 nm × 97 nm. Colors were subjectively attributed to linear elements to mark the actin filaments (reddish); other macromolecular complexes, mostly ribosomes (green), and membranes (blue) are shown. (B, D) Actin anchorage to the membrane is illustrated in a surface-rendering of a cortical region demonstrating the plasma membrane associated with actin filaments ((B) is a rendered volume of 500 nm × 300 nm × 20 nm). The high magnification in (D) shows a kink-like structure close to the filament membrane-associating site, creating a filament-membrane bridge at a nonperpendicular angle. (C, E) Actin filament bundles in the cell cortex, and rendered volumes showing actin lateral connections to the membrane (C) or lateral bridges between two filaments (E), (extracted from (C)). Adapted from Medalia *et al.* (2002).

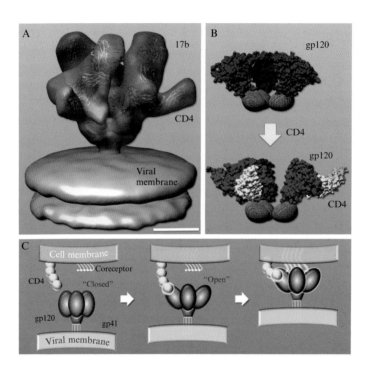

Jun Liu et al., Figure 13.8   3D structure of HIV Env with CD4 and CD4 induced antibody 17b (Liu *et al.*, 2008). The crystal structure from a ternary complex of gp120 with CD4 and 17b (Kwong *et al.*, 1998) was fitted in EM density map as rigid body (A). gp120 was colored in magenta, CD4 in yellow, 17b fab in light green. CD4 induces huge conformational change in the HIV Env spike (B). A model of viral entry was proposed in (C). The scale bar is 5 nm.